T0335829

MATHEMATICAL FEYNMAN PATH INTEGRALS AND THEIR APPLICATIONS

MATHEMATICAL FEYNMAN PATH INTEGRALS AND THEIR APPLICATIONS

Sonia Mazzucchi
University of Trento, Italy

World Scientific
NEW JERSEY • LONDON • SINGAPORE • BEIJING • SHANGHAI • HONG KONG • TAIPEI • CHENNAI

Published by

World Scientific Publishing Co. Pte. Ltd.

5 Toh Tuck Link, Singapore 596224

USA office: 27 Warren Street, Suite 401-402, Hackensack, NJ 07601

UK office: 57 Shelton Street, Covent Garden, London WC2H 9HE

British Library Cataloguing-in-Publication Data
A catalogue record for this book is available from the British Library.

MATHEMATICAL FEYNMAN PATH INTEGRALS AND THEIR APPLICATIONS

Copyright © 2009 by World Scientific Publishing Co. Pte. Ltd.

All rights reserved. This book, or parts thereof, may not be reproduced in any form or by any means, electronic or mechanical, including photocopying, recording or any information storage and retrieval system now known or to be invented, without written permission from the Publisher.

For photocopying of material in this volume, please pay a copying fee through the Copyright Clearance Center, Inc., 222 Rosewood Drive, Danvers, MA 01923, USA. In this case permission to photocopy is not required from the publisher.

ISBN-13 978-981-283-690-8
ISBN-10 981-283-690-X

Printed in Singapore.

Preface

Even if more than 60 years have passed since their first appearance in Feynman's PhD thesis, Feynman path integrals have not lost their fascination yet.

They give a suggestive description of quantum evolution, reintroducing in quantum mechanics the classical concept of trajectory, which had been banned from the traditional formulation of the theory. In fact, they can be recognized as a bridge between the classical description of the physical world and the quantum one. Not only do they provide a quantization method, allowing to associate, at least heuristically a quantum evolution to any classical Lagrangian, but also they make very intuitive the study of the semiclassical limit of quantum mechanics, i.e. the study of the detailed behavior of the wave function when the Planck constant is regarded as a small parameter converging to zero.

Nowadays, the physical applications of Feynman's ideas go beyond non relativistic quantum mechanics and include quantum fields, statistical mechanics, quantum gravity, polymer physics, geometry. Nevertheless, in most cases, Feynman path integrals remain a mathematical challenge as they are not well defined from a mathematical point of view.

Since 1960, a large amount of work has been devoted to the mathematical realization of Feynman path integrals in terms of a well defined functional integral. Despite the several interesting results that have been obtained in the last decades, the feeling that Feynman integrals are only an heuristic tool is still a widespread belief among mathematicians and physicists.

The present book provides a detailed and self-contained description of the rigorous mathematical realization of Feynman path integrals in terms of infinite dimensional oscillatory integrals, a particular kind of functional integrals that can be recognized as the direct generalization of classical oscillatory integrals to the case where the integration is performed on an infinite dimensional space, in particular on a space of paths.

The book describes the mathematical difficulties, the first results obtained in the 70's and the 80's, as well as the more recent development and applications. Special attention has been paid to enlightening the mathematical techniques, including infinite dimensional integration theory, asymptotic expansions and resummation techniques, without losing the connection with the physical interpretation of the theory.

A large amount of references allows the reader to get a deeper knowledge of the most interesting mathematical results as well as of the modern physical applications.

I am grateful to many coworkers, friends and colleagues for fruitful discussions. Special thanks are due to S. Albeverio for his help and support, as well as for reading the manuscript and making lots of useful comments. I am also particularly grateful to G. Greco, V. Moretti, E. Pagani, M. Toller and L. Tubaro.

S. Mazzucchi

Contents

Chapter 1

Introduction

One of the most challenging problems of modern physics is the connection between the macroscopic and the microscopic world, that is between classical and quantum mechanics. In principle a macroscopic system should be described as a collection of microscopic ones, so that classical mechanics should be deduced from quantum theory by means of suitable approximations. At a first glance the solution of the problem is not straightforward: indeed there are deep differences between the classical and the quantum description of the physical world.

In classical mechanics the state of an elementary physical system, for instance, a point particle is given by specifying its position q (a point in its configuration space) and its velocity $\dot{q}$. The time evolution in the time interval $[0, t]$ is described by a path $q(s)_{s\in[0,t]}$ in the configuration space. The dynamics of the particle under the action of a force field described by the real-valued potential V is determined by the classical Lagrangian:

$$\mathcal{L}(q(s), \dot{q}(s)) := \frac{m}{2}\dot{q}^2 - V(q), \tag{1.1}$$

where m is the mass of the particle. By the Hamilton's least action principle, the Euler-Lagrange equation of motion

$$\frac{d}{dt}\frac{\partial \mathcal{L}}{\partial \dot{q}} - \frac{\partial \mathcal{L}}{\partial q} = 0$$

follow by a variational argument. The trajectory of the particle connecting a point x at time t_0 to a point y at time t is the path making stationary the action functional S:

$$\delta S_t(q) = 0, \quad S_t(q) = \int_0^t \mathcal{L}(q(s), \dot{q}(s))ds. \tag{1.2}$$

The quantum description of a point particle appears at a first glance completely different. First of all the concept of trajectory is meaningless.

Heisenberg's uncertainty principle states the existence of "incompatible observables": i.e. the measurement of one of them destroys the information about the measurement of the other one. Position and velocity are typical examples: quantum mechanics forbids the knowledge of the couple $q(s), \dot{q}(s)$ for a time interval $[0,t]$ with a given precision. In other words, from a quantum mechanical point of view, the trajectory of a particle has no physical meaning as there is no way to measure it.

Contrary to classical mechanics, the state of a quantum particle moving in the d-dimensional Euclidean space is described by a unitary vector ψ in the complex separable Hilbert space $L^2(\mathbb{R}^d)$, the so-called "wave function". The physical meaning of the vector ψ is probabilistic. For instance, the probability that the result of the measurement of the position of the particle is contained in a Borel set $A \subset \mathbb{R}^d$, is given by the integral $\int_A |\psi(x)|^2 dx$.

The time evolution is determined by a one-parameter group of unitary evolution operators $U(t)$, whose infinitesimal generator is the quantum Hamiltonian operator H, which is given on vectors ψ belonging to $C_0^\infty(\mathbb{R}^d)$ by

$$H\psi(x) = -\frac{\hbar^2}{2m}\Delta\psi(x) + V(x)\psi(x), \qquad x \in \mathbb{R}^d, \tag{1.3}$$

where $\hbar$ is the reduced Planck constant, Δ the Laplacian. Under suitable assumptions on the potential and on the domain of the operator (see for instance [246]), H is (essentially) self-adjoint and the evolution operator $U(t) = e^{-\frac{i}{\hbar}Ht}$ is uniquely determined by Eq. (1.3). The evolution of the state vector, i.e. the wave function at a given time t, can be described by the Schrödinger equation:

$$\begin{cases} i\hbar\frac{\partial}{\partial t}\psi(t,x) = -\frac{\hbar^2}{2m}\Delta\psi(t,x) + V(x)\psi(t,x) \\ \psi(0,x) = \psi_0(x)\,. \end{cases} \tag{1.4}$$

In 1948, following a suggestion by Dirac [106, 107], R. P. Feynman proposed a new suggestive description of quantum evolution. Feynman's aim was a Lagrangian formulation of quantum mechanics and the introduction of the action functional and of variational arguments in the theory, in analogy to classical mechanics. Feynman developed Dirac's idea that in quantum dynamics the imaginary exponential of the action functional plays a fundamental role. According to Feynman's interpretation, the total transition amplitude $G(0,x;t,y)$ from the point x at time 0 to the point y at time t, i.e. the kernel of the evolution operator $U(t)$ evaluated at the points x, y, should be given by a sum over the contributions of all possible paths $\gamma : [0,t] \to \mathbb{R}^d$ such that $\gamma(0) = x$ and $\gamma(t) = y$:

$$G(0,x;t,y) = \int e^{\frac{i}{\hbar}S_t(\gamma)} D\gamma, \tag{1.5}$$

where $D\gamma$ denotes a Lebesgue-type measure on the space of paths. Analogously, the solution of the Schrödinger equation, i.e. the wave function $\psi(t,x)$ evaluated at the time t in the point $x \in \mathbb{R}^d$, should be given by the integral over the space of paths $\gamma : [0,t] \to \mathbb{R}^d$ such that $\gamma(0) = x$:

$$\psi(t,x) = \int e^{\frac{i}{\hbar}S_t(\gamma)} \psi(0,\gamma(0)) D\gamma. \tag{1.6}$$

In other words, according to Feynman's formulation, the time evolution of a quantum system should be given by a "sum over all possible histories".

Even if more than half a century has passed since Feynman's original paper [122], formulae (1.5) and (1.6) are still astonishing and have not lost their fascination yet. Feynman's approach creates a bridge between the classical Lagrangian description of the physical world and the quantum one, reintroducing in quantum mechanics the classical concept of trajectory, which had been banned by the traditional formulation of the theory. It allows, at least heuristically, to associate a quantum evolution to each classical Lagrangian. Moreover it makes very intuitive the study of the semiclassical limit of quantum mechanics, i.e. the study of the behavior of the wave function when the Planck constant $\hbar$ is regarded as a small parameter which is allowed to converge to 0. In fact, when $\hbar$ becomes small, the integrand $e^{\frac{i}{\hbar}S_t(\gamma)}$ behaves as a strongly oscillatory function and, according to an heuristic application of the stationary phase method (see chapter 4), the main contribution to the integral should come from those paths which make stationary the phase functional S_t. These, by Hamilton's least action principle, are exactly the classical orbits of the system.

An intuitive justification of, at this stage still mysterious, Feynman's formula can be given by means of Trotter product formula [279, 78, 79]. Under suitable assumption on the potential V (see [78, 79, 279, 235] for more details), for instance if V is bounded, the evolution operator $U(t) = e^{-\frac{i}{\hbar}Ht}$ can be written in terms of a strong operator limit:

$$e^{-\frac{i}{\hbar}Ht} = \lim_{n\to\infty} \left(e^{-\frac{it}{\hbar n}H_0} e^{-\frac{it}{\hbar n}V}\right)^n, \tag{1.7}$$

where $H_0 = -\frac{\hbar^2}{2m}\Delta$. In particular, by taking an initial datum $\psi_0 \in \mathcal{S}(\mathbb{R}^d)$, the solution of the Schrödinger equation (1.4) with $V = 0$ can be written as

$$e^{-\frac{i}{\hbar}H_0 t}\psi_0(x) = \left(\frac{2\pi i\hbar t}{m}\right)^{-\frac{d}{2}} \int_{\mathbb{R}^d} e^{im\frac{(x-y)^2}{2\hbar t}} \psi_0(y)dy, \tag{1.8}$$

and Eq. (1.7) gives

$$e^{-\frac{i}{\hbar}Ht}\psi_0(x) = \lim_{n\to\infty}\left(\frac{2\pi i\hbar t}{mn}\right)^{-\frac{nd}{2}}\int_{\mathbb{R}^{nd}} e^{\frac{i}{\hbar}\sum_{j=1}^{n}\left(\frac{m}{2}\frac{(x_j-x_{j-1})^2}{(t/n)^2}-V(x_j)\right)\frac{t}{n}} \psi_0(x_0)dx_0\dots dx_{n-1}. \quad (1.9)$$

If we now divide the time interval $[0,t]$ into n equal parts of amplitude t/n, and if for any path $\gamma:[0,t]\to\mathbb{R}^d$ we consider its approximation by means of a broken line path γ_n:

$$\gamma_n(s) := x_j + \frac{(x_{j+1}-x_j)}{t/n}(s - jt/n), \quad s\in[jt/n,(j+1)t/n], \quad j = 0\dots n-1, \quad (1.10)$$

where $x_j := \gamma(jt/n)$, then the exponent in the integrand of Eq. (1.9) can be regarded as the Riemann approximation of the action functional S_t evaluated along the path γ_n. In other words, Eq. (1.6) can be regarded as a intuitive way to write the limit (1.9).

However, as we have already discussed above, the intuitive power of Feynman's formula goes beyond a simple mnemonic tool to write a limiting procedure. Indeed Feynman extended his approach to the description of the dynamics of more general quantum systems, including the case of quantum fields [124, 123, 125] and producing an heuristic calculus that, from a physical point of view, works even in cases where rigorous arguments fail.

Despite the successfully predicting power of Feynman path integral, it lacks of mathematical rigour. Feynman himself was conscious of this problem:

> [...] one feels like Cavalieri must have felt calculating the volume of a pyramid before the invention of the calculus. [122]

The challenge to give meaning to Feynman's heuristic calculus and to define rigorously oscillatory integrals in infinite dimension, was left to mathematicians.

1.1 Wiener's and Feynman's integration

When we try to interpret the heuristic integral (1.6) we have to face mainly with two mathematical difficulties.

First of all one has to implement a non trivial integration theory on a space of paths, that is on an infinite dimensional space. It is reasonable to assume that the function space containing the "Feynman paths" has a metric or at least a topological structure. A possible candidate is the space of paths with "finite kinetic energy", that is the Hilbert space $\mathcal{H}_t$ of absolutely continuous paths $\gamma : [0,t] \to \mathbb{R}^d$ such that $\gamma(t) = 0$ and $\int_0^t |\dot{\gamma}(s)|^2 ds < \infty$ ($\dot{\gamma}$ denoting the distributional derivative of the path γ) endowed with the inner product

$$\langle \gamma_1, \gamma_2 \rangle = \int_0^t \dot{\gamma}_1(s)\dot{\gamma}_2(s)ds. \tag{1.11}$$

$\mathcal{H}_t$ will be called the *Cameron-Martin space.*

Another possible choice is the Banach space $C([0,t], \mathbb{R}^d)$ of continuous paths $\omega : [0,t] \to \mathbb{R}^d$ such that $\omega(0) = 0$, endowed with the sup-norm $\|\omega\| = \sup_{s\in[0,t]} |\omega(s)|$. In the case where $d = 1$, this space will be denoted by C_t.

In both cases the expression $D\gamma$ in Eq. (1.6), denoting a Lebesgue "flat" measure is meaningless. A rather simple argument shows that a Lebesgue-type measure cannot be defined on infinite dimensional Hilbert spaces. Indeed the assumption of the existence of a σ-additive measure μ which is invariant under rotations and translations and assigns a positive finite measure to all bounded open sets, leads to a contradiction. By taking an orthonormal system $\{e_i\}_{i\in\mathbb{N}}$ in an infinite dimensional Hilbert space $\mathcal{H}$ and by considering the open balls $B_i = \{x \in \mathcal{H}, \|x - e_i\| < 1/2\}$, one has that they are pairwise disjoint and their union is contained in the open ball $B(0,2) = \{x \in \mathcal{H}, \|x\| < 2\}$. By the Euclidean invariance of the Lebesgue-type measure μ one can deduce that $\mu(B_i) = a$, $0 < a < \infty$, for all $i \in \mathbb{N}$. By the σ-additivity one has

$$\mu(B(0,2)) \geq \mu(\cup_i B_i) = \sum_i \mu(B_i) = \infty,$$

but, on the other hand $\mu(B(0,2))$ should be finite as $B(0,2)$ is bounded. An analogous argument holds also for Banach spaces [140] and in particular for the space C_t. In other words, the expression $D\gamma$ in formulae (1.5) and (1.6) is not defined from a mathematical point of view and cannot be used as "reference measure", i.e. the measure with respect to which Feynman's measure has density $e^{i\frac{S_t}{\hbar}}$.

Integration theory on a space of continuous paths was already present at Feynman's time. In particular an example of a non trivial measure on the space C_t had been already provided by N. Wiener [291] in his work on Brownian motion, however there is no mention of Wiener integral in Feynman's paper.

The connection between Feynman's idea and Brownian motion was discovered for the first time by M. Kac [187, 188], who was inspired by

Feynman's lecture at Cornell University. Kac noted that by considering the heat equation

$$\begin{cases} \frac{\partial}{\partial t}u(t,x) = \frac{1}{2}\Delta u(t,x) - Vu(t,x) \\ u(0,x) = u_0(x) \end{cases} \tag{1.12}$$

instead of Eq. (1.4), and by replacing in Feynman's formula (in the simple case $V = 0$ and $\hbar = m = 1$) the oscillatory term $e^{iS_t(\gamma)} = e^{i\int_0^t \frac{\dot\gamma(s)^2}{2}ds}$ with the not oscillatory one $e^{-\int_0^t \frac{\dot\gamma(s)^2}{2}ds}$:

$$\frac{e^{iS_t(\gamma)}D\gamma}{\int e^{iS_t(\gamma)}D\gamma} \longrightarrow \frac{e^{-S_t(\gamma)}D\gamma}{\int e^{-S_t(\gamma)}D\gamma}$$

it is possible to replace the heuristic expression (1.6) with a well defined integral on the space of continuous paths with respect to the Wiener measure W:

$$u(t,x) = \int_{C_t} u_0(\omega(t)+x)dW(\omega). \tag{1.13}$$

In the case $V \neq 0$ Eq. (1.13) becomes:

$$u(t,x) = \int_{C_t} u_0(\omega(t)+x)e^{-\int_0^t V(\omega(s)+x)ds}dW(\omega). \tag{1.14}$$

In other words the solution of the heat equation admits a mathematically rigorous path integral representation which is now called *Feynman-Kac formula.*

We give here a brief description's of Wiener's measure and of Kac's result, in order to explain the underlying ideas and show not only the similarities, but also the differences with Feynman's case.

Wiener measure is a σ-additive, positive Gaussian probability measure on the space $C([0,t],\mathbb{R}^d)$. It can be constructed in several ways (see for instance [195, 263, 235] and section A.2 in the appendix). We have chosen an intuitive approach that shows the analogies between Feynman's and Wiener's integration.

In the following, for notational simplicity, we shall assume that $d = 1$, but the whole discussion can be simply generalized to arbitrary dimension d. Let us consider the *cylindrical sets*, i.e. the subsets of C_t of the form

$$A(t_1,\dots t_k; I_1\dots I_k) := \{\gamma \in C([0,t],\mathbb{R}) \,:\, \gamma(t_1)\in I_1,\dots,\gamma(t_k)\in I_k\}, \tag{1.15}$$

where $0 \le t_1 \le \dots t_k \le t$ and $I_1,\dots,I_k$ are intervals of $\mathbb{R}$.

The Wiener measure of these sets is given by the following formula:

$$W(A(t_1,\dots t_k; I_1\dots I_k)) = \Big(\frac{1}{(2\pi)^k t_1(t_2-t_1)\dots(t_k-t_{k-1})}\Big)^{1/2}$$
$$\int_{I_k}\dots\int_{I_1} e^{-\sum_{j=1}^k \frac{(x_j-x_{j-1})^2}{2(t_j-t_{j-1})}} dx_1\dots dx_k, \quad (1.16)$$

where $t_0 \equiv 0$ and $x_0 \equiv 0$. The collection $\mathcal{I}$ of cylindrical sets is a semi-algebra[1] and W is additive on it. By the Caratheodory extension theorem [83], W can be extended to a σ-additive measure on $\sigma(\mathcal{I})$, the σ-algebra generated by the cylindrical sets, which is equal to $\mathcal{B}(C_t)$, the Borel σ-algebra on C_t.

Let us introduce now a "polygonal path" $\gamma : [0,t] \to \mathbb{R}$, such that $\gamma(t_j) = x_j$ and $\gamma(s)$ for $s \in [t_j, t_{j+1}]$ coincides with the constant velocity path connecting x_j with x_{j+1}:

$$\gamma(s) = \sum_{j=0}^{k-1} \chi_{[t_j,t_{j+1}]}(s)\Big(x_j + \frac{x_{j+1}-x_j}{t_{j+1}-t_j}(s-t_j)\Big), \qquad s\in[0,t],$$

where $\chi_{[t_j,t_{j+1}]}$ is the characteristic function of the interval $[t_i, t_{i+1}]$ and $\Delta t_i := t_{i+1} - t_i$ is its amplitude. Let

$$S_t^\circ(\gamma) \equiv \frac{1}{2}\int_0^t |\dot\gamma(s)|^2 ds \quad (1.17)$$

be the free action, i.e. the time integral of the kinetic energy of the path, that is

$$S_t^\circ(\gamma) \equiv \frac{1}{2}\sum_{j=0}^{k-1}\Big|\frac{\Delta x_j}{\Delta t_j}\Big|^2 \Delta t_j. \quad (1.18)$$

Let us define

$$D\gamma \equiv Z^{-1} \prod_{t\in\{t_1,\dots,t_k\}} d\gamma(t), \quad (1.19)$$

with

$$Z \equiv ((2\pi)^n \Delta t_{n-1}\dots\Delta t_0)^{\frac{1}{2}}, \quad (1.20)$$

---

[1] A collection $\mathcal{C}$ of subsets of a set A is called a semi-algebra if

(1) $\emptyset \in \mathcal{C}$ and $A \in \mathcal{C}$

(2) if $B, C \in \mathcal{C}$ then $B \cap C \in \mathcal{C}$

(3) if $B \in \mathcal{C}$, then $A \setminus B$ is the finite disjoint union of subsets in $\mathcal{C}$.

and $d\gamma(t) = dx_j$ per $t = t_j$. With these notations, the Wiener measure of cylindrical sets can be written in terms of the following formula:

$$W(\gamma_{t_1} \in I_1, \ldots, \gamma_{t_k} \in I_k) = \int e^{-S_t^\circ(\gamma)} D\gamma. \tag{1.21}$$

The right hand side has a meaning as soon as we restrict ourselves to cylindrical sets, as the infinite dimensional Wiener integration can be reduced to a (finite dimensional) integration on $\mathbb{R}^k$. In this case both the "normalized Lebesgue measure" $D\gamma$ and the factor $e^{-S_t^\circ(\gamma)}$ make sense. When the measure is extended on the whole sigma algebra $\mathcal{B}(C_t)$, formula (1.21) has to been interpreted as a finite dimensional approximation: the key point is that, even if the single terms $D\gamma$ and $e^{-S_t^\circ(\gamma)}$ lose a well defined meaning, their combination is still meaningfull and it gives exactly Wiener Gaussian measure.

If we now consider the heat equation (1.12) and, analogously to the Schrödinger case, we write the corresponding heat semigroup in terms of the Trotter product formula [279, 78, 79], we obtain:

$$\begin{aligned} e^{-Ht}u_0(x) &= \lim_{n\to\infty} \left(e^{-\frac{t}{n}H_0} e^{-\frac{t}{n}V}\right)^n \\ &= \lim_{n\to\infty} \left(\frac{2\pi t}{nm}\right)^{-\frac{nd}{2}} \int_{\mathbb{R}^{nd}} e^{-\sum_{j=1}^n \left(\frac{(x_j - x_{j-1})^2}{(t/n)^2} - V(x_j)\right)\frac{t}{n}} \\ &\qquad u_0(x_0)dx_0 \ldots dx_{n-1}. \end{aligned} \tag{1.22}$$

Now the latter line can be recognized as the finite dimensional approximation of the Wiener integral (1.14).

The analogies between Feynman's and Wiener's integral end at this stage because there is a deep difference between them: the presence of oscillations in the heuristic "Feynman measure" $e^{\frac{i}{\hbar}S_t(\gamma)}D\gamma$, as M. Kac himself writes:

> The occurrence of i (which is essential for Quantum Mechanics) makes manipulations with integrals like [formula (1.9)] extremely tricky.
> [189]

The Wiener integration is a Lebesgue type integration, where the absolute convergence is fundamental. On the other hand, physical intuition leads us to stress the importance of the oscillatory behavior of the integrand in Feynman's formula, which describes the concept of coherent superposition and of interference, which is typical of quantum phenomena. In principle the convergence of the integral should be given by the cancellations due to this oscillatory behavior and we should not expect to implement an integration theory in the Lebesgue's traditional way.

This fact was clearly explained by Cameron in 1960 [64], who proved that it is not possible to realize Feynman's measure as a infinite dimensional Gaussian measure with complex covariance (a complex version of Wiener measure) as it would have infinite total variation. We give here a sketch of Cameron's argument as it helps us to understand the core of the problem: the oscillations and the infinite dimensional setting. Cameron's starting point is Wiener measure on $C([0,t],\mathbb{R}^d)$ with a generic covariance $\sigma \in \mathbb{R}^+$. As we have seen above, it is completely determined by its value on cylindrical sets (1.15). In the general case $\sigma \neq 1$ Eq. (1.16) becomes:

$$W_\sigma(A(t_1,\dots t_k; I_1 \dots I_k)) = \left(\frac{1}{(2\pi)^k t_1(t_2-t_1)\dots(t_k-t_{k-1})\sigma^k}\right)^{1/2} \int_{I_1\times\dots\times I_k} e^{-\sum_{j=1}^k \frac{(x_j-x_{j-1})^2}{2\sigma(t_j-t_{j-1})}} dx_1\dots dx_k\,. \quad (1.23)$$

Let us now assume that σ is a complex parameter and try to extend the definition of W_σ to this case. If we try to compute the total variation[2] $|W_{\sigma,k}|$ of the "complex measure" W_σ, when restricted to the cylindrical sets $A(t_1,\dots t_k; I_1 \dots I_k)$ *with* k *fixed*, we find out that it depends on both σ and k in the following way:

$$|W_{\sigma,k}| = \left(|\sigma|\mathrm{Re}(\sigma^{-1})\right)^{-k/2}.$$

By letting now $k \to \infty$ we can conclude that the total variation of the measure W_σ, with $\sigma \in \mathbb{C}\backslash\mathbb{R}^+$, would be infinite. Furthermore, it is possible to see that the complex measure W_σ would have infinite total variation even on bounded sets[3]. In other words there is no σ-additive measure W_σ, with $\sigma \in \mathbb{C}\setminus\mathbb{R}^+$, on C_t such that its value on cylindrical sets is given by Eq. (1.23) and which allows the implementation of an integration theory in the Lebesgue's sense.

It is worthwhile to remark that, in the case $\sigma = i$, the Gaussian measure with covariance σ cannot have finite total variation, even when it is defined on a finite dimensional space. As an example we can consider the Fresnel integral

$$\int_{\mathbb{R}} e^{\frac{i}{2}x^2}dx. \quad (1.24)$$

---

[2]We recall that the total variation of a complex Borel measure μ on a set A is given by

$$|\mu| = \sup\sum_i |\mu(A_i)|,$$

where the supremum is taken over all sequences $\{A_i\}$ of pairwise disjoint Borel subsets of A, such that $\cup_i A_i = A$.

[3]Even the Lebesgue measure on $\mathbb{R}^n$ has infinite total variation, but its total variation on bounded pluri-intervals is finite.

We cannot interpret Eq. (1.24) as an integral with respect to a complex measure $d\mu := e^{\frac{i}{2}x^2}\,dx$, as its total variation would be infinite:

$$|\mu| = \int |e^{\frac{i}{2}x^2}|dx = \int dx = \infty\,.$$

The integral (1.24) has to be defined in an alternative way, for instance as an improper Riemann integral, and its convergence is given by the cancellations due to the oscillatory behavior of the integrand, is such a way that:

$$\int_{\mathbb{R}} e^{\frac{i}{2}x^2}\,dx = \sqrt{2\pi i}.$$

We shall keep in mind this example when in the following chapters we will construct Feynman's integration.

1.2 The Feynman functional

Cameron's result shows that it is impossible to realize Feynman integral as an absolutely convergent (Lebesgue) integral with respect to a "mysterious" σ-additive bounded variation Feynman complex measure μ_F, heuristically given by:

$$\mu_F(\gamma) = \frac{e^{\frac{i}{\hbar}S_t}d\gamma}{\int e^{\frac{i}{\hbar}S_t}d\gamma}, \tag{1.25}$$

as the latter cannot exist. Feynman integration requires an alternative approach.

Let us slightly change our point of view and recall that, by Riesz-Markov theorem (see for instance [245], section IV.4), any complex finite regular Borel measure μ on a locally compact space X can be seen as an element of the dual of $C_\infty(X)$, the space of continuous complex valued functions vanishing at ∞, endowed with the sup-norm:

$$\|f\| = \sup_{x\in X}|f(x)|, \qquad f \in C_\infty(X).$$

In this way, the integral of a function $f \in C_\infty(X)$ with respect to a finite measure μ can be represented as the action on f of the functional $l_\mu \in C_\infty(X)^*$ associated to μ:

$$\int_X f(x)d\mu(x) \equiv l_\mu(f). \tag{1.26}$$

In Feynman's case, as we have seen so far, the left hand side of Eq. (1.26) is not defined, as Feynman's measure does not exists, but one could try to make sense to the right hand side of Eq. (1.26), by slightly changing the functional setting. In other words, one could try to realize the Feynman integral as linear continuous functional on a sufficiently rich Banach algebra of functions, different from $C_\infty(X)$. In order to mirror the features of the heuristic Feynman measure, such a functional should have some properties:

(1) It should behave in a simple way under "translations and rotations in path space", reflecting the fact that $D\gamma$ is a "flat" measure.
(2) It should satisfy a Fubini type theorem, concerning iterated integrations in path space (allowing the construction, in physical applications, of a one-parameter group of unitary operators).
(3) It should be approximated by finite dimensional oscillatory integrals, allowing a sequential approach, close to Feynman's original work.
(4) It should be related to probabilistic integrals with respect to the Wiener measure, allowing an "analytic continuation approach to Feynman path integrals from Wiener type integrals".
(5) It should be sufficiently flexible to allow a rigorous mathematical implementation of an infinite dimensional version of the stationary phase method and the corresponding study of the semiclassical limit of quantum mechanics.

Nowadays several implementations of this program can be found in the physical and in the mathematical literature. One of the first techniques which has been introduced and that was largely developed, also in connection to quantum fields, is the analytic continuation of Wiener Gaussian integrals [64, 235, 180, 282, 191, 108, 218, 230, 81, 277, 278]. The starting point of this approach is the transformation of variable formula for the Wiener integral with covariance σ:

$$\int f(\omega)dW_\sigma(\omega) = \int f(\sqrt{\sigma}\omega)dW(\omega). \tag{1.27}$$

As Cameron proved, the left hand side of Eq. (1.27) is not defined when σ is complex. The leading idea of analytic continuation approach is to give meaning to the right hand side of Eq. (1.27) in the case where $\sigma = i$ for a suitable class of functions f.

Another alternative approach is the realization of Feynman measure as an infinite dimensional distribution. The idea was proposed by C. De Witt-Morette [100]. Its rigorous mathematical realization has been more recently undertaken in the framework of Hida calculus [163]. The latter approach has given particularly interesting results in the applications to Chern-Simons theory [15].

Other possible approaches involve "complex Poisson measures" [222, 60, 86, 207] and non standard analysis [12]. All these approaches will be described in detail in chapter 6.

1.3 Infinite dimensional oscillatory integrals

In this book we shall describe the rigorous mathematical realization of Feynman path integrals in terms of the *infinite dimensional oscillatory integrals.*

The leading idea is the rigorous definition of an infinite dimensional analogue of integral (1.24), in particular of expressions of this form:

$$\int_{\mathcal{H}} f(x) e^{\frac{i}{2\hbar}\|x\|^2} dx, \tag{1.28}$$

where $\mathcal{H}$ is a real separable Hilbert space, $\hbar \in \mathbb{R}^+$ a positive parameter, $f : \mathcal{H} \to \mathbb{C}$ is a suitable function and $e^{\frac{i}{2\hbar}\|x\|^2}$ plays the role of the oscillatory factor, the density of an heuristic complex Gaussian measure.

The roots of this approach can be found in two papers by Ito in the 60's [172, 173] (see also [32] for a discussion of Ito's work), but it was systematic developed by Albeverio and Høegh-Krohn in the 70's [16, 17] in terms of *infinite dimensional Fresnel integrals.*

In Albeverio and Høegh-Krohn's work, the integral (1.28) is defined by dualization. By taking a function f which is the Fourier transform of a complex bounded-variation measure μ_f on $\mathcal{H}$

$$f(x) = \int_{\mathcal{H}} e^{i\langle x,y\rangle} d\mu_f(y), \tag{1.29}$$

the infinite dimensional Fresnel integral of f is defined as:

$$\int_{\mathcal{H}} f(x) e^{\frac{i}{2\hbar}\|x\|^2} dx := \int_{\mathcal{H}} e^{-\frac{i\hbar}{2}\|x\|^2} d\mu_f(x). \tag{1.30}$$

The integral on the right hand side of (1.30) is absolutely convergent and well defined in Lebesgue's sense. The class of functions f of the form (1.29), endowed with a suitable norm, is a Banach algebra, the *Fresnel algebra*, denoted with $\mathcal{F}(\mathcal{H})$. The application

$$I_F : \mathcal{F}(\mathcal{H}) \to \mathbb{C}, \qquad f \mapsto \int_{\mathcal{H}} f(x) e^{\frac{i}{2\hbar}\|x\|^2} dx,$$

is a linear continuous functional.

In the application of this formalism to the Feynman path integral representation of the solution of the Schrödinger equation, by assuming that the potential V is of the following form

$$V(x) = \frac{1}{2}x \cdot \Omega^2 x + V_1(x) \tag{1.31}$$

(where Ω^2 is a positive definite symmetric $d \times d$ matrix and V_1 is the Fourier transform of a complex bounded variation measure on $\mathbb{R}^d$), Albeverio and Høegh-Krohn prove that the solution can be represented as an infinite dimensional Fresnel integral on the Cameron-Martin space (see Eq. (1.11)).

In [16] an infinite dimensional version of the stationary phase method has been developed and applied to the semiclassical limit of quantum mechanics. Some of these results will be described in detail in the next chapters.

The study of oscillatory integrals on infinite dimensional Hilbert spaces was further implemented by D. Elworthy and A. Truman [114], S. Albeverio and Z. Brzeźniak [7, 9]. In [114] a slightly different definition was proposed. The integral (1.28) is defined as the limit of a sequence of finite dimensional approximations. Each term of the sequence is a classical oscillatory integral on a finite dimensional space, which is defined as an improper Riemann integral, by modifying a definition proposed by Hörmander [164, 165]. This new definition of integral (1.28) can be recognized as the infinite dimensional generalization of oscillatory integrals of the type (1.24).

The class of "integrable function" f is in principle different from the Fresnel class $\mathcal{F}(\mathcal{H})$ considered by Albeverio and Høegh-Krohn, as the definition of the infinite dimensional Fresnel integrals and of the infinite dimensional oscillatory integrals are different. However it is possible to prove that any function $f \in \mathcal{F}(\mathcal{H})$ is integrable and its infinite dimensional oscillatory integral, i.e. the limit of a sequence of finite dimensional approximations, exists and is given by Eq. (1.30), that in this new setting has to be interpreted as a theorem instead of a definition.

It is worthwhile to point out that the definition of infinite dimensional oscillatory integrals is rather flexible and allows not only to enlarge the class of integrable functions [27] to sets larger than $\mathcal{F}(\mathcal{H})$, but also to study several applications to quantum mechanics [26].

In this book I shall extensively describe both the theory and the applications, including the most recent ones.

- Chapter 2 contains the theory of finite and infinite dimensional oscillatory integrals.

- Chapter 3 describes the application of infinite dimensional oscillatory integrals to the rigorous mathematical realization of the Feynman path integral representation of the solution of the Schrödinger equation.
- Chapter 4 is devoted to the stationary phase method and its application to the semiclassical limit of quantum mechanics.
- Chapter 5 shows that it is possible to generalize the definition of infinite dimensional oscillatory integrals in order to deal with complex-valued phase functions. Such a functional is applied to the solution of a stochastic Schrödinger equation appearing in the theory of continuous quantum measurement: the Schrödinger-Belavkin equation. A mathematical definition and construction of the Feynman-Vernon influence functional is also given.
- The last chapter is devoted to a brief description of some alternative approaches to the mathematical definition of Feynman path integrals and to their applications.

Chapter 2

Infinite Dimensional Oscillatory Integrals

2.1 Finite dimensional oscillatory integrals

The study of oscillatory integrals of the form

$$\int_{\mathbb{R}^n} e^{\frac{i}{\hbar}\Phi(x)} f(x)dx, \tag{2.1}$$

(where $\Phi : \mathbb{R}^n \to \mathbb{R}$ is a smooth phase function, $f : \mathbb{R}^n \to \mathbb{C}$ is the function which is integrated and $\hbar \in \mathbb{R} \setminus \{0\}$) is a classical topic of investigation, largely developed in connection with several applications in mathematics, such as the theory of Fourier integral operators [164, 165], and in physics (such as in optics).

Well known examples of integrals of the above form are the Fresnel integrals

$$\int e^{ix^2} dx, \tag{2.2}$$

applied in the theory of wave diffraction [268] and the Airy integrals

$$\int e^{ix^3} dx, \tag{2.3}$$

applied in the theory of rainbow.

In the mathematical literature, a particular interest has been devoted to the study of the asymptotic behavior of integrals (2.1) when $\hbar$ is regarded as a small parameter converging to 0. Originally introduced by Stokes and Kelvin and successively developed by several mathematicians, in particular van der Corput, the "stationary phase method" provides a powerful tool to handle the asymptotics of (2.1) as $\hbar \downarrow 0$. According to it, the main contribution to the asymptotic behavior of the integral should come from those points $x_c \in \mathbb{R}^n$ which belong to the critical manifold:

$$\mathcal{C}(\Phi) := \{x \in \mathbb{R}^n, \mid \Phi'(x) = 0\}, \tag{2.4}$$

that is the points which make stationary the phase function Φ.

Beautiful mathematical work on oscillatory integrals and the method of stationary phase is connected with the mathematical classification of singularities of algebraic and geometric structures (Coxeter indices, catastrophe theory), see, e.g. [41, 44, 110]. We shall give more details on this topic in chapter 4.

If the function f in Eq. (2.1) is summable, the integral (2.1) is well defined as a convergent Lebesgue integral, but in several interesting cases, as the examples (2.2) and (2.3) show, this condition is not satisfied. Indeed for several applications it is convenient to introduce a definition of Eq. (2.1) which allows to handle a general class of phases Φ and functions f. As in the case of the convergence of some improper Riemann integrals, the sign of the function which is integrated plays an important role. In particular, in the convergence of oscillatory integrals, the cancellations due to the oscillatory behavior of the integrand $e^{\frac{i}{\hbar}\Phi(x)}$ are fundamental and have to be taken into account in proposing a definition of Eq. (2.1) which allows to handle not only integrals of the form (2.2) and (2.3), but also expression as

$$\int e^{ix^2} x^m dx, \qquad m \in \mathbb{N}.$$

It is worthwhile to recall that this particular feature makes oscillatory integrals the suitable mathematical tool describing the physical concept of coherent superposition, that is of interference.

We give here a definition which is a generalization of the one proposed in [114], which is due to Hörmander [164, 165], and propose a related, more general definition of *oscillatory integral in the Σ-sense.*

Definition 2.1. Let Φ be a continuous real-valued function on $\mathbb{R}^n$. The oscillatory integral on $\mathbb{R}^n$, with $\hbar \in \mathbb{R} \setminus \{0\}$,

$$\int_{\mathbb{R}^n} e^{\frac{i}{\hbar}\Phi(x)} f(x) dx,$$

is well defined if for each test function $\phi \in \mathcal{S}(\mathbb{R}^n)$, such that $\phi(0) = 1$, the limit of the sequence of absolutely convergent integrals

$$\lim_{\epsilon \downarrow 0} \int_{\mathbb{R}^n} e^{\frac{i}{\hbar}\Phi(x)} \phi(\epsilon x) f(x) dx,$$

exists and is independent of ϕ. In this case the limit is denoted by

$$\int_{\mathbb{R}^n}^{o} e^{\frac{i}{\hbar}\Phi(x)} f(x) dx.$$

If the same holds only for ϕ such that $\phi(0) = 1$ and $\phi \in \Sigma$, for some subset Σ of $\mathcal{S}(\mathbb{R}^N)$, we say that the oscillatory integral exists in the Σ-sense and we shall denote it by the same symbol.

Hörmander in his work on Fourier integral operators [164, 165] gives a detailed treatment of oscillatory integrals and describes a large class of integrable functions, called *symbols*.

Definition 2.2. A C^∞ map $f : \mathbb{R}^n \to \mathbb{C}$ belongs to the space of symbols $S^N_\lambda(\mathbb{R}^n)$, where N, λ are two real numbers and $0 < \lambda \leq 1$, if for each $\alpha = (\alpha_1, \dots, \alpha_n) \in \mathbb{N}^n_0$ there exists a constant $C_\alpha \in \mathbb{R}$ such that

$$\left| \frac{d^{\alpha_1}}{dx_1^{\alpha_1}} \cdots \frac{d^{\alpha_n}}{dx_n^{\alpha_n}} f \right| \leq C_\alpha (1 + |x|)^{N - \lambda|\alpha|}, \qquad |x| \to \infty, \tag{2.5}$$

where $|\alpha| = \alpha_1 + \alpha_2 + \cdots + \alpha_n$.

One can prove that S^N_λ is a Fréchet space under the topology defined by taking as seminorms $|f|_\alpha$ the best constants C_α in (2.5) (see [164]). The space increases as N increases and λ decreases. If $f \in S^N_\lambda$ and $g \in S^M_\lambda$, then $fg \in S^{N+M}_\lambda$. We denote

$$\bigcup_N S^N_\lambda \equiv S^\infty_\lambda.$$

It is quite simple to verify that S^N_λ, with $\lambda = 1$, includes for instance the homogeneous polynomials of degree N. The following theorem shows that, under rather general assumptions on the phase functions Φ, if f belong to S^∞_λ for some $\lambda \in (0, 1]$, the oscillatory integral $\int^o e^{\frac{i}{\hbar}\Phi(x)} f(x)dx$ is well defined.

Theorem 2.1. *Let Φ be a real-valued C^2 function on $\mathbb{R}^n$ with the critical set $C(\Phi)$ (defined by (2.4)) being finite. Let us assume that for each $N \in \mathbb{N}$ there exists a $k \in \mathbb{N}$ such that $\frac{|x|^{N+1}}{|\nabla\Phi(x)|^k}$ is bounded for $|x| \to \infty$. Let $f \in S^M_\lambda$, with $M, \lambda \in \mathbb{R}$, $0 < \lambda \leq 1$. Then the oscillatory integral $\int^o e^{\frac{i}{\hbar}\Phi(x)} f(x)dx$ exists for each $\hbar \in \mathbb{R} \setminus \{0\}$.*

Proof. We follow the method of Hörmander [164], see also [114, 7].

Let us assume that the phase function $\Phi(x)$ has l stationary points $c_1, \dots, c_l$, that is:

$$\nabla\Phi(c_i) = 0, \qquad i = 1, \dots, l.$$

Let us choose a suitable partition of unity $1 = \sum_{i=0}^l \chi_i$, where χ_i, $i = 1, \dots, l$, are $C^\infty_0(\mathbb{R}^n)$ functions constant equal to 1 in an open ball centered in the stationary point c_i, respectively, and $\chi_0 = 1 - \sum_{i=1}^{l} \chi_i$. Each of the integrals

$$I_i(f) \equiv \int_{\mathbb{R}^n} e^{\frac{i}{\hbar}\Phi(x)} \chi_i(x) f(x) dx, \qquad i = 1, \ldots, l,$$

is well defined in Lebesgue sense since $f\chi_i \in C_0(\mathbb{R}^n)$.

Let

$$I_0 \equiv \int_{\mathbb{R}^n} e^{\frac{i}{\hbar}\Phi(x)} \chi_0(x) f(x) dx.$$

To see that I_0 is a well defined oscillatory integral, let us introduce the operator L^+ given by

$$L^+ g(x) = -i\hbar \frac{\chi_0(x)}{|\nabla\Phi(x)|^2} \nabla\Phi(x) \nabla g(x),$$

while its adjoint in $L^2(\mathbb{R}^n)$ is given by

$$Lf(x) = i\hbar \frac{\chi_0(x)}{|\nabla\Phi(x)|^2} \nabla\Phi(x)\nabla f(x) + i\hbar \operatorname{div}\left(\frac{\chi_0(x)}{|\nabla\Phi(x)|^2}\nabla\Phi(x)\right) f(x).$$

Let us choose $\phi \in \mathcal{S}(\mathbb{R}^n)$, such that $\phi(0) = 1$. It is easy to see that if $f \in S_\lambda^M$ then f_ϵ, defined as $f_\epsilon(x) := \phi(\epsilon x) f(x)$, belongs to $S_\lambda^{M+1} \cap \mathcal{S}(\mathbb{R}^n)$, for any $\epsilon > 0$. By iterated application of the Stokes formula, we have:

$$\int_{\mathbb{R}^n} e^{\frac{i}{\hbar}\Phi(x)} \phi(\epsilon x) f(x) \chi_0(x) dx = \int_{\mathbb{R}^n} L^+(e^{\frac{i}{\hbar}\Phi(x)}) \phi(\epsilon x) f(x) dx$$

$$= \int_{\mathbb{R}^n} e^{\frac{i}{\hbar}\Phi(x)} L f_\epsilon(x) dx. \quad (2.6)$$

By iterating the procedure k times, for k sufficiently large, one obtains an absolutely convergent integral and it is possible to pass to the limit $\epsilon \to 0$ by the Lebesgue dominated convergence theorem.

Considering $\sum_{i=0}^{l} I_i(f)$ we have, by the existence result proved for I_0 and the additivity property of oscillatory integrals, that $\int_{\mathbb{R}^n}^{o} e^{\frac{i}{\hbar}\Phi(x)} f(x) dx$ is well defined and equal to $\sum_{i=0}^{l} I_i(f)$. □

Remark 2.1. If $\mathcal{C}(\Phi)$ has countably many non accumulating points $\{x_c^i\}_{i\in\mathbb{N}}$, the same method yields $\int_{\mathbb{R}^n}^{o} e^{\frac{i}{\hbar}\Phi(x)} f(x) dx = \sum_{i=0}^{\infty} I_i(f)$ provided this sum converges.

There are partial extensions of the above construction in the case of critical points which form a submanifold in $\mathbb{R}^n$ [110], or are degenerate [41], see also [97].

In particular we have proved the existence for $f \in S_\lambda^\infty$, $0 < \lambda \leq 1$, of the oscillatory integrals $\int e^{ix^M} f(x) dx$, with M arbitrary. For $M = 2$ one has the Fresnel integral of [16, 17], for $M = 3$ one has integrals called, for $n = 1$, Airy integrals [165].

2.2 The Parseval type equality

This section is devoted to the study of the *Fresnel integrals*, a particular class of oscillatory integrals with quadratic phase function:

$$\int_{\mathbb{R}^n}^{o} e^{\frac{i}{\hbar}\|x\|^2} f(x)dx. \tag{2.7}$$

We shall see that in this case it is possible to identify a subclass of integrable functions f, such that the corresponding Fresnel integral can be explicitly computed in terms of a Parseval type equality. This property is particular important because, as we shall see below, it allows the generalization of the definition to an infinite dimensional setting.

In the following we shall denote by $\mathcal{H}$ a (finite or infinite dimensional) real separable Hilbert space. The norm will be denoted by $\| \ \|$ and the inner product with $\langle \ , \ \rangle$. If the dimension of $\mathcal{H}$ is finite, $\dim(\mathcal{H}) = n$, we shall identify it with $\mathbb{R}^n$.

Let us consider the space $\mathcal{M}(\mathcal{H})$ of complex bounded variation measures on $\mathcal{H}$ endowed with the total variation norm [254]:

$$\|\mu\| = \sup \sum_i |\mu(A_i)|,$$

where the supremum is taken over all sequences $\{A_i\}$ of pairwise disjoint Borel subsets $\mathcal{B}(\mathcal{H})$ of $\mathcal{H}$, such that $\cup_i A_i = \mathcal{H}$. $\mathcal{M}(\mathcal{H})$ is a Banach algebra [254], where the product of two measures $\mu * \nu$ is by definition their convolution:

$$\mu * \nu(A) = \int_{\mathcal{H}} \mu(A - x)d\nu(x), \qquad \mu, \nu \in \mathcal{M}(\mathcal{H}), A \in \mathcal{B}(\mathcal{H})$$

and the unit element is the Dirac measure δ_0.

Let us denote with $\mathcal{F}(\mathcal{H})$ the space of functions $f : \mathcal{H} \to \mathbb{C}$ which are the Fourier transforms of complex bounded variation measures $\mu_f \in \mathcal{M}(\mathcal{H})$:

$$f(x) = \hat{\mu}_f(x) := \int_{\mathcal{H}} e^{i\langle y,x\rangle} d\mu_f(y), \qquad \mu_f \in \mathcal{M}(\mathcal{H}), x \in \mathcal{H}.$$

$\mathcal{F}(\mathcal{H})$ is a Banach algebra of functions, the *Fresnel algebra*, where the norm is the total variation of the corresponding measure:

$$\|f\| := \|\mu_f\|, \qquad f = \hat{\mu}_f,$$

the multiplication is the pointwise one

$$fg(x) = f(x)g(x),$$

and the unity is the constant function

$$f(x) = 1 \qquad \forall x \in \mathcal{H}.$$

A complete characterizations of the functions belonging to $\mathcal{F}(\mathcal{H})$ is not easy. The elements in $\mathcal{F}(\mathcal{H})$ are bounded uniformly continuous functions on $\mathcal{H}$ [254], and the following inequality holds:

$$|f|_\infty \leq \|f\|, \qquad f \in \mathcal{F}(\mathcal{H}),$$

where $|f|_\infty$ denotes the sup-norm. Moreover, since each $\mu \in \mathcal{M}(\mathcal{H})$ is a finite linear complex combination of positive measures, every $f \in \mathcal{F}(\mathcal{H})$ is a corresponding linear combination of functions of positives type[1] [211]. In the finite dimensional case $\mathcal{H} = \mathbb{R}^n$, one has the inclusion $\mathcal{S}(\mathbb{R}^n) \subset \mathcal{F}(\mathbb{R}^n)$.

For the applications that will follow, it is convenient to introduce a particular definition of finite dimensional Fresnel integrals, which differs from the general definition 2.1 for the presence of a normalization constant.

Definition 2.3. A function $f : \mathbb{R}^n \to \mathbb{C}$ is called Fresnel integrable if for each $\phi \in \mathcal{S}(\mathbb{R}^n)$ such that $\phi(0) = 1$ the limit

$$\lim_{\epsilon \to 0} (2\pi i\hbar)^{-n/2} \int e^{\frac{i}{2\hbar}\langle x,x\rangle} f(x)\phi(\epsilon x)dx \tag{2.8}$$

exists and is independent of ϕ. In this case the limit is called the Fresnel integral of f and denoted by

$$\widetilde{\int} e^{\frac{i}{2\hbar}\langle x,x\rangle} f(x)dx. \tag{2.9}$$

The factor $(2\pi i\hbar)^{-n/2}$ in the definition of the Fresnel integral plays the role of a normalization constant. Indeed if $f = 1$, it is easy to verify that

$$\widetilde{\int} e^{\frac{i}{2\hbar}\langle x,x\rangle} f(x)dx = 1.$$

The description of the full class of Fresnel integrable functions is still an open problem, but the following theorem shows that it includes $\mathcal{F}(\mathbb{R}^n)$.

[1] A function $f : \mathcal{H} \to \mathbb{C}$ is called *positive definite* if for any $x_1, \ldots x_n \in \mathcal{H}$ and for any $c_1, \ldots c_n \in \mathbb{C}$, the following inequality holds

$$\sum_{j,k=1}^{n} c_j f(x_j - x_k)\bar{c}_k \geq 0.$$

Theorem 2.2. *Let $L : \mathbb{R}^n \to \mathbb{R}^n$ be a self-adjoint linear operator such that $I - L$ is invertible (I being the identity operator on $\mathbb{R}^n$) and let $f \in \mathcal{F}(\mathbb{R}^n)$. Then the Fresnel integral of the function*

$$x \mapsto e^{-\frac{i}{2\hbar}\langle x, Lx\rangle} f(x), \qquad x \in \mathbb{R}^n$$

is well defined and is given by the following Parseval-type equality:

$$\widetilde{\int} e^{\frac{i}{2\hbar}\langle x,(I-L)x\rangle} f(x)dx$$
$$= e^{-\frac{\pi i}{2}\mathrm{Ind}(I-L)}|\det(I-L)|^{-1/2}\int_{\mathbb{R}^n} e^{-\frac{i\hbar}{2}\langle x,(I-L)^{-1}x\rangle}d\mu_f(x), \quad (2.10)$$

where $\mathrm{Ind}(I-L)$ *is the index of the operator* $I - L$, *that is the number of negative eigenvalues.*

Proof. We present here the proof first presented in [114].

Let $\phi \in \mathcal{S}(\mathbb{R}^n)$ and $\epsilon > 0$ and let us consider the integral

$$(2\pi i\hbar)^{-n/2}\int_{\mathbb{R}^n} e^{\frac{i}{2\hbar}\langle x,(I-L)x\rangle} f(x)\phi(\epsilon x)dx\,. \quad (2.11)$$

We claim that for any $f \in \mathcal{F}(\mathbb{R}^n)$ and $g \in \mathcal{S}(\mathbb{R}^n)$, the following holds:

$$(2\pi i\hbar)^{-n/2}\int_{\mathbb{R}^n} e^{\frac{i}{2\hbar}\langle x,(I-L)x\rangle} f(x)g(x)dx$$
$$= e^{-\frac{\pi i}{2}\mathrm{Ind}(I-L)}|\det(I-L)|^{-1/2}\int_{\mathbb{R}^n\times\mathbb{R}^n} e^{-\frac{i\hbar}{2}\langle x,(I-L)^{-1}x\rangle}\tilde{g}(x-y)d\mu_f(y)dx. \quad (2.12)$$

By Eq. (2.12) the integral (2.11) is equal to

$$(2\pi i\hbar)^{-n/2}\int_{\mathbb{R}^n} e^{\frac{i}{2\hbar}\langle x,(I-L)x\rangle} f(x)\phi(\epsilon x)dx$$
$$= e^{-\frac{\pi i}{2}\mathrm{Ind}(I-L)}|\det(I-L)|^{-1/2}\int_{\mathbb{R}^n\times\mathbb{R}^n} e^{-\frac{i\hbar}{2}\langle x,(I-L)^{-1}x\rangle}\epsilon^{-n}\tilde{\phi}(\frac{x-y}{\epsilon})d\mu_f(y)dx$$
$$= e^{-\frac{\pi i}{2}\mathrm{Ind}(I-L)}|\det(I-L)|^{-1/2}\int_{\mathbb{R}^n\times\mathbb{R}^n} e^{-\frac{i\hbar}{2}\langle y+\epsilon x,(I-L)^{-1}(y+\epsilon x)\rangle}\tilde{\phi}(x)d\mu_f(y)dx\,.$$

By letting $\epsilon \downarrow 0$, taking into account that $\int\tilde{\phi}(x)dx = 1$, we get Eq. (2.10).

Let us now prove Eq. (2.12). First of all, let us assume that $f = 1$.

(1) If $I - L$ is positive definite, by a change of variable Eq. (2.12) is equivalent to

$$(2\pi i\hbar)^{-n/2}\int_{\mathbb{R}^n} e^{\frac{i}{2\hbar}\langle x,x\rangle}g(x)dx = \int_{\mathbb{R}^n} e^{-\frac{i\hbar}{2}\langle x,x\rangle}\tilde{g}(x)dx.$$

Indeed, as $(2\pi i\hbar)^{-n/2}e^{\frac{i}{2\hbar}\langle x,x\rangle}$ is a bounded continuous function, it has a Fourier transform in the sense of tempered distributions

$$(2\pi i\hbar)^{-n/2}\int_{\mathbb{R}^n} e^{\frac{i}{2\hbar}\langle x,x\rangle}e^{i\langle x,y\rangle}dx = e^{-\frac{i\hbar}{2}\langle y,y\rangle}.$$

(2) If $I - L$ is negative definite, $\mathrm{Ind}(I - L) = n$, the result follows by replacing $\hbar$ with $-\hbar$ and from the fact that

$$(-2\pi i\hbar)^{-n/2} = (2\pi i\hbar)^{-n/2}e^{-\frac{\pi i n}{2}}.$$

(3) For general $I - L$, it is possible to decompose $\mathbb{R}^n$ as the product of the positive and negative eigenspaces, i.e.

$$\mathbb{R}^n = E^+ \times E^-,$$

and analogously $I - L$ is the product of its positive and negative parts

$$I - L = (I - L)^+ \times (I - L)^-,$$

where $(I - L)^+ : E^+ \to E^+$ and $(I - L)^- : E^+ \to E^-$, $\dim(E^-) = \mathrm{Ind}(I - L)$. The result follows from the points (1) and (2) in the case of a function $g : E^+ \times E^- \to \mathbb{C}$ of the form

$$g(x_+, x_-) = g_+(x_+)g_-(x_-), \qquad g_+ : E^+ \to \mathbb{C}, \quad g_- : E^- \to \mathbb{C}. \tag{2.13}$$

The same can be deduced for g that are finite linear combinations of factorisable functions of the form (2.13). For general g it is sufficient to note that $\mathcal{S}(E^+) \times \mathcal{S}(E^-)$ is dense in $\mathcal{S}(\mathbb{R}^n)$ and the result follows by a continuity argument.

Let us now prove Eq. (2.12) for $f \in \mathcal{F}(\mathbb{R}^n)$ and $g \in \mathcal{S}(\mathbb{R}^n)$. Let us denote with I_{rs}, with $r + s = n$, the linear operator on $\mathbb{R}^n$ having r eigenvalues equal to -1 and s eigenvalues equal to $+1$. By substituting in the left hand side of Eq. (2.12) $f(x) = \int e^{i\langle x,y\rangle}d\mu_f(y)$ and by applying Fubini theorem, we get:

$$\begin{aligned}(2\pi i\hbar)^{-n/2}&\int_{\mathbb{R}^n} e^{\frac{i}{2\hbar}\langle x,I_{rs}x\rangle}f(x)g(x)dx \\ &= (2\pi i\hbar)^{-n/2}\int_{\mathbb{R}^n} d\mu_f(y)\int_{\mathbb{R}^n} e^{\frac{i}{2\hbar}\langle x,I_{rs}x\rangle}e^{i\langle x,y\rangle}g(x)dx.\end{aligned}$$

By a change of variable the latter is equal to

$$(2\pi i\hbar)^{-n/2}\int_{\mathbb{R}^n} d\mu_f(y)e^{-\frac{i\hbar}{2}\langle y,I_{rs}y\rangle}\int_{\mathbb{R}^n} e^{\frac{i}{2\hbar}\langle x,I_{rs}x\rangle}g(x-I_{rs}\hbar y)dx$$
$$= e^{-\frac{i\pi r}{2}}\int_{\mathbb{R}^n} d\mu_f(y)\int_{\mathbb{R}^n}\tilde{g}(x-y)e^{-\frac{i\hbar}{2}\langle x,I_{rs}x\rangle}dx.$$

By a change of scale the latter equality is equivalent to

$$(2\pi i\hbar)^{-n/2}\int_{\mathbb{R}^n} e^{\frac{i}{2\hbar}\langle x,(I-L)x\rangle}f(x)g(x)dx$$
$$= e^{-\frac{\pi i}{2}\mathrm{Ind}(I-L)}|\det(I-L)|^{-1/2}\int_{\mathbb{R}^n\times\mathbb{R}^n} e^{-\frac{i\hbar}{2}\langle x,(I-L)^{-1}x\rangle}\tilde{g}(x-y)d\mu_f(y)dx.$$

□

2.3 Generalized Fresnel integrals

In the present section, following [30], we shall generalize the result of the previous section to more general phase functions Φ, in particular those given by an even polynomial $P(x)$ in the variables $x_1,\dots,x_n$:

$$P(x) = A_{2M}(x,\dots,x) + A_{2M-1}(x,\dots,x) + \dots + A_1(x) + A_0, \qquad (2.14)$$

where A_k are k_{th}-order covariant tensors on $\mathbb{R}^n$:

$$A_k : \underbrace{\mathbb{R}^n\times\mathbb{R}^n\times\dots\times\mathbb{R}^n}_{k-times}\to\mathbb{R}$$

and the leading term, namely $A_{2M}(x,\dots,x)$, is a $2M_{th}$-order completely symmetric positive covariant tensor on $\mathbb{R}^n$.

By theorem 2.1 the *generalized Fresnel integral*

$$\int^{o}_{\mathbb{R}^n} e^{\frac{i}{\hbar}P(x)}f(x)dx \qquad (2.15)$$

is well defined for P of the form (2.14) and f belonging to the class of symbols (see definition 2.2).

In the case where $\hbar\in\mathbb{C}$ with $Im(\hbar)<0$ and if Φ is of the form (2.14), then the generalized Fresnel integral (2.15) also exists, even in Lebesgue sense, as an analytic function in $\hbar$, as easily seen by the fact that the integrand is bounded by $|f|\exp(\frac{Im(\hbar)}{|\hbar|^2}\Phi)$.

The key tool for the generalization of Parseval-type equality (2.10) is an estimate of the Fourier transform of the function $x\mapsto e^{\frac{i}{\hbar}P(x)}$, $x\in\mathbb{R}^n$.

Lemma 2.1. *Let $P : \mathbb{R}^n \to \mathbb{R}$ be given by (2.14) and let $\hbar \in \mathbb{C}$, with $Im(\hbar) \leq 0$. Then the Fourier transform of the distribution $e^{\frac{i}{\hbar}P(x)}$:*

$$\tilde{F}(k) = \int_{\mathbb{R}^n} e^{ik\cdot x} e^{\frac{i}{\hbar}P(x)} dx, \qquad \hbar \in \mathbb{R}\setminus\{0\} \tag{2.16}$$

is an entire bounded function and admits in the case where $\hbar \in \mathbb{R}$ the following representation:

$$\tilde{F}(k) = e^{in\pi/4M} \int_{\mathbb{R}^n} e^{ie^{i\pi/4M}k\cdot x} e^{\frac{i}{\hbar}P(e^{i\pi/4M}x)} dx, \qquad \hbar > 0, \tag{2.17}$$

or

$$\tilde{F}(k) = e^{-in\pi/4M} \int_{\mathbb{R}^n} e^{ie^{-i\pi/4M}k\cdot x} e^{\frac{i}{\hbar}P(e^{-i\pi/4M}x)} dx, \qquad \hbar < 0. \tag{2.18}$$

Remark 2.2. The integral on the right hand side of (2.17) is absolutely convergent as

$$e^{\frac{i}{\hbar}P(e^{i\pi/4M}x)} = e^{-\frac{1}{\hbar}A_{2M}(x,\ldots,x)} e^{\frac{i}{\hbar}(A_{2M-1}(e^{i\pi/4M}x,\ldots,e^{i\pi/4M}x)+\ldots+A_1(xe^{i\pi/4M})+A_0)}.$$

A similar calculation shows the absolute convergence of the integral on the right hand side of (2.18).

Proof. [Proof of lemma 2.1] The proof is divided into three steps.

(1) Let us prove first of all formulas (2.17) and (2.18) by using the analyticity of $e^{ikz+\frac{i}{\hbar}P(z)}$, $z \in \mathbb{C}$, and a change of integration contour.
Let us denote D the region of the complex plane:

$$D \subset \mathbb{C}, \qquad D \equiv \{z \in \mathbb{C} \mid Im(z) < 0\}$$

Let us assume $\hbar$ is a complex variable belonging to the region $\bar{D}\setminus\{0\}$.
Let us introduce the polar coordinates in $\mathbb{R}^n$:

$$\int_{\mathbb{R}^n} e^{ik\cdot x} e^{\frac{i}{\hbar}P(x)} dx$$
$$= \int_{S_{n-1}} \left(\int_0^{+\infty} e^{i|k|rf_k(\phi_1,\ldots,\phi_{n-1})} e^{\frac{i}{\hbar}P_{(\phi_1,\ldots,\phi_{n-1})}(r)} r^{n-1} dr \right) d\Omega_{n-1} \tag{2.19}$$

where instead of n Cartesian coordinates we use $n-1$ angular coordinates $(\phi_1, \ldots, \phi_{n-1})$ and the variable $r = |x|$. S_{n-1} denotes the $(n-1)$-dimensional spherical surface, $d\Omega_{n-1}$ is the measure on it,

$\mathcal{P}_{(\phi_1,\dots,\phi_{n-1})}(r)$ is a $2M_{th}$ order polynomial in the variable r with coefficients depending on the $n-1$ angular variables $(\phi_1,\dots,\phi_{n-1})$, namely:

$$\begin{aligned} P(x) &= r^{2M}A_{2M}\Big(\frac{x}{|x|},\dots,\frac{x}{|x|}\Big) + r^{2M-1}A_{2M-1}\Big(\frac{x}{|x|},\dots,\frac{x}{|x|}\Big) + \dots \\ &\qquad + \dots + rA_1\Big(\frac{x}{|x|}\Big) + A_0 \\ &= a_{2M}(\phi_1,\dots,\phi_{n-1})r^{2M} + a_{2M-1}(\phi_1,\dots,\phi_{n-1})r^{2M-1} \\ &\qquad + \dots + a_1(\phi_1,\dots,\phi_{n-1})r + a_0 \\ &\qquad\qquad = \mathcal{P}_{(\phi_1,\dots,\phi_{n-1})}(r) \end{aligned}$$

where $a_{2M}(\phi_1,\dots,\phi_{n-1}) > 0$ for all $(\phi_1,\dots,\phi_{n-1}) \in S_{n-1}$. The function $f_k : S_{n-1} \to [-1,1]$ is defined by

$$\frac{k}{|k|}\frac{x}{|x|} = f_k(\phi_1,\dots,\phi_{n-1}).$$

Let us focus on the integral

$$\int_0^{+\infty} e^{i|k|rf_k(\phi_1,\dots,\phi_{n-1})} e^{\frac{i}{\hbar}\mathcal{P}_{(\phi_1,\dots,\phi_{n-1})}(r)} r^{n-1}dr, \tag{2.20}$$

which can be interpreted as the Fourier transform of the distribution on the real line

$$F(r) = \Theta(r)r^{n-1}e^{\frac{i}{\hbar}\mathcal{P}_{(\phi_1,\dots,\phi_{n-1})}(r)},$$

with $\Theta(r) = 1$ for $r \geq 0$ and $\Theta(r) = 0$ for $r < 0$. Let us introduce the notation $k' \equiv |k|f_k(\phi_1,\dots,\phi_{n-1})$, $a_j \equiv a_j(\phi_1,\dots,\phi_{n-1})$, $j = 0,\dots,2M$, $P'(r) = \sum_{j=0}^{2M} a_j r^j$ and $\hbar \in \mathbb{C}$, $\hbar = |\hbar|e^{i\phi}$, with $-\pi \leq \phi \leq 0$.

Let us consider the complex plane and set $z = \rho e^{i\theta}$. If $Im(\hbar) < 0$ the integral (2.20) is absolutely convergent, while if $\hbar \in \mathbb{R}\setminus\{0\}$ it needs a regularization. If $\hbar \in \mathbb{R}$, $\hbar > 0$ we have

$$\int_0^{+\infty} e^{ik'r}e^{\frac{i}{\hbar}P'(r)}r^{n-1}dr = \lim_{\epsilon\downarrow 0}\int_{z=\rho e^{i\epsilon}} e^{ik'z}e^{\frac{i}{\hbar}P'(z)}z^{n-1}dz$$

while if $\hbar < 0$

$$\int_0^{+\infty} e^{ik'r}e^{\frac{i}{\hbar}P'(r)}r^{n-1}dr = \lim_{\epsilon\downarrow 0}\int_{z=\rho e^{-i\epsilon}} e^{ik'z}e^{\frac{i}{\hbar}P'(z)}z^{n-1}dz. \tag{2.21}$$

We deal first of all with the case $\hbar \in \mathbb{R}$, $\hbar > 0$ (the case $\hbar < 0$ can be handled in a completely similar way). Let

$$\gamma_1(R) = \{z = \rho e^{i\theta} \in \mathbb{C} \;\mid\; 0 \leq \rho \leq R, \quad \theta = \epsilon\},$$

$$\gamma_2(R) = \{z = \rho e^{i\theta} \in \mathbb{C} \mid \rho = R, \quad \epsilon \le \theta \le \pi/4M\},$$

$$\gamma_3(R) = \{z = \rho e^{i\theta} \in \mathbb{C} \mid 0 \le \rho \le R, \quad \theta = \pi/4M\}.$$

From the analyticity of the integrand and the Cauchy theorem we have

$$\int_{\gamma_1(R)\cup\gamma_2(R)\cup\gamma_3(R)} e^{ik'z} e^{\frac{i}{\hbar}P'(z)} z^{n-1} dz = 0.$$

In particular:

$$\begin{aligned}\Big|\int_{\gamma_2(R)} e^{ik'z}e^{\frac{i}{\hbar}P'(z)}z^{n-1}dz\Big| &= R^n\Big|\int_\epsilon^{\pi/4M} e^{ik'Re^{i\theta}}e^{\frac{i}{\hbar}P'(Re^{i\theta})}e^{in\theta}d\theta\Big| \\ &\le R^n\int_\epsilon^{\pi/4M} e^{-k'R\sin(\theta)}e^{-\frac{1}{\hbar}\sum_{j=1}^{2M}a_jR^j\sin(j\theta)}d\theta \\ &\le R^n\int_\epsilon^{\pi/4M} e^{-k''R\theta}e^{-a_{2M}\frac{4M}{\hbar\pi}R^{2M}\theta}e^{-\sum_{j=1}^{2M-1}a'_jR^j\theta}d\theta,\end{aligned}$$

where k'', a'_k for $k = 1, \dots, 2M-1$ are suitable constants. We have used the fact that if $\alpha \in [0, \pi/2]$ then $\frac{2}{\pi}\alpha \le \sin(\alpha) \le \alpha$. The latter integral can be explicitly computed and gives:

$$R^n\Big(\frac{e^{-\epsilon(a_{2M}\frac{4M}{\hbar\pi}R^{2M}+k''R+\sum_{j=1}^{2M-1}a'_jR^j)} - e^{-\frac{\pi}{4M}(a_{2M}\frac{4M}{\hbar\pi}R^{2M}+k''R+\sum_{j=1}^{2M-1}a'_jR^j)}}{a_{2M}\frac{4M}{\hbar\pi}R^{2M}+k''R+\sum_{j=1}^{2M-1}a'_jR^j}\Big),$$

which converges to 0 as $R \to \infty$. We get

$$\int_{z=\rho e^{i\epsilon}} e^{ik'z}e^{\frac{i}{\hbar}P'(z)}z^{n-1}dz = \int_{z=\rho e^{i(\pi/4M)}} e^{ik'z}e^{\frac{i}{\hbar}P'(z)}z^{n-1}dz.$$

By taking the limit as $\epsilon \downarrow 0$ of both sides one gets:

$$\int_0^{+\infty} e^{ik'r}e^{\frac{i}{\hbar}P'(r)}r^{n-1}dr = e^{in\pi/4M}\int_0^{+\infty} e^{ik\rho e^{i\pi/4M}}e^{\frac{i}{\hbar}P'(re^{i\pi/4M})}\rho^{n-1}d\rho.$$

By substituting into (2.19) we get the final result:

$$\tilde{F}(k) = \int_{\mathbb{R}^n} e^{ik\cdot x}e^{\frac{i}{\hbar}P(x)}dx = e^{in\pi/4M}\int_{\mathbb{R}^n} e^{ie^{i\pi/4M}k\cdot x}e^{\frac{i}{\hbar}P(e^{i\pi/4M}x)}dx. \tag{2.22}$$

In the case $\hbar < 0$ an analogous reasoning gives:

$$\tilde{F}(k) = \int_{\mathbb{R}^n} e^{ik\cdot x}e^{\frac{i}{\hbar}P(x)}dx = e^{-in\pi/4M}\int_{\mathbb{R}^n} e^{ie^{-i\pi/4M}k\cdot x}e^{\frac{i}{\hbar}P(e^{-i\pi/4M}x)}dx. \tag{2.23}$$

(2) The analyticity of $\tilde{F}(k)$ is trivial in the case $Im(\hbar) < 0$, and follows from equations (2.22) and (2.23) when $\hbar \in \mathbb{R} \setminus \{0\}$.

(3) Let us now prove that $\tilde{F}$

$$\tilde{F}(k) = \int_{\mathbb{R}^n} e^{ikx} e^{\frac{i}{\hbar}P(x)} dx$$

is bounded as a function of k by studying its asymptotic behavior as $|k| \to \infty$.

Let us focus on the case $\hbar \in \mathbb{R} \setminus \{0\}$ (in the case $Im(\hbar) < 0$ $|\tilde{F}|$ is trivially bounded by

$$\int_{\mathbb{R}^n} |e^{\frac{i}{\hbar}P(x)}| dx = \int_{\mathbb{R}^n} e^{\frac{Im(\hbar)}{|\hbar|^2}P(x)} dx < +\infty.$$

Let us assume for notational simplicity that $\hbar = 1$, the general case can be handled in a completely similar way. In order to study $\int_{\mathbb{R}^N} e^{ikx} e^{iP(x)} dx$ one has to introduce a suitable regularization. Chosen $\psi \in \mathcal{S}(\mathbb{R}^n)$, such that $\psi(0) = 1$ we have

$$e^{iP(x)}\psi(\epsilon x) \to e^{iP(x)}, \qquad \text{in } \mathcal{S}'(\mathbb{R}^n) \text{ as } \epsilon \to 0,$$

$$\tilde{F}(k) = \lim_{\epsilon \to 0} \int_{\mathbb{R}^n} e^{ikx} e^{iP(x)} \psi(\epsilon x) dx.$$

Let us consider first of all the case $n = 1$ and $P(x) = x^{2M}/2M$. The unique real stationary point of the phase function $\Phi(x) = kx + x^{2M}/2M$ is $c_k = (-k)^{\frac{1}{2M-1}}$. Let χ_1 be a positive C^∞ function such that $\chi_1(x) = 1$ if $|x - c_k| \leq 1/2$, $\chi_1(x) = 0$ if $|x - c_k| \geq 1$ and $0 \leq \chi_1(x) \leq 1$ if $1/2 \leq |x - c_k| \leq 1$. Let $\chi_0 \equiv 1 - \chi_1$. Then $\tilde{F}(k) = I_1(k) + I_0(k)$, where

$$I_0(k) = \lim_{\epsilon \to 0} \int e^{ikx} e^{ix^{2M}/2M} \chi_0(x) \psi(\epsilon x) dx,$$

$$I_1(k) = \int e^{ikx} e^{ix^{2M}/2M} \chi_1(x) dx.$$

For the study of the boundedness of $|\tilde{F}(k)|$ as $|k| \to \infty$ it is enough to look at I_0, since one has, by the choice of χ_1, that $|I_1| \leq 2$. By repeating the same reasoning used in the proof of theorem 2.1, I_0 can be computed by means of Stokes formula:

$$\lim_{\epsilon \to 0} \int e^{ikx} e^{ix^{2M}/2M} \chi_0(x) \psi(\epsilon x) dx = i \lim_{\epsilon \to 0} \epsilon \int e^{ikx} e^{i\frac{x^{2M}}{2M}} \frac{\chi_0(x)\psi'(\epsilon x)}{k + x^{2M-1}} dx$$
$$+ i \lim_{\epsilon \to 0} \int e^{ikx} e^{ix^{2M}/2M} \frac{d}{dx}\Big(\frac{\chi_0(x)}{k + x^{2M-1}}\Big) \psi(\epsilon x) dx.$$

Both integrals are absolutely convergent and, by dominated convergence, we can take the limit $\epsilon \to 0$, so that

$$\begin{aligned} I_0(k) &= i \int e^{ikx} e^{ix^{2M}/2M} \frac{d}{dx}\Big(\frac{\chi_0(x)}{k + x^{2M-1}} \Big) dx \\ &= i \int e^{ikx} e^{ix^{2M}/2M} \Big(\frac{\chi_0'(x)}{k + x^{2M-1}} \Big) dx \\ &\quad - i \int e^{ikx} e^{ix^{2M}/2M} \Big(\frac{(2M-1)\chi_0(x) x^{2M-2}}{(k + x^{2M-1})^2} \Big) dx. \end{aligned}$$

Thus:

$$|I_0(k)| \leq \sup|\chi_0'| \int_{c_k-1}^{c_k-1/2} \Big| \frac{1}{k + x^{2M-1}} \Big| dx + \sup|\chi_0'| \int_{c_k+1/2}^{c_k+1} \Big| \frac{1}{k + x^{2M-1}} \Big| dx$$

$$+ (2M-1) \int_{-\infty}^{c_k-1/2} \Big| \frac{x^{2M-2}}{(k + x^{2M-1})^2} \Big| dx + (2M-1) \int_{c_k+1/2}^{+\infty} \Big| \frac{x^{2M-2}}{(k + x^{2M-1})^2} \Big| dx\,.$$

By a change of variables it is possible to see that both integrals remain bounded as $|k| \to \infty$ (see [30] for more details) By such considerations we can deduce that $|\tilde{F}(k)|$ is bounded as $|k| \to \infty$.

A similar reasoning holds also in the case $n = 1$ and $P(x) = \sum_{i=1}^{2M} a_i x^i$ is a generic polynomial. Indeed for $|k|$ sufficiently large the derivative of the phase function $\Phi'(x) = k + P'(x)$ has only one simple real root, denoted by c_k. One can repeat the same reasoning valid for the case $P(x) = x^{2M}/2M$ and prove that for $|k| \to \infty$ one has $|\int e^{ikx+iP(x)}dx| \leq C$ (where C is a function of the coefficients a_i of P at most with polynomial growth).

The general case $\mathbb{R}^n$ can also be essentially reduced to the one-dimensional case. Indeed let us consider a generic vector $k \in \mathbb{R}^n$, $k = |k|u_1$, and study the behavior of $\tilde{F}(k)$ as $|k| \to \infty$. By choosing as orthonormal base $u_1, \ldots, u_n$ of $\mathbb{R}^n$, where $u_1 = k/|k|$, we have

$$\begin{aligned} \tilde{F}(k) = \lim_{\epsilon \to 0} \int_{\mathbb{R}^{n-1}} & e^{iQ(x_2,\ldots,x_n)} \psi(\epsilon x_2) \cdots \psi(\epsilon x_n) \\ & \Big(\int_{\mathbb{R}} e^{i|k|x_1} e^{iP_{x_2,\ldots,x_n}(x_1)} \psi(\epsilon x_1) dx_1 \Big) dx_2 \ldots dx_n \end{aligned} \tag{2.24}$$

where $\psi \in \mathcal{S}(\mathbb{R})$, $\psi(0) = 1$; $x_i = x \cdot u_i$, $P_{x_2,\ldots,x_n}(x_1)$ is the polynomial in the variable x_1 with coefficients depending on powers of the remaining $n-1$ variables $x_2, \ldots, x_n$, obtained by considering in the initial polynomial $P(x_1, x_2, \ldots, x_n)$ all the terms containing some power of x_1. The polynomial Q in the $n-1$ variables $x_2, \ldots, x_n$ is given by

$P(x_1, x_2, \dots, x_n) - P_{x_2,\dots,x_n}(x_1)$.
Let us set

$$I^\epsilon(k, x_2, \dots, x_n) \equiv \int_{\mathbb{R}} e^{i|k|x_1} e^{iP_{x_2,\dots,x_n}(x_1)} \psi(\epsilon x_1) dx_1.$$

By the previous considerations we know that, for each $\epsilon \geq 0$, $|I^\epsilon(k, x_2, \dots, x_n)|$ is bounded by a function of $G(x_2, \dots, x_n)$ of polynomial growth. By the same reasoning as in the proof of theorem 2.1 we can deduce that the oscillatory integral (2.24) is a well defined bounded function of k.

□

Remark 2.3. A representation similar to (2.17) holds also in the more general case $\hbar \in \mathbb{C}$, $Im(\hbar) < 0$, $\hbar \neq 0$. By setting $\hbar \equiv |\hbar| e^{i\phi}$, $\phi \in [-\pi, 0]$ one has:

$$\tilde{F}(k) = \int_{\mathbb{R}^n} e^{ik\cdot x} e^{\frac{i}{\hbar}P(x)} dx$$
$$= e^{in(\pi/4M+\phi/2M)} \int_{\mathbb{R}^n} e^{ie^{i(\pi/4M+\phi/2M)} k\cdot x} e^{\frac{i}{\hbar}P(e^{i(\pi/4M+\phi/2M)}x)} dx. \quad (2.25)$$

By mimicking the proof of equation (2.17), one can prove in the case $\hbar > 0$ the following result (a similar one holds also in the case $\hbar < 0$):

Theorem 2.3. *Let us denote by Λ_M the subset of the complex plane*

$$\Lambda_M = \{z \in \mathbb{C} \mid 0 < \arg(z) < \pi/4M\} \subset \mathbb{C}, \quad (2.26)$$

and let $\bar{\Lambda}_M$ be its closure. Let $f : \mathbb{R}^n \to \mathbb{C}$ be a Borel function defined for all y of the form $y = \lambda x$, where $\lambda \in \bar{\Lambda}_M$ and $x \in \mathbb{R}^n$, with the following properties:

(1) the function $\lambda \mapsto f(\lambda x)$ is analytic in Λ_M and continuous in $\bar{\Lambda}_M$ for each $x \in \mathbb{R}^n$, $|x| = 1$,
(2) for all $x \in \mathbb{R}^n$ and all $\theta \in (0, \pi/4M)$

$$|f(e^{i\theta}x)| \leq AG(x),$$

where $A \in \mathbb{R}$ and $G : \mathbb{R}^n \to \mathbb{R}$ is a positive function satisfying bound (a) or (b) respectively:

(a) if P is as in the general case defined by (2.14)

$$G(x) \leq e^{B|x|^{2M-1}}, \quad B > 0,$$

(b) if P is homogeneous, i.e. $P(x) = A_{2M}(x, \dots, x)$,

$$G(x) \leq e^{\frac{\sin(2M\theta)}{\hbar} A_{2M}(x,x,\dots,x)} g(|x|),$$

where $g(t) = O(t^{-(n+\delta)})$, $\delta > 0$, as $t \to \infty$.

Then the limit of regularized integrals:

$$\lim_{\epsilon\downarrow 0}\int e^{\frac{i}{\hbar}P(xe^{i\epsilon})}f(xe^{i\epsilon})dx, \qquad 0<\epsilon<\pi/4M, \quad \hbar>0$$

is given by:

$$e^{in\pi/4M}\int_{\mathbb{R}^n} e^{\frac{i}{\hbar}P(e^{i\pi/4M}x)}f(e^{i\pi/4M}x)dx\,. \tag{2.27}$$

The latter integral is absolutely convergent and it is understood in Lebesgue sense.

The class of functions satisfying conditions (1) and (2) in theorem 2.3 includes for instance the polynomials of any degree and the exponentials. In the case $f\in S^N_\lambda$ for some N,λ, one is tempted to interpret expression (2.27) as an explicit formula for the evaluation of the generalized Fresnel integral $\int e^{\frac{i}{\hbar}P(x)}f(x)dx$, $\hbar>0$, whose existence is assured by theorem 2.1. This is, however, not necessarily true for all $f\in S^\infty_\lambda$ satisfying (1) and (2). Indeed the definition 2.1 of oscillatory integral requires that the limit of the sequence of regularized integrals exists and is independent on the regularization. The identity

$$\lim_{\epsilon\to 0}\int_{\mathbb{R}^n} e^{\frac{i}{\hbar}P(x)}f(x)\psi(\epsilon x)dx = e^{in\pi/4M}\int_{\mathbb{R}^n} e^{\frac{i}{\hbar}P(e^{i\pi/4M}x)}f(e^{i\pi/4M}x)dx,$$

can be proved only by choosing regularizing functions $\psi\in\mathcal{S}(\mathbb{R}^n)$ with $\psi(0)=1$ and ψ in the class Σ consisting of all $\psi\in\mathcal{S}(\mathbb{R}^n)$ which satisfy assumption (1) of theorem 2.3 and are such that $|\psi(e^{i\theta}x)|$ is bounded as $|x|\to\infty$ for each $\theta\in(0,\pi/4M)$. In fact we will prove that expression (2.27) coincides with the oscillatory integral (2.15), i.e. one can take $\Sigma=\mathcal{S}(\mathbb{R}^N)$, by imposing stronger assumptions on the function f. First of all we show that the representation (2.17) (resp. (2.18)) for the Fourier transform of $e^{\frac{i}{\hbar}P(x)}$ allows a generalization of equation (2.10). Let us denote by $\bar{D}\subset\mathbb{C}$ the lower semiplane in the complex plane

$$\bar{D}\equiv\{z\in\mathbb{C}\mid Im(z)\le 0\}\,. \tag{2.28}$$

Theorem 2.4. *Let $f\in\mathcal{F}(\mathbb{R}^n)$, $f=\hat{\mu}_f$. Then the generalized Fresnel integral*

$$I(f)\equiv\int_{\mathbb{R}^n} e^{\frac{i}{\hbar}P(x)}f(x)dx, \qquad \hbar\in\bar{D}\setminus\{0\}$$

is well defined and it is given by the formula of Parseval's type:

$$\int_{\mathbb{R}^n} e^{\frac{i}{\hbar}P(x)}f(x)dx = \int_{\mathbb{R}^n}\tilde{F}(k)\mu_f(dk), \tag{2.29}$$

where $\tilde{F}(k)$ is given by (2.25) *(see lemma 2.1 and remark 2.3)*

$$\tilde{F}(k) = \int_{\mathbb{R}^n} e^{ikx} e^{\frac{i}{\hbar}P(x)} dx.$$

The integral on the right hand side of (2.29) *is absolutely convergent (hence it can be understood in Lebesgue sense).*

Proof. Let us choose a test function $\psi \in \mathcal{S}(\mathbb{R}^n)$, such that $\psi(0) = 1$ and let us compute the limit

$$I(f) \equiv \lim_{\epsilon \downarrow 0} \int_{\mathbb{R}^n} e^{\frac{i}{\hbar}P(x)} \psi(\epsilon x) f(x) dx.$$

By hypothesis $f(x) = \int_{\mathbb{R}^n} e^{ikx} \mu_f(dk)$, $x \in \mathbb{R}^N$, and substituting in the previous expression we get:

$$I(f) = \lim_{\epsilon \downarrow 0} \int_{\mathbb{R}^n} e^{\frac{i}{\hbar}P(x)} \psi(\epsilon x) \Big(\int_{\mathbb{R}^n} e^{ikx} \mu_f(dk) \Big) dx.$$

By Fubini theorem (which applies for any $\epsilon > 0$ since the integrand is bounded by $|\psi(\epsilon x)|$ which is dx-integrable, and μ_f is a bounded measure) the right hand side is

$$= \lim_{\epsilon \downarrow 0} \int_{\mathbb{R}^n} \Big(\int_{\mathbb{R}^n} e^{\frac{i}{\hbar}P(x)} \psi(\epsilon x) e^{ikx} dx \Big) \mu_f(dk)$$

$$= \frac{1}{(2\pi)^n} \lim_{\epsilon \downarrow 0} \int_{\mathbb{R}^n} \int_{\mathbb{R}^n} \tilde{F}(k - \alpha\epsilon) \tilde{\psi}(\alpha) d\alpha \mu_f(dk) \quad (2.30)$$

(here we have used the fact that the integral with respect to x is the Fourier transform of $e^{\frac{iP(x)}{\hbar}} \psi(\epsilon x)$ and the inverse Fourier transform of a product is a convolution). Now we can pass to the limit using the Lebesgue bounded convergence theorem and get the desired result:

$$\lim_{\epsilon \downarrow 0} \int_{\mathbb{R}^n} e^{\frac{i}{\hbar}P(x)} \psi(\epsilon x) f(x) dx = \int_{\mathbb{R}^n} \tilde{F}(k) \mu_f(dk),$$

where we have used that $\int \tilde{\psi}(\alpha) d\alpha = (2\pi)^n \psi(0)$ and lemma 2.1, which assures the boundedness of $\tilde{F}(k)$. □

Corollary 2.1. *Let $\hbar = |\hbar| e^{i\phi}$, $\phi \in [-\pi, 0]$, $\hbar \neq 0$, $f \in \mathcal{F}(\mathbb{R}^n)$, $f = \hat{\mu}_f$ be such that $\forall x \in \mathbb{R}^n$*

$$\int_{\mathbb{R}^n} e^{-kx \sin(\pi/4M + \phi/2M)} |\mu_f|(dk) \leq AG(x), \quad (2.31)$$

where $A \in \mathbb{R}$ and $G : \mathbb{R}^n \to \mathbb{R}$ is a positive function satisfying bound (1) or (2) respectively:

(1) if P is defined by (2.14),

$$G(x) \leq e^{B|x|^{2M-1}}, \quad B > 0,$$

(2) if P is homogeneous, i.e. $P(x) = A_{2M}(x, \ldots, x)$:

$$G(x) \leq e^{\frac{1}{\hbar} A_{2M}(x,x,\ldots,x)} g(|x|),$$

where $g(t) = O(t^{-(n+\delta)})$, $\delta > 0$, as $t \to \infty$.

Then f extends to an analytic function on $\mathbb{C}^n$ and its generalized Fresnel integral (2.15) is well defined and it is given by

$$\int_{\mathbb{R}^n} e^{\frac{i}{\hbar}P(x)} f(x)dx$$
$$= e^{in(\pi/4M+\phi/2M)} \int_{\mathbb{R}^n} e^{\frac{i}{\hbar}P(e^{i(\pi/4M+\phi/2M)}x)} f(e^{i(\pi/4M+\phi/2M)}x)dx.$$

Proof. By bound (2.31) it follows that the Laplace transform $f^L : \mathbb{C}^n \to \mathbb{C}$, $f^L(z) = \int_{\mathbb{R}^n} e^{kz} \mu_f(dk)$, of μ_f is a well defined entire function such that, for $x \in \mathbb{R}^n$, $f^L(ix) = f(x)$. By theorem 2.4 the generalized Fresnel integral can be computed by means of the Parseval type equality

$$\int_{\mathbb{R}^n} e^{\frac{i}{\hbar}P(x)} f(x)dx = \int_{\mathbb{R}^n} \tilde{F}(k)\mu_f(dk)$$
$$= e^{in(\pi/4M+\phi/2M)} \int_{\mathbb{R}^n} \left(\int_{\mathbb{R}^n} e^{ikxe^{i(\pi/4M+\phi/2M)}} e^{\frac{i}{\hbar}P(e^{i(\pi/4M+\phi/2M)}x)} dx \right) \mu_f(dk)$$

By Fubini theorem, which applies given the assumptions on the measure μ_f, this is equal to

$$e^{in(\pi/4M+\phi/2M)} \int_{\mathbb{R}^n} e^{\frac{i}{\hbar}P(e^{i(\pi/4M+\phi/2M)}x)} \int_{\mathbb{R}^n} e^{ikxe^{i(\pi/4M+\phi/2M)}} \mu_f(dk)dx$$
$$= e^{in(\pi/4M+\phi/2M)} \int_{\mathbb{R}^n} e^{\frac{i}{\hbar}P(e^{i(\pi/4M+\phi/2M)}x)} f^L(ie^{i(\pi/4M+\phi/2M)}x)dx$$
$$= e^{in(\pi/4M+\phi/2M)} \int_{\mathbb{R}^n} e^{\frac{i}{\hbar}P(e^{i(\pi/4M+\phi/2M)}x)} f(e^{i(\pi/4M+\phi/2M)}x)dx$$

and the conclusion follows. □

2.4 Infinite dimensional oscillatory integrals

The definition 2.3 of the (finite dimensional) Fresnel integral can be generalized to the case the integration is performed on an infinite dimensional real separable Hilbert space $(\mathcal{H}, \langle\ ,\ \rangle)$. An *infinite dimensional oscillatory integral* is defined as the limit of a sequence of finite dimensional approximations. This approach was proposed in [114] and further developed in [7] in connection with the study of an infinite dimensional version of the stationary phase method. It is also related to previous work by K. Ito [172, 173] and S. Albeverio and R. Høegh-Krohn [16, 17].

Definition 2.4. A Borel measurable function $f : \mathcal{H} \to \mathbb{C}$ is called Fresnel integrable if for each sequence $\{P_n\}_{n\in\mathbb{N}}$ of projectors onto n-dimensional subspaces of $\mathcal{H}$, such that

- $P_n \leq P_{n+1}$ (i.e. $P_n(\mathcal{H}) \subset P_{n+1}(\mathcal{H})$),
- $P_n \to I$ strongly as $n \to \infty$ (I being the identity operator in $\mathcal{H}$),

the finite dimensional approximations of the oscillatory integral of f, suitably normalized

$$\int^{\circ}_{P_n\mathcal{H}} e^{\frac{i}{2\hbar}\|P_nx\|^2} f(P_nx)d(P_nx)\Big(\int^{\circ}_{P_n\mathcal{H}} e^{\frac{i}{2\hbar}\|P_nx\|^2} d(P_nx)\Big)^{-1},$$

are well defined (in the sense of definition 2.1) and the limit

$$\lim_{n\to\infty}\int^{\circ}_{P_n\mathcal{H}} e^{\frac{i}{2\hbar}\|P_nx\|^2} f(P_nx)d(P_nx)\Big(\int^{\circ}_{P_n\mathcal{H}} e^{\frac{i}{2\hbar}\|P_nx\|^2} d(P_nx)\Big)^{-1},$$

exists and is independent on the sequence $\{P_n\}$.
In this case the limit is called the infinite dimensional oscillatory integral of f and is denoted by

$$\widetilde{\int_{\mathcal{H}}} e^{\frac{i}{2\hbar}\|x\|^2} f(x)dx.$$

Analogously to definition 2.3, also in the definition 2.4 of the infinite dimensional oscillatory integral a normalization constant

$$\Big(\int^{\circ}_{P_n\mathcal{H}} e^{\frac{i}{2\hbar}\|P_nx\|^2} d(P_nx)\Big)^{-1} = (2\pi i\hbar)^{-n/2}$$

is present. Indeed if $f : \mathcal{H} \to \mathbb{C}$ is the identity function, i.e. $f(x) = 1$ $\forall x \in \mathcal{H}$, it is simple to verify that

$$\widetilde{\int_{\mathcal{H}}} e^{\frac{i}{2\hbar}\|x\|^2} f(x)dx = 1.$$

Moreover the presence of the normalization constant makes definition 2.4 the direct generalization of definition 2.3. Indeed if $f : \mathcal{H} \to \mathbb{C}$ is a finite based function depending on a finite number of variables, i.e. if $f(x) = f(P_nx)$ for a finite dimensional projection operator P_n in $\mathcal{H}$, it is possible to see that the infinite dimensional oscillatory integral of f on $\mathcal{H}$ coincides with the finite dimensional Fresnel integral on the finite dimensional subspace $P_n(\mathcal{H})$ of the restriction f_n of the function f on $P_n(\mathcal{H})$:

$$f_n : P_n\mathcal{H} \to \mathbb{C}, \qquad f_n(x) := f(P_n x), \qquad x \in P_n(\mathcal{H}),$$

$$\widetilde{\int_{\mathcal{H}}} e^{\frac{i}{2\hbar}\|x\|^2} f(x)dx = \widetilde{\int_{P_n(\mathcal{H})}} e^{\frac{i}{2\hbar}\|x\|^2} f_n(x)dx.$$

The "concrete" description of the class of all Fresnel integrable functions is still an open problem of harmonic analysis, even when $dim(\mathcal{H}) < \infty$ (as already discussed in sections 2.1, 2.2 and 2.3). However, analogously to the finite dimensional case, it is possible to prove that this class includes $\mathcal{F}(\mathcal{H})$, the Banach algebra of functions that are Fourier transform of bounded variation measures on $\mathcal{H}$, and a Parseval-type formula analogous to Eq. (2.10) holds. In order that all the terms on the right hand side of Eq. (2.10) make sense even in an infinite dimensional setting, we have to impose some conditions to the operator L.

In the following we shall consider a self-adjoint trace class operator $L : \mathcal{H} \to \mathcal{H}$, such that $I - L$ is invertible. As L is compact, it has a complete set of eigenvectors, with eigenvalues of finite multiplicity and with 0 as their only possible limit point. As $I - L$ is invertible by assumption, the index of $(I - L)$, i.e. the number of negatives eigenvalues, counted with their multiplicity, is finite. The following lemma is taken from [114].

Lemma 2.2. *Let $\{P_n\}_{n\in\mathbb{N}}$ be a sequence of of projectors onto n-dimensional subspaces of $\mathcal{H}$, convergent strongly to the identity when $n \to \infty$. For any compact self-adjoint operator L the following holds:*

$$\lim_{n\to\infty} \text{Ind}(I - P_n L P_n) = \text{Ind}(I - L)\,. \tag{2.32}$$

Proof. Let $\text{Ind}(I - L) = p$ and let $\lambda_1, \ldots, \lambda_p$ be the eigenvalues of L that are greater than 1, with corresponding orthonormal vectors $e_1, \ldots, e_p$. We have

$$\langle e_i, (I - L)e_i\rangle = 1 - \lambda_i < 0,$$

and by the strong convergence of the sequence $\{P_n\}_{n\in\mathbb{N}}$ to the identity,

$$\lim_n \langle P_n e_i, (I - L)P_n e_i\rangle = 1 - \lambda_i.$$

In particular for sufficiently large n we have

$$\langle P_n e_i, (I - L)P_n e_i\rangle < 0.$$

As $\langle P_n e_i, (I_n - P_n L P_n) P_n e_i \rangle = \langle P_n e_i, (I - L) P_n e_i \rangle$ (I_n being the identity operator on $P_n \mathcal{H}$), it is possible to conclude that the vectors $P_n e_1, \ldots, P_n e_p$ lie in a subspace of $P_n \mathcal{H}$ in which $(I_n - P_n L P_n)$ is negative definite. Since $P_n e_1, \ldots, P_n e_p$ are linearly independent for sufficiently large n, as

$$\lim_n \langle P_n e_i, P_n e_j \rangle = \delta_{ij},$$

we can conclude that

$$\underline{\lim}_n \text{Ind}(I_n - P_n L P_n) \geq \text{Ind}(I - L).$$

In order to prove the converse inequality, it is sufficient to note that, if V_n is the subspace of $P_n \mathcal{H}$ spanned by the negative eigenvectors of $I_n - P_n L P_n$, then for $v \in V_n$ we have

$$\langle v, (I - L) v \rangle = \langle P_n v, (I - L) P_n v \rangle = \langle v, (I_n - P_n L P_n) v \rangle < 0.$$

This gives, for sufficiently large n,

$$\text{Ind}(I - L) \geq \text{Ind}(I_n - P_n L P_n).$$ □

For a self-adjoint trace class operator $L : \mathcal{H} \to \mathcal{H}$, it is possible to define the Fredholm determinant of the operator $I - L$ [262], given by the infinite product of its eigenvalues

$$\Pi_n (1 - \lambda_n).$$

It is denoted with $\det(I - L)$ and given by

$$\det(I - L) = |\det(I - L)| e^{-\pi i \text{ Ind } (I-L)}$$

where $|\det(I - L)|$ is its absolute value and $\text{Ind}((I - L))$ is the number of negative eigenvalues of the operator $(I - L)$, counted with their multiplicity.

We can now state the fundamental theorem.

Theorem 2.5. *Let $L : \mathcal{H} \to \mathcal{H}$ be a self-adjoint trace class operator such that $(I - L)$ is invertible (I being the identity operator in $\mathcal{H}$). Let us assume that $f \in \mathcal{F}(\mathcal{H})$. Then the function $g : \mathcal{H} \to \mathbb{C}$ given by*

$$g(x) = e^{-\frac{i}{2\hbar}\langle x, Lx \rangle} f(x), \qquad x \in \mathcal{H}$$

is Fresnel integrable and the corresponding infinite dimensional oscillatory integral is given by the following Parseval-type formula:

$$\widetilde{\int_{\mathcal{H}}} e^{\frac{i}{2\hbar}\langle x, (I-L)x \rangle} f(x) dx = (\det(I - L))^{-1/2} \int_{\mathcal{H}} e^{-\frac{i\hbar}{2}\langle x, (I-L)^{-1} x \rangle} d\mu_f(x) \tag{2.33}$$

where

$$\det(I - L) = |\det(I - L)| e^{-\pi i \text{ Ind } (I-L)}$$

is the Fredholm determinant of the operator $(I - L)$.

Proof. Given a sequence $\{P_n\}_{n\in\mathbb{N}}$ of projectors onto n-dimensional subspaces of $\mathcal{H}$, such that $P_n \leq P_{n+1}$ and $P_n \to I$ strongly as $n \to \infty$ (I being the identity operator in $\mathcal{H}$), the finite dimensional approximations of the infinite dimensional oscillatory integral $\widetilde{\int_{\mathcal{H}}} e^{\frac{i}{2\hbar}\langle x,(I-L)x\rangle} f(x)dx$ are equal to

$$\int^{\circ}_{P_n\mathcal{H}} e^{\frac{i}{2\hbar}\|P_nx\|^2} e^{-\frac{i}{2\hbar}\langle P_nx, LP_nx\rangle} f_n(P_nx)d(P_nx)\Big(\int^{\circ}_{P_n\mathcal{H}} e^{\frac{i}{2\hbar}\|P_nx\|^2} d(P_nx)\Big)^{-1}, \tag{2.34}$$

where $f_n : P_n\mathcal{H} \to \mathbb{C}$ is given by

$$f_n(x) := f(P_nx), \qquad x \in P_n\mathcal{H}.$$

The function f_n belongs to $\mathcal{F}(P_n\mathcal{H})$, indeed for $x \in P_n\mathcal{H}$ we have:

$$f_n(x) = \int_{\mathcal{H}} e^{i\langle y,x\rangle} d\mu_f(y) = \int_{\mathcal{H}} e^{i\langle P_ny,P_nx\rangle} d\mu_f(y) = \int_{\mathcal{H}} e^{i\langle y,P_nx\rangle} d\mu_{f,n}(y)$$

where $\mu_{f,n}$ given by

$$\int_{\mathcal{H}} g(y)d\mu_{f,n}(y) = \int_{\mathcal{H}} g(P_ny)d\mu_f(y)$$

is the restriction of the measure μ_f on $P_n\mathcal{H}$. It can be expressed in terms of a bounded variation measure $\bar{\mu}_{f,n}$, given by

$$\int_{P_n\mathcal{H}} g(y)d\bar{\mu}_{f,n}(y) = \int_{\mathcal{H}} g(P_ny)d\mu_f(y), \qquad g : P_n\mathcal{H} \to \mathbb{C}.$$

Let $L_n : P_n\mathcal{H} \to P_n\mathcal{H}$ be the operator on $P_n\mathcal{H}$ given by $L_n := P_nLP_n$. As $(I-L)$ is invertible, it is easy to see that for n sufficiently large the operator $(I_n - L_n)$ on $P_n\mathcal{H}$ (I_n being the identity operator in $P_n\mathcal{H}$) is invertible. By Parseval's formula (2.10) the expression (2.34) is equal to

$$\begin{aligned}(\det(I_n - L_n))^{-1/2} &\int_{P_n\mathcal{H}} e^{-\frac{i\hbar}{2}\langle P_nx,(I_n-L_n)^{-1}P_nx\rangle} d\bar{\mu}_{f,n}(P_nx) \\ &= (\det(I_n - L_n))^{-1/2} \int_{\mathcal{H}} e^{-\frac{i\hbar}{2}\langle P_nx,(I-L_n)^{-1}P_nx\rangle} d\mu_f(x). \end{aligned} \tag{2.35}$$

By letting $n \to \infty$, $L_n \to L$ in trace norm and expression (2.35) converges to the right hand side of (2.33). □

A fundamental property of infinite dimensional oscillatory integrals is the covariance under translations of vectors belonging to $\mathcal{H}$, more precisely the following holds [114]:

Theorem 2.6. *Let $a \in \mathcal{H}$ and $f \in \mathcal{F}(\mathcal{H})$. Let us define $f_a : \mathcal{H} \to \mathbb{C}$ by $f_a(x) := f(x+a)$. Then the function $g : \mathcal{H} \to \mathbb{C}$, given by*

$$g(x) := e^{\frac{i}{\hbar}\langle a,x\rangle} f_a(x), \qquad x \in \mathcal{H} \tag{2.36}$$

is Fresnel integrable and the corresponding infinite dimensional oscillatory integral is given by

$$\widetilde{\int_{\mathcal{H}}} e^{\frac{i}{2\hbar}\langle x,x\rangle} e^{\frac{i}{\hbar}\langle a,x\rangle} f_a(x)dx = e^{-\frac{i}{2\hbar}\|a\|^2} \widetilde{\int_{\mathcal{H}}} e^{\frac{i}{2\hbar}\langle x,x\rangle} f(x)dx. \tag{2.37}$$

Proof. It is easy to verify that the function g belongs to $\mathcal{F}(\mathcal{H})$, indeed it is the Fourier transform of the measure μ_g, whose action on a Borel bounded function $h : \mathcal{H} \to \mathbb{C}$ is given by:

$$\int_{\mathcal{H}} h(x)d\mu_g(x) = \int_{\mathcal{H}\times\mathcal{H}} h(x+y)\delta_{a/\hbar}(x)e^{i\langle y,a\rangle} d\mu_f(y).$$

By theorem 2.5, we have:

$$\begin{aligned}
&\widetilde{\int_{\mathcal{H}}} e^{\frac{i}{2\hbar}\langle x,x\rangle} e^{\frac{i}{\hbar}\langle a,x\rangle} f_a(x)dx \\
&= \int_{\mathcal{H}\times\mathcal{H}} e^{-\frac{i\hbar}{2}\langle x+y,x+y\rangle} \delta_{a/\hbar}(x)e^{i\langle y,a\rangle} d\mu_f(y) \\
&= e^{-\frac{i}{2\hbar}\|a\|^2} \widetilde{\int_{\mathcal{H}}} e^{-\frac{i\hbar}{2}\langle y,y\rangle} d\mu_f(y) \\
&= e^{-\frac{i}{2\hbar}\|a\|^2} \widetilde{\int_{\mathcal{H}}} e^{\frac{i}{2\hbar}\langle x,x\rangle} f(x)dx.
\end{aligned}$$

□

The Parseval type equality (2.33) allows also the proof of the following Fubini theorem [7]:

Theorem 2.7. *Let $T : \mathcal{H} \to \mathcal{H}$, $T := I - L$ with L self-adjoint trace class, a linear invertible operator. Let $\mathcal{H} = Y + Z$ be the direct sum of two closed subspaces, with Z being finite dimensional. Assume that*

$$\langle Ty, z\rangle = 0 \qquad \forall y \in Y, z \in Z.$$

Let

$$T_1 y = (P_Y \circ T)(y), \qquad y \in Y,$$

$$T_2 z = (P_Z \circ T)(z), \qquad z \in Z,$$

where P_Y and P_Z are respectively orthogonal projections onto Y and Z. Assume that both T_1 and T_2 are isomorphisms of Y and Z, respectively. Then, if $f \in \mathcal{F}(\mathcal{H})$:

$$\widetilde{\int_{\mathcal{H}}} e^{\frac{i}{2\hbar}\langle x,Tx\rangle} f(x)dx = C_T \widetilde{\int_Z} e^{\frac{i}{2\hbar}\langle z,T_2 z\rangle} \widetilde{\int_Y} e^{\frac{i}{2\hbar}\langle y,T_1 y\rangle} f(y+z)dydz, \tag{2.38}$$

with

$$C_T = (\det T)^{-1/2}(\det T_1)^{1/2}(\det T_2)^{1/2}.$$

Proof. For the proof of Eq. (2.38) it is convenient to introduce a different notation. Let us define the bilinear form ((,)) on $\mathcal{H} \times \mathcal{H}$:

$$((x_1, x_2)) = \langle x_1, Tx_2 \rangle.$$

Since $f \in \mathcal{F}(\mathcal{H})$, there exists a bounded variation measure $\mu \in \mathcal{M}(\mathcal{H})$ such that

$$f(x) = \int_{\mathcal{H}} e^{i\langle Tx, \xi \rangle} d\mu(\xi) = \int_{\mathcal{H}} e^{i((x,\xi))} d\mu(\xi).$$

The measure μ is given by:

$$\int_{\mathcal{H}} h(x) d\mu(x) = \int_{\mathcal{H}} h(T^{-1}x) d\mu_f(x), \qquad f(x) = \int_{\mathcal{H}} e^{i\langle x, \xi \rangle} d\mu_f(\xi).$$

By Parseval type equality (2.33), we have

$$\widetilde{\int_{\mathcal{H}}} e^{\frac{i}{2\hbar}\langle x, Tx \rangle} f(x) dx = (\det T)^{-1/2} \int_{\mathcal{H}} e^{-\frac{i\hbar}{2}((x,x))} d\mu(x).$$

By considering $y + z, \eta + \xi \in Y \oplus Z$, we have that the function $f_z : Y \to \mathbb{C}$, given by

$$f_z(y) = f(y + z)$$

is of the form

$$f_z(y) = \int_Y e^{i((y,\eta))} d\mu_z(\eta),$$

indeed

$$\begin{aligned} f(y+z) &= \int_{\mathcal{H}} e^{i\langle T(y+z), \eta + \zeta \rangle} d\mu(\eta + \zeta) \\ &= \int_{\mathcal{H}} e^{i\langle Ty, \eta \rangle} e^{i\langle Tz, \zeta \rangle} d\mu(\eta + \zeta) \\ &= \int_Y e^{i\langle Ty, \eta \rangle} d\mu_z(\eta) = \int_Y e^{i((y,\eta))} d\mu_z(\eta). \end{aligned}$$

The measure μ_z is defined by

$$\int_Y h(\eta) d\mu_z(\eta) = \int_{\mathcal{H}} h(\eta) e^{i\langle Tz, \zeta \rangle} d\mu(\eta + \zeta),$$

for any Borel bounded function $h : Y \to \mathbb{C}$.
Since

$$((y, y)) = \langle y, T_1 y \rangle, \quad ((z, z)) = \langle z, T_2 z \rangle, \qquad y \in Y, z \in Z,$$

and

$$\widetilde{\int_Y} e^{\frac{i}{2\hbar}\langle y, T_1 y \rangle} f(y + z) dy = (\det T_1)^{-1/2} \int_Y e^{-\frac{i\hbar}{2}((y,y))} d\mu_z(y), \qquad (2.39)$$

we have that the function $I : Z \to \mathbb{C}$ given by

$$I(z) := \widetilde{\int_Y} e^{\frac{i}{2\hbar}\langle y, T_1 y\rangle} f(y+z)dy$$

is of the form $I(z) = \int_Z e^{i((z,\zeta))} d\tilde{\mu}(\zeta)$, with $\tilde{\mu} \in \mathcal{M}(Z)$. Indeed

$$\begin{aligned}
&(\det T_1)^{-1/2} \int_Y e^{-\frac{i\hbar}{2}((y,y))} d\mu_z(y) \\
&= (\det T_1)^{-1/2} \int_{\mathcal{H}} e^{-\frac{i\hbar}{2}((\eta,\eta))} e^{i((z,\zeta))} d\mu(\eta+\zeta) \\
&= \int_Z e^{i((z,\zeta))} d\tilde{\mu}(\zeta)
\end{aligned}$$

with $\tilde{\mu}$ defined by

$$\int_Z h(z)\tilde{\mu}(z) = (\det T_1)^{-1/2} \int_{\mathcal{H}} h(z) e^{-\frac{i\hbar}{2}((\eta,\eta))} d\mu(\eta+\zeta),$$

for any Borel bounded function $h : Z \to \mathbb{C}$.
Analogously to Eq. (2.39) we have

$$\widetilde{\int_Z} e^{\frac{i}{2\hbar}\langle z, T_2 z\rangle} I(z)dz = (\det T_2)^{-1/2} \int_Z e^{-\frac{i\hbar}{2}((z,z))} d\tilde{\mu}(z),$$

and by a sequence of calculations we have

$$\begin{aligned}
&\widetilde{\int_Z} e^{\frac{i}{2\hbar}\langle z, T_2 z\rangle} \widetilde{\int_Y} e^{\frac{i}{2\hbar}\langle y, T_1 y\rangle} f(y+z)dydz \\
&= (\det T_2)^{-1/2}(\det T_1)^{-1/2} \int_{\mathcal{H}} e^{-\frac{i\hbar}{2}((\zeta,\zeta))} e^{-\frac{i\hbar}{2}((\eta,\eta))} d\mu(\eta+\zeta) \\
&= (\det T_2)^{-1/2}(\det T_1)^{-1/2} \int_{\mathcal{H}} e^{-\frac{i\hbar}{2}((\zeta+\eta,\zeta+\eta))} d\mu(\eta+\zeta) \\
&= (\det T_2)^{-1/2}(\det T_1)^{-1/2}(\det T)^{1/2} \widetilde{\int_{\mathcal{H}}} e^{\frac{i}{2\hbar}\langle x, Tx\rangle} f(x)dx.
\end{aligned}$$

□

As we have already remarked above, the normalization constant in the definition 2.4 plays a crucial role. Other alternative definitions of infinite dimensional oscillatory integrals can be considered. They can be obtained by introducing in the finite dimensional approximations different normalization constants.

Given a a self-adjoint invertible operator B on $\mathcal{H}$ (we do not impose any assumption on its trace), we can consider the definition of the *normalized infinite dimensional oscillatory integral with respect to B*.

Definition 2.5. A Borel function $f : \mathcal{H} \to \mathbb{C}$ is called $\mathcal{F}^\hbar_B$ integrable if for each sequence $\{P_n\}_{n\in\mathbb{N}}$ of projectors onto n-dimensional subspaces of $\mathcal{H}$, such that $P_n \leq P_{n+1}$ and $P_n \to I$ strongly as $n \to \infty$ (I being the identity operator in $\mathcal{H}$) the finite dimensional approximations

$$\widetilde{\int_{P_n\mathcal{H}}} e^{\frac{i}{2\hbar}\langle P_nx, BP_nx\rangle} f(P_nx)d(P_nx),$$

are well defined (in the sense of definition 2.3) and the limit

$$\lim_{n\to\infty} (\det P_nBP_n)^{\frac{1}{2}} \widetilde{\int_{P_n\mathcal{H}}} e^{\frac{i}{2\hbar}\langle P_nx, BP_nx\rangle} f(P_nx)d(P_nx) \tag{2.40}$$

exists and is independent on the sequence $\{P_n\}$.
In this case the limit is called the normalized oscillatory integral of f with respect to B and is denoted by

$$\widetilde{\int_{\mathcal{H}}^B} e^{\frac{i}{2\hbar}\langle x, Bx\rangle} f(x)dx.$$

Again, given a function $f \in \mathcal{F}(\mathcal{H})$, it is possible to prove that f is $\mathcal{F}^\hbar_B$ integrable and the corresponding normalized infinite dimensional oscillatory integral can be computed by means of a formula similar to (2.33):

Theorem 2.8. *Let us assume that $f \in \mathcal{F}(\mathcal{H})$. Then f is $\mathcal{F}^\hbar_B$ integrable and the corresponding normalized oscillatory integral is given by the following Parseval-type formula:*

$$\widetilde{\int_{\mathcal{H}}^B} e^{\frac{i}{2\hbar}\langle x, Bx\rangle} f(x)dx = \int_{\mathcal{H}} e^{-\frac{i\hbar}{2}\langle x, B^{-1}x\rangle} d\mu_f(x). \tag{2.41}$$

Proof. The proof is completely similar to the proof of theorem 2.5 and we left the details to the reader. □

Remark 2.4. Formula (2.41) has already been discussed in the first part of [17].

The difference between definitions 2.4 and 2.5 is the normalization constant. In fact definition 2.5 can be seen as a generalization of definition 2.4, which can be obtained by setting $B = I$, the identity operator.

The integral $\widetilde{\int_{\mathcal{H}}^B} e^{\frac{i}{2\hbar}\langle x, Bx\rangle} f(x)dx$ is called "normalized" because if we substitute into Eq. (2.41) the function $f = 1$, we have

$$\widetilde{\int_{\mathcal{H}}^B} e^{\frac{i}{2\hbar}\langle x, Bx\rangle} f(x)dx = 1.$$

The importance of the normalization constant in the finite dimensional approximations is highlighted by theorems 2.5 and 2.8. Indeed theorem 2.8 makes sense even if the operator $L := I - B$ is not trace class (in that case the Fredholm determinant $\det(I - B)$ cannot be defined).

In fact it is possible to introduce different normalization constants in the finite dimensional approximations and the properties of the corresponding infinite dimensional oscillatory integrals are related to the trace properties of the operator associated to the quadratic part of the phase function [9]. More precisely, for any $p \in \mathbb{N}$, let us consider the Schatten class $\mathcal{T}_p(\mathcal{H})$ of bounded linear operators L in $\mathcal{H}$ such that

$$\|L\|_p = (\mathrm{Tr}(L^+ L)^{p/2})^{1/p}$$

is finite. $(\mathcal{T}_p(\mathcal{H}), \|\cdot\|_p)$ is a Banach space (see [262]). For any $p \in \mathbb{N}$, $p \geq 2$ and $L \in \mathcal{T}_p(\mathcal{H})$ one defines the regularized Fredholm determinant $\det_{(p)} : I + \mathcal{T}_p(\mathcal{H}) \to \mathbb{R}$:

$$\det_{(p)}(I + L) = \det\left((I + L)\exp\sum_{j=1}^{p-1} \frac{(-1)^j}{j} L^j\right), \qquad L \in \mathcal{T}_p(\mathcal{H}),$$

where det denotes the usual Fredholm determinant, which is well defined as it is possible to prove that the operator $(I + L)\exp\sum_{j=1}^{p-1}\frac{(-1)^j}{j}L^j - I$ is trace class [262]. In particular $\det_{(2)}$ is called Carleman determinant.

For $p \in \mathbb{N}$, $p \geq 2$, $L \in \mathcal{T}_1(\mathcal{H})$, let us define the normalized quadratic form on $\mathcal{H}$:

$$N_p(L)(x) = (x, Lx) - i\hbar \mathrm{Tr}\sum_{j=1}^{p-1}\frac{L^j}{j}, \qquad x \in \mathcal{H}. \tag{2.42}$$

Again, for $p \in \mathbb{N}$, $p \geq 2$, let us define the *class p normalized oscillatory integral*:

Definition 2.6. Let $p \in \mathbb{N}$, $p \geq 2$, L a bounded linear operator in $\mathcal{H}$, $f : \mathcal{H} \to \mathbb{C}$ a Borel measurable function. The class p normalized oscillatory integral of the function f with respect to the operator L is well defined if for each sequence $\{P_n\}_{n\in\mathbb{N}}$ of projectors onto n-dimensional subspaces of $\mathcal{H}$, such that $P_n \leq P_{n+1}$ and $P_n \to I$ strongly as $n \to \infty$ (I being the identity operator in $\mathcal{H}$) the finite dimensional approximations

$$\widetilde{\int_{P_n\mathcal{H}}^{\circ}} e^{\frac{i}{2\hbar}\|x\|^2} e^{-\frac{i}{2\hbar}N_p(P_nLP_n)(P_nx)} f(P_nx)d(P_nx), \tag{2.43}$$

are well defined and the limit

$$\lim_{n\to\infty}\widetilde{\int_{P_n\mathcal{H}}^{\circ}} e^{\frac{i}{2\hbar}\|x\|^2} e^{-\frac{i}{2\hbar}N_p(P_nLP_n)(P_nx)} f(P_nx)d(P_nx) \tag{2.44}$$

exists and is independent on the sequence $\{P_n\}$.
In this case the limit is denoted by

$$\mathcal{I}_{p,L}(f) = \widetilde{\int_{\mathcal{H}}^{p}} e^{\frac{i}{2\hbar}\|x\|^2} e^{-\frac{i}{2\hbar}\langle x, Lx\rangle} f(x)dx.$$

If L is not a trace class operator, then the quadratic form (2.42) is not well defined. Nevertheless expression (2.43) still makes sense thanks to the fact that all the functions under the integral are restricted to finite dimensional subspaces. Moreover the limit (2.44) can make sense, as the following result shows [9].

Theorem 2.9. *Let us assume that $f \in \mathcal{F}(\mathcal{H})$ and let L be a self-adjoint operator such that $L \in \mathcal{T}_p(\mathcal{H})$ and $\det_{(p)}(I - L) \neq 0$. Then the class-p normalized oscillatory integral of the function f with respect to the operator L exists and is given by the following Parseval-type formula:*

$$\widetilde{\int_{\mathcal{H}}^{p}} e^{\frac{i}{2\hbar}\|x\|^2} e^{-\frac{i}{2\hbar}\langle x, Lx\rangle} f(x)dx = [\det_{(p)}(I - L)]^{-1/2} \int_{\mathcal{H}} e^{-\frac{i\hbar}{2}\langle x,(I-L)^{-1}x\rangle} d\mu_f(x). \tag{2.45}$$

Proof. See [9]. □

2.5 Polynomial phase functions

As we have seen in the previous section, the possible generalizations of the definition of finite dimensional oscillatory integrals to an infinite dimensional Hilbert space $\mathcal{H}$

$$\widetilde{\int_{\mathcal{H}}} e^{\frac{i}{\hbar}\Phi(x)} dx$$

concerns only quadratic phase functions $\Phi : \mathcal{H} \to \mathbb{C}$ of the form $\Phi(x) = \langle x, x\rangle/2$ or, more generally $\Phi(x) = \langle x, Bx\rangle/2$, with $B : \mathcal{H} \to \mathcal{H}$ a linear operator satisfying suitable assumptions.

As one can understand by a careful reading of theorem 2.5 and analogously theorems 2.8 and 2.9), the key tool allowing the extension of the results of section 2.2 is the Parseval type equality (2.10). Indeed given a function $f \in \mathcal{F}(\mathbb{R}^n)$, its Fresnel integral is given by:

$$\widetilde{\int} e^{\frac{i}{2\hbar}\langle x,(I-L)x\rangle} f(x)dx = (\det(I - L))^{-1/2} \int_{\mathbb{R}^n} e^{-\frac{i\hbar}{2}\langle x,(I-L)^{-1}x\rangle} d\mu_f(x), \tag{2.46}$$

with $L : \mathbb{R}^n \to \mathbb{R}^n$ self-adjoint and $I - L$ invertible.

Equation (2.46) admits a generalization on an infinite dimensional Hilbert space $\mathcal{H}$ because, even if the left hand side loses any meaning when $n \to \infty$, the right hand side can have a well defined meaning also when $\mathbb{R}^n$ is replaced with $\mathcal{H}$, provided the operator L satisfies suitable assumptions (as the existence of the inverse and of the Fredholm determinant of the operator $I - L$).

In the case where the quadratic phase function is replaced with an higher degree polynomial function, as in section 2.3, the situation becomes more involved. Theorem 2.4 generalizes a Parseval type equality to the case of oscillatory integrals of the form:

$$\int_{\mathbb{R}^n}^{o} e^{\frac{i}{\hbar}P(x)} f(x)dx, \qquad f \in \mathcal{F}(\mathbb{R}^n),$$

where $P : \mathbb{R}^n \to \mathbb{R}$ is a $2M$ degree polynomial with positive leading coefficient[2]. More precisely the following formula holds:

$$\int_{\mathbb{R}^n} e^{\frac{i}{\hbar}P(x)} f(x)dx = \int_{\mathbb{R}^n} \tilde{F}(k)\mu_f(dk), \tag{2.47}$$

with the function $\tilde{F} : \mathbb{R}^n \to \mathbb{C}$ is a smooth function, represented by an absolutely convergent integral

$$\tilde{F}(k) = e^{in\pi/4M} \int_{\mathbb{R}^n} e^{ie^{i\pi/4M}kx} e^{\frac{i}{\hbar}P(e^{i\pi/4M}x)} dx. \tag{2.48}$$

The generalization of formulae (2.47) and (2.48) to the infinite dimensional case is not straitforward, since neither of them makes sense in the limit $n \to \infty$.

The quadratic phase functions are much simpler to deal with, because when the polynomial P is of the form

$$P(x) = \frac{1}{2}\langle x, x\rangle,$$

the function $\tilde{F}$, i.e. the Fourier transform of the imaginary exponential of the phase function, can be explicitly computed:

$$\tilde{F}(k) = (2\pi i\hbar)^{-n/2} \int_{\mathbb{R}^n} e^{ikx} e^{\frac{i}{2\hbar}\langle x,x\rangle} dx = e^{-\frac{i\hbar}{2}\langle x,x\rangle}. \tag{2.49}$$

If the left hand side of Eq. (2.49) loses any meaning in the limit $n \to \infty$ because of the non convergence of the constant $(2\pi i\hbar)^{-n/2}$ and the non existence of the Lebesgue measure dx, the right hand side is still meaningful, even when x belongs to an infinite dimensional Hilbert space.

[2]An analogous result is obtained in the case where the leading coefficient is negative.

In the following we shall present the case of a polynomial phase function with quartic growth, where an infinite dimensional generalization of Parseval type equality (2.47) is allowed.

Let us deal first of all with the finite dimensional case, i.e. $dim(\mathcal{H}) = n$. Let $A : \mathcal{H}\times\mathcal{H}\times\mathcal{H}\times\mathcal{H} \to \mathbb{R}$ be a completely symmetric and positive fourth order covariant tensor on $\mathcal{H}$. After the introduction of an orthonormal basis in $\mathcal{H}$, the elements $x \in \mathcal{H}$ can be identified with $n-$ple of real numbers, i.e. $x = (x_1, \dots, x_n)$, and the action of the tensor A on the 4-ple (x, x, x, x) is represented by an homogeneous fourth order polynomial in the variables $x_1, \dots, x_n$:

$$P(x) = A(x,x,x,x) = \sum_{j,k,l,m} a_{j,k,l,m} x_j x_k x_l x_m \tag{2.50}$$

with $a_{j,k,l,m} \in \mathbb{R}$.
We are going to define the following generalized Fresnel integral:

$$\widetilde{\int} e^{\frac{i}{2\hbar}x\cdot(I-B)x} e^{\frac{-i\lambda}{\hbar}P(x)} f(x)dx \tag{2.51}$$

where I, B are $n\times n$ matrices, I being the identity, $\lambda \in \mathbb{R}$, $f \in \mathcal{F}(\mathbb{R}^n)$ and $\hbar > 0$.

Lemma 2.3. [27] *Let $P : \mathbb{R}^n \to \mathbb{R}$ be given by (2.50). Then the Fourier transform of the distribution $\frac{e^{\frac{i}{2\hbar}x\cdot(I-B)x}}{(2\pi i\hbar)^{n/2}} e^{\frac{-i\lambda}{\hbar}P(x)}$:*

$$\tilde{F}(k) = \int_{\mathbb{R}^n} e^{ik\cdot x} \frac{e^{\frac{i}{2\hbar}x\cdot(I-B)x}}{(2\pi i\hbar)^{n/2}} e^{\frac{-i\lambda}{\hbar}P(x)} dx \tag{2.52}$$

is a bounded complex-valued entire function on $\mathbb{R}^n$ admitting, if A is strictly positive, the following representations

$$\tilde{F}(k) = \begin{cases} e^{in\pi/8} \int_{\mathbb{R}^n} e^{ie^{i\pi/8}k\cdot x} \frac{e^{\frac{ie^{i\pi/4}}{2\hbar}x\cdot(I-B)x}}{(2\pi i\hbar)^{N/2}} e^{\frac{\lambda}{\hbar}P(x)} dx & \lambda < 0 \\ e^{-in\pi/8} \int_{\mathbb{R}^n} e^{ie^{-i\pi/8}k\cdot x} \frac{e^{\frac{ie^{-i\pi/4}}{2\hbar}x\cdot(I-B)x}}{(2\pi i\hbar)^{n/2}} e^{-\frac{\lambda}{\hbar}P(x)} dx & \lambda > 0\,. \end{cases} \tag{2.53}$$

Moreover, for general $A \geq 0$, if $\lambda \leq 0$ and $(I - B)$ is symmetric strictly positive then $\tilde{F}(k)$ can also be represented by

$$\tilde{F}(k) = \int_{\mathbb{R}^n} e^{ie^{i\pi/4}k\cdot x} \frac{e^{-\frac{1}{2\hbar}x\cdot(I-B)x}}{(2\pi\hbar)^{n/2}} e^{\frac{i\lambda}{\hbar}P(x)} dx = \mathbb{E}[e^{ie^{i\pi/4}k\cdot x} e^{\frac{i\lambda}{\hbar}P(x)} e^{\frac{1}{2\hbar}x\cdot Bx}] \tag{2.54}$$

where $\mathbb{E}$ denotes the expectation value with respect to the centered Gaussian measure on $\mathbb{R}^n$ with covariance operator $\hbar I$.

Proof. Representation (2.53) and the boundedness of $\tilde{F}$ follow from lemma 2.1, where a more general case is handled. From the representations (2.53) and (2.54) the analyticity of $\tilde{F}(k)$, $k \in \mathbb{C}$ follows immediately.

Let us now prove representation (2.54) for the Fourier transform

$$\tilde{F}(k) = \int_{\mathbb{R}^n} e^{ik\cdot x} \frac{e^{\frac{i}{2\hbar}x\cdot(I-B)x}}{(2\pi i\hbar)^{n/2}} e^{\frac{-i\lambda}{\hbar}P(x)} dx,$$

by mimicking the same procedure used in the proof of lemma 2.1. Without loss of generality we can assume that the quadratic form $x\cdot(I-B)x$ is equal to $x \cdot x$, as it can always be reduced to this form by a change of coordinates.

Let us compute the $n-$dimensional integral defining $\tilde{F}(k)$ by introducing the polar coordinates in $\mathbb{R}^n$:

$$\int_{\mathbb{R}^n} e^{ik\cdot x} \frac{e^{\frac{i}{2\hbar}x\cdot x}}{(2\pi i\hbar)^{n/2}} e^{\frac{-i\lambda}{\hbar}P(x)} dx$$
$$= \int_{S_{n-1}} \Big(\int_0^{+\infty} e^{i|k|rf(\phi_1,\dots,\phi_{n-1})} \frac{e^{\frac{i}{2\hbar}r^2}}{(2\pi i\hbar)^{n/2}} e^{\frac{-i\lambda}{\hbar}P(r)} r^{n-1} dr \Big) d\Omega_{n-1} \qquad (2.55)$$

where instead of n Cartesian coordinates we use $n-1$ angular coordinates $(\phi_1, \dots, \phi_{n-1})$ and the variable $r = |x|$. S_{n-1} denotes the $(n-1)$-dimensional spherical surface, $d\Omega_{n-1}$ is the Haar measure on it, $f(\phi_1, \dots, \phi_{n-1}) = (k \cdot x)/|k|r$, $P(r)$ is a fourth order polynomial in the variable r with coefficients depending on the $n-1$ angular variables $(\phi_1, \dots, \phi_{n-1})$, namely:

$$P(r) = r^4 A\Big(\frac{x}{|x|}, \frac{x}{|x|}, \frac{x}{|x|}, \frac{x}{|x|}\Big) = r^4 a(\phi_1, \dots, \phi_{n-1}), \qquad (2.56)$$

where $a(\phi_1, \dots, \phi_{n-1}) > 0$ for all $(\phi_1, \dots, \phi_{n-1}) \in S_{n-1}$. Let us focus on the integral

$$\int_0^{+\infty} e^{i|k|rf(\phi_1,\dots,\phi_{n-1})} \frac{e^{\frac{i}{2\hbar}r^2}}{(2\pi i\hbar)^{n/2}} e^{\frac{-i\lambda}{\hbar}P(r)} r^{n-1} dr.$$

This can be interpreted as the Fourier transform of the distribution on the real line

$$F(r) = \theta(r) r^{n-1} \frac{e^{\frac{i}{2\hbar}r^2}}{(2\pi i\hbar)^{n/2}} e^{\frac{-i\lambda}{\hbar}P(r)},$$

with $\theta(r) = 1$ if $r \geq 0$ and $\theta(r) = 0$ otherwise, $\lambda < 0$ and $P(r) = ar^4$, $a > 0$:

$$\int_0^{+\infty} e^{ikr} \frac{e^{\frac{i}{2\hbar}r^2}}{(2\pi i\hbar)^{n/2}} e^{\frac{-i\lambda}{\hbar}P(r)} r^{n-1} dr. \qquad (2.57)$$

Let us consider the complex plane and set $z = re^{i\theta}$. We have

$$
\begin{aligned}
&\int_0^{+\infty} e^{ikr} \frac{e^{\frac{i}{2\hbar}r^2}}{(2\pi i\hbar)^{n/2}} e^{\frac{-i\lambda}{\hbar}P(r)} r^{n-1} dr \\
&= \lim_{\epsilon \downarrow 0} \int_{z=\rho e^{i\epsilon}} e^{ikz} \frac{e^{\frac{i}{2\hbar}z^2}}{(2\pi i\hbar)^{n/2}} e^{\frac{-i\lambda}{\hbar}P(z)} z^{n-1} dz \\
&= \lim_{\epsilon \downarrow 0} \lim_{R \to +\infty} \int_0^R e^{ik\rho e^{i\epsilon}} \frac{e^{\frac{i}{2\hbar}\rho^2 e^{2i\epsilon}}}{(2\pi i\hbar)^{n/2}} e^{\frac{-i\lambda}{\hbar}P(\rho e^{i\epsilon})} \rho^{n-1} e^{ni\epsilon} d\rho .
\end{aligned}
$$

Given:

$$
\begin{aligned}
\gamma_1(R) &= \{z \in \mathbb{C} \mid 0 \le \rho \le R, \quad \theta = \epsilon\} \\
\gamma_2(R) &= \{z \in \mathbb{C} \mid \rho = R, \epsilon \le \theta \le \pi/4 - \epsilon\} \\
\gamma_3(R) &= \{z \in \mathbb{C} \mid 0 \le \rho \le R, \quad \theta = \pi/4 - \epsilon\}
\end{aligned}
$$

with $\epsilon > 0$ small, from the analyticity of the integrand and the Cauchy theorem we have

$$
\int_{\gamma_1(R)\cup\gamma_2(R)\cup\gamma_3(R)} e^{ikz} \frac{e^{\frac{i}{2\hbar}z^2}}{(2\pi i\hbar)^{n/2}} e^{\frac{-i\lambda}{\hbar}P(z)} z^{n-1} dz = 0.
$$

In particular:

$$
\begin{aligned}
&\Big| \int_{\gamma_2(R)} e^{ikz} \frac{e^{\frac{i}{2\hbar}z^2}}{(2\pi i\hbar)^{n/2}} e^{\frac{-i\lambda}{\hbar}P(z)} z^{n-1} dz \Big| \\
&= R^n \Big| \int_\epsilon^{\pi/4-\epsilon} e^{ikRe^{i\theta}} \frac{e^{\frac{ie^{i2\theta}}{2\hbar}R^2}}{(2\pi i\hbar)^{n/2}} e^{\frac{-i\lambda}{\hbar}P(Re^{i\theta})} e^{in\theta} d\theta \Big| \\
&\le R^n \int_\epsilon^{\pi/4-\epsilon} e^{-kR\sin(\theta)} \frac{e^{\frac{-\sin(2\theta)}{2\hbar}R^2}}{(2\pi\hbar)^{n/2}} e^{\frac{\lambda}{\hbar}(aR^4\sin(4\theta))} d\theta \\
&\le R^n \int_\epsilon^{\pi/8} e^{-k'R\theta} \frac{e^{\frac{-2\theta}{\pi\hbar}R^2}}{(2\pi\hbar)^{n/2}} e^{\frac{\lambda}{\hbar}(aR^4\frac{8}{\pi})\theta} d\theta \\
&\quad + R^n e^{\frac{\lambda}{\hbar}2aR^4} \int_{\pi/8}^{\pi/4-\epsilon} e^{-k'R\theta} \frac{e^{\frac{-2\theta}{\pi\hbar}R^2}}{(2\pi\hbar)^{n/2}} e^{\frac{\lambda}{\hbar}(-aR^4\frac{8}{\pi})\theta} d\theta \\
&= \frac{R^n}{(2\pi\hbar)^{n/2}} \Big\{ \Big(\frac{e^{(\frac{8a\lambda}{\pi\hbar}R^4 - \frac{2}{\pi\hbar}R^2 - k'R)\pi/8} - e^{(\frac{8a\lambda}{\pi\hbar}R^4 - \frac{2}{\pi\hbar}R^2 - k'R)\epsilon}}{(\frac{8a\lambda}{\pi\hbar}R^4 - \frac{2}{\pi\hbar}R^2 - k'R)} \Big) \\
&\quad + \Big(\frac{e^{\frac{8\epsilon a\lambda}{\pi\hbar}R^4} e^{(-\frac{2}{\pi\hbar}R^2 - k'R)(\pi/4-\epsilon)} - e^{\frac{a\lambda}{\hbar}R^4} e^{(-\frac{2}{\pi\hbar})R^2 - k'R)\pi/8}}{(\frac{-8a\lambda}{\pi\hbar}R^4 - \frac{2}{\pi\hbar}R^2 + -k'R)} \Big) \Big\} \qquad (2.58)
\end{aligned}
$$

where $k' \in \mathbb{R}$ is a suitable constant. We have used the fact that if $\alpha \in [0, \pi/2]$ then $\frac{2}{\pi}\alpha \le \sin(\alpha) \le \alpha$, while if $\alpha \in [\pi/2, \pi]$ then $\sin(\alpha) \ge 2 - \frac{2}{\pi}\alpha$. From the last line one can deduce that

$$
\Big| \int_{\gamma_2(R)} e^{ikz} \frac{e^{\frac{i}{2\hbar}z^2}}{(2\pi i\hbar)^{n/2}} e^{\frac{-i\lambda}{\hbar}P(z)} z^{n-1} dz \Big| \to 0, \qquad R \to \infty,
$$

so that

$$\int_{z=\rho e^{i\epsilon}} e^{ikz}\frac{e^{\frac{i}{2\hbar}z^2}}{(2\pi i\hbar)^{n/2}}e^{\frac{-i\lambda}{\hbar}P(z)}z^{n-1}dz$$

$$= \int_{z=\rho e^{i(\pi/4-\epsilon)}} e^{ikz}\frac{e^{\frac{i}{2\hbar}z^2}}{(2\pi i\hbar)^{n/2}}e^{\frac{-i\lambda}{\hbar}P(z)}z^{n-1}dz.$$

By taking the limit as $\epsilon \downarrow 0$ of both sides one gets:

$$\int_0^{+\infty} e^{ikr}\frac{e^{\frac{i}{2\hbar}r^2}}{(2\pi i\hbar)^{n/2}}e^{\frac{-i\lambda}{\hbar}P(r)}r^{n-1}dr$$

$$= \int_0^{+\infty} e^{ik\rho e^{i\pi/4}}\frac{e^{\frac{-\rho^2}{2\hbar}}}{(2\pi\hbar)^{n/2}}e^{\frac{-i\lambda}{\hbar}P(\rho e^{i\pi/4})}\rho^{n-1}d\rho \qquad (2.59)$$

By substituting into (2.55) we get the final result:

$$\int_{\mathbb{R}^n} e^{ik\cdot x}\frac{e^{\frac{i}{2\hbar}x\cdot x}}{(2\pi i\hbar)^{n/2}}e^{\frac{-i\lambda}{\hbar}P(x)}dx$$

$$= \int_{\mathbb{R}^n} e^{ie^{i\pi/4}k\cdot x}\frac{e^{\frac{-x\cdot x}{2\hbar}}}{(2\pi\hbar)^{n/2}}e^{\frac{-i\lambda}{\hbar}P(xe^{i\pi/4})}dx$$

$$= \mathbb{E}[e^{ie^{i\pi/4}k\cdot x}e^{\frac{-i\lambda}{\hbar}P(xe^{i\pi/4})}].$$

□

Remark 2.5. A careful reading of this proof shows that the second part of the statement, that is representation (2.54), is valid if and only if the degree of P is 4, but cannot be generalized to polynomial functions of higher (even) degree. In fact the proof is based on the analyticity of the integrand and on a deformation of the contour of integration into a region of the complex plane in which the real part of the leading term of the polynomial, that is $Re(-i\lambda a z^4)$, is negative, where $\lambda < 0$, $a > 0$. By setting $z = \rho e^{i\theta}$ one can immediately verify that this condition is satisfied if and only if $0 \leq \theta \leq \pi/4$. By considering a polynomial of higher even degree $2M$ this condition becomes $0 \leq \theta \leq \pi/2M$ and if $M > 2$ the angle $\theta = \pi/4$ is no longer included. This angle is fundamental as the oscillatory function $\frac{e^{\frac{i}{2\hbar}z^2}}{(2\pi i\hbar)^{1/2}}$ evaluated in $z = \rho e^{i\pi/4}$ gives $e^{-i\pi/4}\frac{e^{\frac{-\rho^2}{2\hbar}}}{(2\pi\hbar)^{1/2}}$, that is the density of the normal distribution with mean zero and variance $\hbar^2$, multiplied by the factor $e^{-i\pi/4}$. These considerations also show the necessity of considering $\lambda \leq 0$.

Remark 2.6. We note that to have $\lambda = 0$ is equivalent to take $P = 0$. In this case one has immediately:

$$\int_{\mathbb{R}^n} e^{ik\cdot x} \frac{e^{\frac{i}{2\hbar}x\cdot x}}{(2\pi i\hbar)^{n/2}} dx$$
$$= \int_{\mathbb{R}^n} e^{ik\cdot xe^{i\pi/4}} \frac{e^{\frac{-x\cdot x}{2\hbar}}}{(2\pi\hbar)^{n/2}} dx = \mathbb{E}[e^{ik\cdot xe^{i\pi/4}}] = e^{\frac{-i\hbar}{2}k\cdot k}. \quad (2.60)$$

These results can now be applied to the definition of a generalized Fresnel integral of the following form

$$I(f) \equiv \widetilde{\int} e^{\frac{i}{2\hbar}x\cdot(I-B)x} e^{\frac{-i\lambda}{\hbar}P(x)} f(x)dx\,. \quad (2.61)$$

Theorem 2.10. *("Parseval equality") Let $f \in \mathcal{F}(\mathbb{R}^n)$, $f = \hat{\mu}_f$. Then the generalized Fresnel integral (2.61) is well defined and it is given by:*

$$\widetilde{\int} e^{\frac{i}{2\hbar}x\cdot(I-B)x} e^{\frac{-i\lambda}{\hbar}P(x)} f(x)dx = \int \tilde{F}(k)\mu_f(dk), \quad (2.62)$$

where $\tilde{F}(k)$ is given by Eq. (2.53) if A in (2.50) is strictly positive, or by Eq. (2.54) if $A \geq 0$, $\lambda \leq 0$ and $(I - B)$ is symmetric strictly positive. The integral on the right hand side of (2.62) is absolutely convergent (hence it can be understood in Lebesgue sense).

Proof. The proof is completely analogous to the proof of theorem 2.4. □

Corollary 2.2. *Let $(I - B)$ be symmetric and strictly positive, $\lambda \leq 0$ and $f \in \mathcal{F}(\mathbb{R}^n)$, $f = \hat{\mu}_f$, such that $\forall x \in \mathbb{R}^n$ the integral $\int e^{-\frac{\sqrt{2}}{2}kx}|\mu_f|(dk)$ is convergent and the positive function $g : \mathbb{R}^n \to \mathbb{R}$, defined by*

$$g(x) = e^{\frac{1}{2\hbar}x\cdot Bx} \int e^{-\frac{\sqrt{2}}{2}kx}|\mu_f|(dk), \qquad x \in \mathbb{R}^n$$

is summable with respect to the centered Gaussian measure on $\mathbb{R}^n$ with covariance $\hbar I$.

Then f extends to an analytic function on $\mathbb{C}^n$ and the corresponding generalized Fresnel integral is well defined and it is given by

$$\widetilde{\int_{\mathbb{R}^n}} \frac{e^{\frac{i}{2\hbar}x\cdot(I-B)x}}{(2\pi i\hbar)^{n/2}} e^{\frac{-i\lambda}{\hbar}P(x)} f(x)dx = \mathbb{E}[e^{\frac{i\lambda}{\hbar}P(x)} e^{\frac{1}{2\hbar}x\cdot Bx} f(e^{i\pi/4}x)]. \quad (2.63)$$

Proof. By the assumption on the measure μ_f, it follows that its Laplace transform $f^L : \mathbb{C}^n \to \mathbb{C}$,

$$f^L(z) = \int_{\mathbb{R}^n} e^{kz}\mu_f(dk),$$

is a well defined entire function such that $f^L(ix) = f(x)$, $x \in \mathbb{R}^n$. By theorem 2.10 the generalized Fresnel integral can be computed by means of Parseval equality

$$\widetilde{\int_{\mathbb{R}^n}} \frac{e^{\frac{i}{2\hbar}x\cdot(I-B)x}}{(2\pi i\hbar)^{n/2}} e^{\frac{-i\lambda}{\hbar}P(x)} f(x)dx = \int_{\mathbb{R}^n} \tilde{F}(k)\mu_f(dk)$$
$$= \int_{\mathbb{R}^n} \mathbb{E}[e^{ikxe^{i\pi/4}} e^{\frac{1}{2\hbar}x\cdot Bx} e^{\frac{i\lambda}{\hbar}P(x)}]\mu_f(dk).$$

By Fubini theorem, which applies given the assumptions on the measure μ_f, this is equal to

$$\mathbb{E}[e^{\frac{1}{2\hbar}x\cdot Bx} e^{\frac{i\lambda}{\hbar}P(x)} \int_{\mathbb{R}^n} e^{ikxe^{i\pi/4}} \mu_f(dk)] = \mathbb{E}[e^{\frac{1}{2\hbar}x\cdot Bx} e^{\frac{i\lambda}{\hbar}P(x)} f^L(ie^{i\pi/4}x)]$$
$$= \mathbb{E}[e^{\frac{1}{2\hbar}x\cdot Bx} e^{\frac{i\lambda}{\hbar}P(x)} f(e^{i\pi/4}x)] \quad (2.64)$$

and the conclusion follows. □

Remark 2.7. The latter theorem shows that, under suitable assumptions on the function f, the generalized Fresnel integral (2.61) can be explicitly computed by means of a Gaussian integral. By mimicking the proof of lemma 2.3 one can be tempted to generalize Eq. (2.63) to a larger class of functions, that are analytic in a suitable region of $\mathbb{C}^n$, but do not belong to $\mathcal{F}(\mathbb{R}^n)$. In fact this is not possible, as the definition 2.1 of oscillatory integral requires that the limit of the sequence of regularized integrals exists and is independent of the regularization. Let us consider the subset of the complex plane

$$\Lambda = \{z \in \mathbb{C} \mid 0 < \arg(z) < \pi/4\} \subset \mathbb{C}, \quad (2.65)$$

and let $\bar{\Lambda}$ be its closure. The identity

$$\lim_{\epsilon\to 0} \int\int_{\mathbb{R}^n} \frac{e^{\frac{i}{2\hbar}x\cdot(I-B)x}}{(2\pi i\hbar)^{n/2}} e^{\frac{-i\lambda}{\hbar}P(x)} f(x)\psi(\epsilon x)dx = \mathbb{E}[e^{\frac{1}{2\hbar}x\cdot Bx} e^{\frac{i\lambda}{\hbar}P(x)} f(e^{i\pi/4}x)]$$

(with $(I-B)$ symmetric strictly positive and $\lambda \leq 0$) can only be proved by choosing a regularizing function $\psi \in \mathcal{S}$, $\psi(0) = 1$, such that the function $z \mapsto \psi(zx)$ is analytic for $z \in \Lambda$ and continuous for $z \in \bar{\Lambda}$ for each $x \in \mathbb{R}^n$. Moreover one has to assume that $|\psi(e^{i\theta}x)|$ is bounded as $|x| \to \infty$ for each $\theta \in (0, \pi/4)$.

We can now generalize these results to the case the generalized Fresnel integral is defined on a real separable infinite dimensional Hilbert space $(\mathcal{H}, \langle\ ,\ \rangle, \|\ \|)$.

Let us consider the *abstract Wiener space* built on $\mathcal{H}$ (see [149, 211] and appendix). Indeed, let ν be the finitely additive cylinder measure on $\mathcal{H}$, defined by its characteristic functional $\hat{\nu}(x) = e^{-\frac{\hbar}{2}\|x\|^2}$. Let $|\ |$ be a "measurable" norm on $\mathcal{H}$, that is $|\ |$ is such that for every $\epsilon > 0$ there exist a finite-dimensional projection $P_\epsilon : \mathcal{H} \to \mathcal{H}$, such that for all $P \perp P_\epsilon$ one has

$$\nu(\{x \in \mathcal{H} |\ |P(x)| > \epsilon\}) < \epsilon,$$

where P and P_ϵ are called orthogonal ($P \perp P_\epsilon$) if their ranges are orthogonal in $(\mathcal{H}, \langle\ ,\ \rangle)$. One can easily verify that $|\ |$ is weaker than $\|\ \|$. Denoted by $\mathcal{B}$ the completion of $\mathcal{H}$ in the $|\ |$-norm and by i the continuous inclusion of $\mathcal{H}$ in $\mathcal{B}$, one can prove that $\mu \equiv \nu \circ i^{-1}$ is a countably additive Gaussian measure on the Borel subsets of $\mathcal{B}$ (see theorems A.2 and A.3). The triple $(i, \mathcal{H}, \mathcal{B})$ is called abstract Wiener space.

Given $y \in \mathcal{B}^*$ one can easily verify that the restriction of y to $\mathcal{H}$ is continuous on $\mathcal{H}$, so that one can identify $\mathcal{B}^*$ as a subset of $\mathcal{H}$. Moreover $\mathcal{B}^*$ is dense in $\mathcal{H}$ and we have the dense continuous inclusions $\mathcal{B}^* \subset \mathcal{H} \subset \mathcal{B}$. Each element $y \in \mathcal{B}^*$ can be regarded as a random variable $n(y)$ on $(\mathcal{B}, \mu)$. A direct computation shows that $n(y)$ is normally distributed, with covariance $|y|^2$ (see theorem A.2 and Eq. (A.4)). The density of $\mathcal{B}^*$ in $\mathcal{H}$ and Eq. (A.4) allow the extension of the map $n : \mathcal{B}^* \to L^2(\mathcal{B}, \mu)$ to the whole space $\mathcal{H}$ (the extended map will be denoted with the same symbol with an abuse of notation).

Given an orthogonal projection P in $\mathcal{H}$, with

$$P(x) = \sum_{i=1}^{n} \langle e_i, x \rangle e_i$$

for some orthonormal $e_1, \ldots, e_n \in \mathcal{H}$, the stochastic extension $\tilde{P}$ of P on $\mathcal{B}$ is defined by

$$\tilde{P}(\cdot) = \sum_{i=1}^{n} n(e_i)(\cdot) e_i.$$

Analogously, given a function $f : \mathcal{H} \to \mathcal{B}_1$, where $(\mathcal{B}_1, |\ |_{\mathcal{B}_1})$ is another real separable Banach space, the stochastic extension $\tilde{f}$ of f to $\mathcal{B}$ exists if the functions $f \circ \tilde{P} : \mathcal{B} \to \mathcal{B}_1$ converges to $\tilde{f}$ in probability with respect to μ as P converges strongly to the identity in $\mathcal{H}$.

Let $A : \mathcal{H}\times\mathcal{H}\times\mathcal{H}\times\mathcal{H} \to \mathbb{R}$ be a completely symmetric positive covariant tensor operator on $\mathcal{H}$ such that the map $V : \mathcal{H} \to \mathbb{R}^+$,

$$x \mapsto V(x) \equiv A(x, x, x, x), \qquad x \in \mathcal{H},$$

is continuous in the measurable norm $|\ |$. As a consequence V is continuous in the Hilbert norm $\|\ \|$, moreover it can be extended by continuity to a random variable $\bar{V}$ on $\mathcal{B}$, with $\bar{V}|_{\mathcal{H}} = V$. By theorem A.4, the stochastic extension $\tilde{V}$ of V exists and coincides with $\bar{V} : \mathcal{B} \to \mathbb{R}$ $\mu-$a.e. Moreover for any increasing sequence of $n-$dimensional projectors P_n in $\mathcal{H}$, the family of bounded random variables on $(\mathcal{B}, \mu)$

$$e^{i\frac{\lambda}{\hbar}V\circ\tilde{P}_n(\,\cdot\,)} \equiv e^{i\frac{\lambda}{\hbar}V^n(\,\cdot\,)}$$

converges $\mu-$a.e. to

$$e^{i\frac{\lambda}{\hbar}\bar{V}(\,\cdot\,)}.$$

Furthermore for any $h \in \mathcal{H}$ the sequence of random variables

$$\sum_{i=1}^{n} h_i n(e_i), \qquad h_i = \langle e_i, h\rangle$$

converges in $L^2(\mathcal{B}, \mu)$, and by subsequences almost everywhere, to the random variable $n(h)$. Analogously, given a self-adjoint trace class operator $B : \mathcal{H} \to \mathcal{H}$, the quadratic form on $\mathcal{H} \times \mathcal{H}$:

$$x \in \mathcal{H} \mapsto \langle x, Bx\rangle$$

can be extended to a random variable on $\mathcal{B}$, denoted again by $\langle\,\cdot\,, B\,\cdot\,\rangle$ (see appendix).

Let us assume that the largest eigenvalue of B is strictly less than 1 (or, in other words, that the operator $(I - B)$ is strictly positive) and let $y \in \mathcal{H}$. Then one can prove (see theorem A.5) that the sequence of random variables $f_n : \mathcal{B} \to \mathbb{R}$

$$f_n(\,\cdot\,) = e^{\sum_{i=1}^{n} y_i n(e_i)(\,\cdot\,)} e^{\frac{1}{2\hbar}\sum_{i=1}^{n} b_i([n(e_i)(\,\cdot\,)]^2},$$

where $y_i = \langle y, e_i\rangle$, converges $\mu-$a.e. as n goes to ∞ to the random variable

$$f(\,\cdot\,) = e^{n(y)(\,\cdot\,)} e^{\frac{1}{2\hbar}\langle\,\cdot\,, B\,\cdot\,\rangle}$$

and that

$$\int f_n d\mu \to \int f d\mu = (\det(I - B))^{-1/2} e^{\frac{\hbar}{2}\langle y, (I-B)^{-1}y\rangle}. \tag{2.66}$$

In this setting it is possible to prove the following result:

Lemma 2.4. *Let $B : \mathcal{H} \to \mathcal{H}$ be a self adjoint and trace class operator such that $I - B$ is strictly positive, let $k \in \mathcal{H}$ and $\lambda \leq 0$. Then for any increasing* *sequence P_n of projectors onto n-dimensional subspaces of $\mathcal{H}$ such that $P_n \uparrow I$ strongly as $n \to \infty$, the following sequence of finite dimensional integrals:*

$$F_n(k) \equiv (2\pi i\hbar)^{-n/2} \int_{P_n\mathcal{H}} e^{i\langle P_n k, P_n x\rangle} e^{\frac{i}{2\hbar}\langle P_n x,(I-B)P_n x\rangle} e^{-i\frac{\lambda}{\hbar}V(P_n x)} d(P_n x)$$

converges, as $n \to \infty$, to the Gaussian integral on $\mathcal{B}$:

$$F(k) \equiv \mathbb{E}[e^{in(k)(\,\cdot\,)e^{i\pi/4}} e^{\frac{1}{2\hbar}\langle\,\cdot\,,B\,\cdot\,\rangle} e^{i\frac{\lambda}{\hbar}\bar{V}(\,\cdot\,)}] \tag{2.67}$$

($\mathbb{E}$ being the expectation with respect to μ on $\mathcal{B}$).

Proof. By lemma 2.3 and Eq. (2.54) one has

$$(2\pi i\hbar)^{-n/2} \int_{P_n\mathcal{H}} e^{i\langle P_n k, P_n x\rangle} e^{\frac{i}{2\hbar}\langle P_n x,(I-B)P_n x\rangle} e^{-i\frac{\lambda}{\hbar}V(P_n x)} d(P_n x) =$$

$$(2\pi\hbar)^{-n/2} \int_{P_n\mathcal{H}} e^{i\langle P_n k, P_n x\rangle e^{i\pi/4}} e^{-\frac{1}{2\hbar}\langle P_n x, P_n x\rangle} e^{\frac{1}{2\hbar}\langle P_n x, BP_n x\rangle} e^{i\frac{\lambda}{\hbar}V(P_n x)} d(P_n x)\,.$$

Let us introduce an orthonormal base $\{e_i\}$ of $\mathcal{H}$ such that P_n is the projector onto the span of the first n vectors. Each element $P_n x \in P_n\mathcal{H}$ can be represented as an $n-$ple of real numbers $(x_1, \ldots, x_n)$, where $x_i = \langle x, e_i\rangle$. The latter integral can be written in the following form:

$$(2\pi\hbar)^{-n/2} \int_{\mathbb{R}^n} e^{i\sum_{i=1}^n k_i x_i e^{i\pi/4}} e^{-\frac{1}{2\hbar}\sum_{i=1}^n x_i^2} e^{\frac{1}{2\hbar}\sum_{i,j=1}^n B_{ij}x_i x_j}$$

$$e^{i\frac{\lambda}{\hbar}\sum_{ij,k,h=1}^n A_{ijkh}x_i x_j x_k x_h} dx_1 \ldots dx_n$$

where $B_{ij} = \langle e_i, Be_j\rangle$ and $A_{ijkh} = A(e_i, e_j, e_k, e_h)$.
On the other hand, this coincides with the Gaussian integral on $(\mathcal{B}, \mu)$:

$$\mathbb{E}[e^{i\sum_{i=1}^n k_i n(e_i)(\,\cdot\,)e^{i\pi/4}} e^{\frac{1}{2\hbar}\sum_{i,j=1}^n \langle e_i, Be_j\rangle n(e_i)(\,\cdot\,)n(e_j)(\,\cdot\,)} e^{\frac{\lambda}{\hbar}V\circ\tilde{P}_n(\,\cdot\,)}].$$

By Lebesgue's dominated convergence theorem (which holds because of the assumption on the strict positivity of the operator $I - B$) this converges as $n \to \infty$ to

$$\mathbb{E}[e^{in(k)(\,\cdot\,)e^{i\pi/4}} e^{\frac{1}{2\hbar}\langle\,\cdot\,,B\,\cdot\,\rangle} e^{i\frac{\lambda}{\hbar}\bar{V}(\,\cdot\,)}]$$

and the conclusion follows. □

The above result allows the following generalization of theorem 2.10 to the infinite dimensional case.

Theorem 2.11. [27, 25] *Let $B : \mathcal{H} \to \mathcal{H}$ be self-adjoint trace class, $(I-B)$ strictly positive, $\lambda \le 0$ and $f \in \mathcal{F}(\mathcal{H})$, $f \equiv \hat{\mu}_f$, and let us assume that the bounded variation measure μ_f satisfies the following condition:*

$$\int_{\mathcal{H}} e^{\frac{\hbar}{4}\langle k,(I-B)^{-1}k\rangle} |\mu_f|(dk) < +\infty. \tag{2.68}$$

Then the infinite dimensional oscillatory integral

$$\widetilde{\int_{\mathcal{H}}} e^{\frac{i}{2\hbar}\langle x,(I-B)x\rangle} e^{-i\frac{\lambda}{\hbar}A(x,x,x,x)} f(x)dx \tag{2.69}$$

exists and is given by:

$$\int_{\mathcal{H}} \mathbb{E}[e^{in(k)(\,\cdot\,)e^{i\pi/4}} e^{\frac{1}{2\hbar}\langle\,\cdot\,,B\,\cdot\,\rangle} e^{i\frac{\lambda}{\hbar}\bar{V}(\,\cdot\,)}]\mu_f(dk).$$

Proof. By definition, choosing an increasing sequence of finite dimensional projectors P_n on $\mathcal{H}$, with $P_n \uparrow I$ strongly as $n \to \infty$, the oscillatory integral (2.69) is given by:

$$\lim_{n\to\infty}(2\pi i\hbar)^{-n/2} \int_{P_n\mathcal{H}} e^{\frac{i}{2\hbar}\langle P_n x,(I-B)P_n x\rangle} e^{-i\frac{\lambda}{\hbar}A(P_n x,P_n x,P_n x,P_n x)} f(P_n x)dP_n x. \tag{2.70}$$

Let $f^n : P_n\mathcal{H} \to \mathbb{C}$ be the function defined by

$$f^n(y) \equiv f(y), \qquad y \in P_n\mathcal{H}.$$

One can easily verify that $f^n \in \mathcal{F}(P_n\mathcal{H})$, $f^n = \hat{\mu}^n_f$, where μ^n_f is the bounded variation measure on $P_n\mathcal{H}$ defined by

$$\mu^n_f(I) = \mu_f(P_n^{-1}I),$$

I being a Borel subset of $P_n\mathcal{H}$, indeed:

$$\begin{aligned} f^n(y) = f(y) &= \int_{\mathcal{H}} e^{i\langle y,k\rangle}\mu_f(dk) \\ &= \int_{\mathcal{H}} e^{i\langle P_n y,P_n k\rangle}\mu_f(dk) = \int_{P_n\mathcal{H}} e^{i\langle y,P_n k\rangle}\mu^n_f(dP_n k) \end{aligned} \tag{2.71}$$

where $y = P_n y$. By theorem 2.10 the limit (2.70) is equal to

$$\lim_{n\to\infty}\int_{P_n\mathcal{H}} G_n(P_n k)\mu^n_f(dP_n k), \tag{2.72}$$

where $G_n : P_n\mathcal{H} \to \mathbb{C}$ is given by:

$$G_n(P_n k) = (2\pi\hbar)^{-n/2}\int_{P_n\mathcal{H}} e^{i\langle P_n k,P_n x\rangle e^{i\pi/4}} e^{-\frac{1}{2\hbar}\langle P_n x,(I-B)P_n x\rangle} e^{i\frac{\lambda}{\hbar}A(P_n x,P_n x,P_n x,P_n x)}dP_n x.$$

This, on the other hand (see the proof of lemma 2.4) is equal to

$$\mathbb{E}[e^{in(P_n k)(\,\cdot\,)e^{i\pi/4}} e^{\frac{1}{2\hbar}\sum_{i,j=1}^n B_{ij}n(e_i)(\,\cdot\,)n(e_j)(\,\cdot\,)} e^{i\frac{\lambda}{\hbar}V^n(\,\cdot\,)}],$$

where $V^n = V \circ \tilde{P}_n$. By substituting the latter expression into (2.72) we have

$$\lim_{n\to\infty} \int_{P_n\mathcal{H}} \mathbb{E}[e^{in(P_nk)(\cdot)e^{i\pi/4}} e^{\frac{1}{2\hbar}\sum_{i,j=1}^n B_{ij}n(e_i)(\cdot)n(e_j)(\cdot)} e^{i\frac{\lambda}{\hbar}V^n(\cdot)}]\mu_f^n(dP_nk)$$

$$= \lim_{n\to\infty} \int_{\mathcal{H}} \mathbb{E}[e^{in(P_nk)(\cdot)e^{i\pi/4}} e^{\frac{1}{2\hbar}\sum_{i,j=1}^n B_{ij}n(e_i)(\cdot)n(e_j)(\cdot)} e^{i\frac{\lambda}{\hbar}V^n(\cdot)}]\mu_f(dk)$$

$$= \lim_{n\to\infty} \int_{\mathcal{H}} F_n(k)\mu_f(dk) .$$

By lemma 2.4 and the dominated convergence theorem, applicable to the integral with respect to μ_f, due to assumption (2.68), we then get

$$\int_{\mathcal{H}} F(k)\mu_f(dk) = \int_{\mathcal{H}} \mathbb{E}[e^{in(k)(\cdot)e^{i\pi/4}} e^{\frac{1}{2\hbar}\langle \cdot, B\cdot\rangle} e^{i\frac{\lambda}{\hbar}\bar{V}(\cdot)}]\mu_f(dk),$$

and the conclusion follows. □

Also corollary 2.2 can be generalized to the infinite dimensional case. Indeed due to the assumption (2.68) the function f on the real Hilbert space $\mathcal{H}$ can be extended to those vectors $y \in \mathcal{H}^{\mathbb{C}}$ in the complex Hilbert space $\mathcal{H}^{\mathbb{C}}$ of the form $y = zx$, $x \in \mathcal{H}$, $z \in \mathbb{C}$, as the integral

$$\int_{\mathcal{H}} e^{iz\langle x,k\rangle}\mu_f(dk)$$

is absolutely convergent. Moreover the latter can be uniquely extended to a random variable on $\mathcal{B}$, denoted again by f, by

$$f^z(\cdot) \equiv f(z\cdot) \equiv \int_{\mathcal{H}} e^{izn(k)(\cdot)}\mu_f(dk). \tag{2.73}$$

Moreover the random variable

$$e^{\frac{1}{2\hbar}\langle \cdot, B\cdot\rangle} f^z(\cdot)$$

belongs to $L^1(\mathcal{B}, \mu)$ if $Im(z)^2 \leq 1/2$.

Theorem 2.12. [27] *Let $B : \mathcal{H} \to \mathcal{H}$ be self-adjoint trace class, $I - B$ strictly positive, $\lambda \leq 0$ and $f \in \mathcal{F}(\mathcal{H})$ be the Fourier transform of a bounded variation measure μ_f satisfying assumption* (2.68).

Then the infinite dimensional oscillatory integral (2.69) *is well defined and it is given by:*

$$\widetilde{\int_{\mathcal{H}}} e^{\frac{i}{2\hbar}\langle x,(I-B)x\rangle} e^{-i\frac{\lambda}{\hbar}A(x,x,x,x)} f(x)dx = \mathbb{E}[e^{i\frac{\lambda}{\hbar}\bar{V}(\cdot)} e^{\frac{1}{2\hbar}\langle \cdot, B\cdot\rangle} f(e^{i\pi/4}\cdot)] \tag{2.74}$$

Proof. By theorem 2.11 the infinite dimensional oscillatory integral (2.69) can be computed by means of the Parseval-type formula:

$$\widetilde{\int_{\mathcal{H}}} e^{\frac{i}{2\hbar}\langle x,(I-B)x\rangle} e^{-i\frac{\lambda}{\hbar}A(x,x,x,x)} f(x)dx$$

$$= \int_{\mathcal{H}} \mathbb{E}[e^{in(k)(\,\cdot\,)e^{i\pi/4}} e^{\frac{1}{2\hbar}\langle\,\cdot\,,B\,\cdot\,\rangle} e^{i\frac{\lambda}{\hbar}\bar{V}(\,\cdot\,)}]\mu_f(dk) \quad (2.75)$$

By Fubini theorem, which can be applied under the assumption (2.68), the integral on the right hand side of (2.75) is equal to

$$\mathbb{E}[e^{i\frac{\lambda}{\hbar}\bar{V}(\,\cdot\,)} e^{\frac{1}{2\hbar}\langle\,\cdot\,,B\,\cdot\,\rangle} \int_{\mathcal{H}} e^{in(k)(\,\cdot\,)e^{i\pi/4}} \mu_f(dk)]$$

$$= \mathbb{E}[e^{i\frac{\lambda}{\hbar}\bar{V}(\,\cdot\,)} e^{\frac{1}{2\hbar}\langle\,\cdot\,,B\,\cdot\,\rangle} f^{e^{i\pi/4}}(\,\cdot\,)] = \mathbb{E}[e^{i\frac{\lambda}{\hbar}\bar{V}(\,\cdot\,)} e^{\frac{1}{2\hbar}\langle\,\cdot\,,B\,\cdot\,\rangle} f(e^{i\pi/4}\,\cdot\,)]\,.$$

The integral on the right hand side is absolutely convergent as $|e^{i\frac{\lambda}{\hbar}\bar{V}}| = 1$ and $e^{\frac{1}{2\hbar}\langle\,\cdot\,,B\,\cdot\,\rangle} f^{e^{i\pi/4}} \in L^1(\mathcal{B},\mu)$ as $Im(e^{i\pi/4}) = 1/\sqrt{2}$ (see Fernique's theorem in [211]). □

Remark 2.8. In the simpler case $\lambda = 0$, under the above assumptions on the function f and the operator B, the infinite dimensional oscillatory integral (given by (2.74) with $V = 0$) can also be explicitly computed by means of the absolutely convergent integrals:

$$\widetilde{\int_{\mathcal{H}}} e^{\frac{i}{2\hbar}\langle x,(I-B)x\rangle} f(x)dx = \frac{1}{\sqrt{\det(I-B)}} \int_{\mathcal{H}} e^{-\frac{i\hbar}{2}\langle k,(I-B)^{-1}k\rangle} \mu_f(dk)\,. \quad (2.76)$$

In fact, by means of different methods (see section 2.4), Eq. (2.76) can be proved even without the assumption on the positivity of the operator $(I - B)$ (it is sufficient that $(I - B)$ is invertible).

Remark 2.9. So far we have proved, under suitable assumptions on the function $f : \mathcal{H} \to \mathbb{C}$ and the operator B, that if $\lambda \leq 0$ the infinite dimensional generalized Fresnel integral (2.69)

$$I^F(\lambda) \equiv \widetilde{\int_{\mathcal{H}}} e^{\frac{i}{2\hbar}\langle x,(I-B)x\rangle} e^{-i\frac{\lambda}{\hbar}A(x,x,x,x)} f(x)dx$$

on the Hilbert space $\mathcal{H}$ is exactly equal to a Gaussian integral on $\mathcal{B}$:

$$I^G(\lambda) \equiv \int_{\mathcal{H}} \mathbb{E}[e^{in(k)(\,\cdot\,)e^{i\pi/4}} e^{\frac{1}{2\hbar}\langle\,\cdot\,,B\,\cdot\,\rangle} e^{i\frac{\lambda}{\hbar}\bar{V}(\,\cdot\,)}]\mu_f(dk)$$

(theorem 2.11), and to

$$I^A(\lambda) \equiv \mathbb{E}[e^{i\frac{\lambda}{\hbar}\bar{V}(\,\cdot\,)} e^{\frac{1}{2\hbar}\langle\,\cdot\,,B\,\cdot\,\rangle} f(e^{i\pi/4}\,\cdot\,)]$$

(theorem 2.12). One can easily verify that I^G and I^A are analytic functions of the complex variable λ in the region of the complex λ plane $\{Im(\lambda) > 0\}$, while they are continuous in $\{Im(\lambda) = 0\}$ and coincide with I^F in $\{Im(\lambda) = 0, Re(\lambda) \leq 0\}$.

The generalization of these techniques to infinite dimensional oscillatory integrals with polynomial phase function of higher degree is not straightforward. Some partial result concerning particular complex phase function of higher arbitrary degree has been obtained in [34].

Chapter 3

Feynman Path Integrals and the Schrödinger Equation

3.1 The anharmonic oscillator with a bounded anharmonic potential

Let us consider the Schrödinger equation in $L^2(\mathbb{R}^d)$

$$i\hbar\frac{\partial}{\partial t}\psi = H\psi \tag{3.1}$$

with initial datum $\psi_{|t=0} = \psi_0 \in L^2(\mathbb{R}^d)$ and quantum mechanical Hamiltonian, given on the smooth vectors $\psi \in \mathcal{S}(\mathbb{R}^d)$ by

$$H\psi(x) = -\frac{\hbar^2}{2m}\Delta\psi(x) + \frac{1}{2}x\Omega^2 x\psi(x) + V(x)\psi(x), \qquad x \in \mathbb{R}^d,$$

where $m > 0$ is the mass, $\Omega^2 \geq 0$ is a positive $d \times d$ matrix, and V is a bounded continuous real function on $\mathbb{R}^d$. In the following we shall put for notational simplicity $m \equiv 1$, but the whole discussion can be repeated for arbitrary values of the mass parameter.

Let us denote by H_0 the harmonic oscillator Hamiltonian, given on $\psi \in \mathcal{S}(\mathbb{R}^d)$ by

$$H_0\psi(x) = -\frac{\hbar^2}{2}\Delta\psi(x) + \frac{1}{2}x\Omega^2 x\psi(x), \qquad x \in \mathbb{R}^d.$$

Both H_0 and H are self-adjoint operators on $L^2(\mathbb{R}^d)$, on the natural domain of definition of H_0, which can be easily described by writing the vectors $\psi \in L^2(\mathbb{R}^d)$ as linear combination of Hermite functions (see [245, 246]).

H and H_0 generate unitary groups in $L^2(\mathbb{R}^d)$, denoted by $U(t) = e^{-\frac{i}{\hbar}Ht}$ and $U_0(t) = e^{-\frac{i}{\hbar}H_0 t}$. The solution of the Schrödinger equation (3.1) with initial datum $\psi_{|t=0} = \psi_0 \in L^2(\mathbb{R}^d)$, is given by

$$\psi(t,x) = (e^{-\frac{i}{\hbar}Ht}\psi_0)(x). \tag{3.2}$$

In the following we shall show how the Feynman path integral representation for the solution (3.2)

$$\psi(t,x) = \int_{\gamma(t)=x} e^{\frac{i}{\hbar}S_t(\gamma)}\psi_0(\gamma(0))d\gamma, \tag{3.3}$$

can be mathematically realized in terms of a well defined infinite dimensional oscillatory integral on a suitable Hilbert space of paths:

$$\psi(t,x) = \int_{\gamma(t)=0} e^{\frac{i}{\hbar}\int_0^t \left(\dot\gamma(s)^2/2-\gamma(s)\Omega^2\gamma(s)/2-V(\gamma(s)+x)\right)ds}\psi_0(\gamma(0)+x)d\gamma, \tag{3.4}$$

where we performed the formal change of variable $\gamma \mapsto \gamma + x$ in order to deal with the homogeneous condition $\gamma(t) = 0$, which is preserved by linear combinations.

For notational simplicity, in the following we shall put $m = 1$, but the whole discussion can be easily generalized to the case of arbitrary values of the mass parameter.

Let us consider the *Cameron-Martin space* $\mathcal{H}_t$, that is the Sobolev space of absolutely continuous functions $\gamma : [0,t] \to \mathbb{R}^d$, such that $\gamma(t) = 0$, and with square integrable weak derivative $\dot\gamma$:

$$\int_0^t |\dot\gamma(s)|^2 ds < \infty,$$

endowed with the inner product

$$\langle\gamma_1,\gamma_2\rangle = \int_0^t \dot\gamma_1(s)\cdot\dot\gamma_2(s)ds.$$

(This space was already introduced in (1.11) in the introduction). From a physical point of view, $\mathcal{H}_t$ represents a space of Feynman paths with finite kinetic energy.

Let us consider the linear operator $L : \mathcal{H}_t \to \mathcal{H}_t$ given by

$$(L\gamma)(s) := \int_s^t ds' \int_0^{s'} (\Omega^2\gamma)(s'')ds''. \tag{3.5}$$

One can easily verify that

$$\langle\gamma_1, L\gamma_2\rangle = \int_0^t \gamma_1(s)\Omega^2\gamma_2(s)ds,$$

and conclude that L is self-adjoint on $\mathcal{H}_t$ and positive:

$$\langle\gamma, L\gamma\rangle = \int_0^t \gamma(s)\Omega^2\gamma(s)ds \ge 0, \qquad \forall\gamma\in\mathcal{H}_t.$$

Let us now investigate the conditions for the existence of the inverse of the operator $I - L$.

Lemma 3.1. *Let Ω_j, with $j = 1, \ldots, d$, be the eigenvalues of the matrix Ω. If*

$$t \neq \left(n + \frac{1}{2}\right)\frac{\pi}{\Omega_j}, \qquad n \in \mathbb{N}, j = 1, \ldots, d,$$

then the operator $I - L$ is invertible and its inverse is given by:

$$(I-L)^{-1}\gamma(s) = \gamma(s) - \Omega \int_s^t \sin[\Omega(s-s')]\gamma(s')ds' \\ + \sin[\Omega(t-s)] \int_0^t (\cos \Omega t)^{-1}\Omega \cos(\Omega s')\gamma(s')ds'. \quad (3.6)$$

Proof. By equation (3.5), given a vector $\eta \in \mathcal{H}_t$, the vector $\gamma \in \mathcal{H}_t$ is of the form

$$\eta = (I-L)\gamma \qquad \Longleftrightarrow \qquad \gamma = (I-L)^{-1}\eta$$

if it satisfies the integral equation:

$$\eta(s) = \gamma(s) - \int_s^t ds' \int_0^{s'} (\Omega^2\gamma)(s'')ds''. \quad (3.7)$$

If η is sufficiently smooth, by differentiating twice, (3.7) is equivalent to the differential equation:

$$\ddot{\gamma}(s) + \Omega^2\gamma(s) = \ddot{\eta}(s)$$

with conditions $\gamma(t) = 0$ and $\dot{\gamma}(0) = \dot{\eta}(0)$. By standard technique the solution is given by

$$(I-L)^{-1}\eta(s) = \eta(s) - \Omega \int_s^t \sin[\Omega(s-s')]\eta(s')ds' \\ + \sin[\Omega(t-s)] \int_0^t (\cos \Omega t)^{-1}\Omega \cos(\Omega s')\eta(s')ds'. \quad (3.8)$$

It is easy to verify that the latter formula is valid for general $\eta \in \mathcal{H}_t$, and the conclusion follows. □

The following lemma gives a characterization of the spectrum of L [114].

Lemma 3.2. *Under the assumptions of lemma 3.1, the self-adjoint operator $L : \mathcal{H}_t \to \mathcal{H}_t$ given by Eq. (3.5) is trace class and*

$$\mathrm{Ind}(I-L) = \sum_{j=1}^{d} \left[\frac{\Omega_j t}{\pi} + \frac{1}{2}\right],$$

where $[\cdot]$ *denotes the integer part and* $\Omega_1, \ldots, \Omega_d$ *are the eigenvalues of the matrix* Ω*, counted with their multiplicity.*
Moreover the Fredholm determinant of $I - L$ *is given by:*

$$\det(I - L) = \det(\cos(\Omega t)).$$

Proof. We follow here the method by Elworthy and Truman [114]. Let us assume, without loss of generality, that the matrix Ω is diagonal.

Because of the self-adjointness and the positivity of L, we look for eigenvalues of the form p^2, with $p \in \mathbb{R}$. The real positive number p^2 is an eigenvalue of L, if there exists a $\gamma \in \mathcal{H}_t$ such that

$$L\gamma = p^2\gamma,$$

$$\int_s^t ds' \int_0^{s'} (\Omega^2\gamma)(s'')ds'' = p^2\gamma(s).$$

By differentiating twice, we obtain the differential equation

$$p^2\ddot{\gamma}(s) + \Omega^2\gamma(s) = 0,$$

with the conditions $\gamma(t) = 0$ and $\dot{\gamma}(0) = 0$. The solutions are given by the vectors

$$\gamma(s) = (\gamma^1(s), \ldots, \gamma^d(s)),$$

where $\gamma^j(s) = A_j \sin[\Omega_j(s - \phi_j)/p]$, for constant A_j, ϕ_j, $j = 1, \ldots, d$, satisfying the conditions:

$$\Omega_j\phi_j/p = \left(m_j + \frac{1}{2}\right)\pi, \qquad m_j \in \mathbb{Z},$$

$$\Omega_j t/p = \left(n_j + \frac{1}{2}\right)\pi, \qquad n_j \in \mathbb{Z}.$$

The only possible values for p are of the form

$$p = \frac{\Omega_j t}{\left(n_j + \frac{1}{2}\right)\pi}, \qquad n_j \in \mathbb{Z}.$$

Since $\sum_{n=0}^{\infty}(n+\frac{1}{2})^{-2} < \infty$, one can conclude that the operator $L : \mathcal{H}_t \to \mathcal{H}_t$ is trace class.

For the calculation of the index of the operator $I - L$, we have to take into account the eigenvalues of the operator L of the form

$$p^2 = \frac{\Omega_j t}{\left(n_j + \frac{1}{2}\right)\pi},$$

with $p > 1$, that is

$$n_j = 0, 1, \dots, \left[\frac{\Omega_j t}{\pi} - \frac{1}{2}\right],$$

[ ] denoting the integer part. They correspond to the following eigenvectors

$$\gamma_{n_j}(s) = \Big(0, 0, \dots, 0, \cos\left[\Big(n_j + \frac{1}{2}\Big)\frac{s\pi}{t}\right], 0, \dots, 0\Big),$$

where the jth entry being non-zero.

The proof of the second part of the lemma follows from the equality:

$$\cos(x) = \prod_{n=0}^{\infty}\left[1 - \frac{x^2}{\left(n + \frac{1}{2}\right)^2 \pi^2}\right].$$

□

The next lemma gives sufficient conditions for the definition of the infinite dimensional oscillatory integral (3.4) and the application of theorem 2.5.

Lemma 3.3. *Let $\psi_0, V \in \mathcal{F}(\mathbb{R}^d)$. Then the function $f : \mathcal{H}_t \to \mathbb{C}$, given by*

$$f(\gamma) = e^{-\frac{i}{\hbar}\int_0^t V(\gamma(s)+x)ds}\psi_0(\gamma(0) + x), \qquad \gamma \in \mathcal{H}_t$$

belongs to $\mathcal{F}(\mathcal{H}_t)$.

Proof. We give here the proof in the case where $d = 1$ in order to simplify the notation, but the whole reasoning can be easily extended to the case of general dimension d.

Let us denote by γ_τ, $\tau \in [0, t]$, the vector in $\mathcal{H}$ defined by:

$$\gamma_\tau(s) = t - \tau \vee s, \qquad s \in [0, t].$$

One can easily verify that for an $\gamma \in \mathcal{H}_t$

$$\langle \gamma_\tau, \gamma \rangle = \gamma(\tau).$$

If $\psi_0 = \hat{\mu}_0$, with $\mu_0 \in \mathcal{M}(\mathbb{R})$, we have:

$$\begin{aligned}
\psi_0(\gamma(0) + x) &= \int_{\mathbb{R}} e^{ik(\gamma(0)+x)} d\mu_0(k) \\
&= \int_{\mathbb{R}} e^{i\langle \gamma, k\gamma_0 \rangle} e^{ikx} d\mu_0(k) \qquad (3.9) \\
&= \int_{\mathbb{R}}\int_{\mathcal{H}} e^{i\langle \gamma, \eta \rangle} \delta_{k\gamma_0}(d\eta) e^{ikx} d\mu_0(k). \qquad (3.10)
\end{aligned}$$

Analogously, if $V \in \mathcal{F}(\mathbb{R})$, $V = \hat{\mu}_V$, with $\mu_v \in \mathcal{M}(\mathbb{R})$, we have

$$\begin{aligned}
\int_0^t V(\gamma(s) + x)ds &= \int_0^t \int_{\mathbb{R}} e^{ik(\gamma(s)+x)} d\mu_v(k)ds \\
&= \int_0^t \int_{\mathbb{R}}\int_{\mathcal{H}} e^{i\langle \gamma, \eta \rangle} \delta_{k\gamma_s}(d\eta) e^{ikx} d\mu_v(k)ds. \qquad (3.11)
\end{aligned}$$

It follows that both functions on $\mathcal{H}_t$, i.e. $\gamma \mapsto \psi_0(\gamma(0)+x)$ and $\gamma \mapsto \int_0^t V(\gamma(s)+x)ds$, belong to $\mathcal{F}(\mathcal{H}_t)$. As $\mathcal{F}(\mathcal{H}_t)$ is a Banach algebra by pointwise multiplications (see section 2.2), it is possible to conclude that also the function

$$\gamma \mapsto e^{-\frac{i}{\hbar}\int_0^t V(\gamma(s)+x)ds}\psi_0(\gamma(0)+x), \qquad \gamma \in \mathcal{H}_t,$$

is an element of $\mathcal{F}(\mathcal{H}_t)$. □

The Parseval type equality for infinite dimensional oscillatory integrals (theorem 2.5) and the previous lemmas allow us to define and compute the Feynman path integral representation for the solution of the Schrödinger equation with the harmonic oscillator Hamiltonian:

$$\begin{cases} i\hbar\frac{\partial}{\partial t}\psi(t,x) = -\frac{\hbar^2}{2}\Delta\psi(t,x) + \frac{1}{2}x\Omega^2 x\psi(t,x) \\ \psi(0,x) = \psi_0(x) \end{cases} \tag{3.12}$$

that is:

$$\psi(t,x) = \int_{\gamma(t)=0} e^{\frac{i}{2\hbar}\int_0^t \left(\dot\gamma(s)^2 - (\gamma(s)+x)\Omega^2(\gamma(s)+x)\right)ds}\psi_0(\gamma(0)+x)d\gamma, \tag{3.13}$$

Theorem 3.1. *Let $\psi_0 \in \mathcal{S}(\mathbb{R}^d)$ and $t \neq \left(n+\frac{1}{2}\right)\frac{\pi}{\Omega_j}$, $n \in \mathbb{N}$. Then the solution of the Schrödinger equation* (3.12) *is given by the infinite dimensional oscillatory integral on the Cameron-Martin space*

$$\begin{aligned}\psi(t,x) &= \widetilde{\int_{\mathcal{H}_t}} e^{\frac{i}{2\hbar}\int_0^t \dot\gamma(s)^2 ds} e^{-\frac{i}{2\hbar}\int_0^t (\gamma(s)+x)\Omega^2(\gamma(s)+x)ds}\psi_0(\gamma(0)+x)d\gamma \\ &= \widetilde{\int_{\mathcal{H}_t}} e^{\frac{i}{2\hbar}\langle\gamma,(I-L)\gamma\rangle} e^{-\frac{i}{2\hbar}x\Omega^2 xt} e^{-\frac{i}{\hbar}\int_0^t \gamma(s)\Omega^2 x ds}\psi_0(\gamma(0)+x)d\gamma\,. \end{aligned} \tag{3.14}$$

Proof. We give here the proof in the case where $d=1$ in order to simplify the notation, but the whole reasoning can be easily extended to the case of general dimension d.

First of all we can see that the function on $f : \mathcal{H}_t \to \mathbb{C}$ given by

$$f(\gamma) := e^{-\frac{i}{2\hbar}\int_0^t (\gamma(s)+x)\Omega^2(\gamma(s)+x)ds}\psi_0(\gamma(0)+x), \qquad \gamma \in \mathcal{H}_t$$

is Fresnel integrable. Indeed it is of the form

$$f(\gamma) = e^{-\frac{i}{2\hbar}\langle\gamma,L\gamma\rangle}g(\gamma), \qquad \gamma \in \mathcal{H}_t,$$

with $L : \mathcal{H}_t \to \mathcal{H}_t$ is the self-adjoint operator (3.5), that by lemmas 3.1 and 3.2 is trace class with $I-L$ being invertible, and $g : \mathcal{H}_t \to \mathbb{C}$, given by

$$g(\gamma) = e^{-\frac{i}{2\hbar}x\Omega^2 xt} e^{-\frac{i}{\hbar}\int_0^t \gamma(s)\Omega^2 x ds}\psi_0(\gamma(0)+x), \quad \gamma \in \mathcal{H}_t$$

belongs to $\mathcal{F}(\mathcal{H}_t)$. Indeed, as $\mathcal{S}(\mathbb{R}^d) \subset \mathcal{F}(\mathbb{R}^d)$, by lemma 3.3 the function $\gamma \mapsto \psi_0(\gamma(0)+x)$ belongs to $\mathcal{F}(\mathcal{H}_t)$. Moreover the function $\gamma \mapsto e^{-\frac{i}{\hbar}\int_0^t \gamma(s)\Omega^2 x ds}$ is of the form

$$e^{-\frac{i}{\hbar}\int_0^t \gamma(s)\Omega^2 x ds} = \int_{\mathcal{H}_t} e^{i\langle\gamma,\eta\rangle}\delta_{\eta_x}(d\eta), \qquad \gamma \in \mathcal{H}_t,$$

where η_x is the vector of $\mathcal{H}_t$ given by

$$\eta_x(s) = \frac{\Omega^2 x}{\hbar}\Big(\frac{s^2}{2} - \frac{t^2}{2}\Big), \qquad s \in [0,t].$$

By theorem 2.5 the infinite dimensional oscillatory integral

$$\widetilde{\int_{\mathcal{H}_t}} e^{\frac{i}{2\hbar}\langle\gamma,(I-L)\gamma\rangle} e^{-\frac{i}{2\hbar}x\Omega^2 xt} e^{-\frac{i}{\hbar}\int_0^t \gamma(s)\Omega^2 x ds}\psi_0(\gamma(0)+x)d\gamma$$

is well defined and is given by

$$\det(I-L)^{-1/2}\int_{\mathcal{H}_t} e^{-\frac{i\hbar}{2}\langle\gamma,(I-L)^{-1}\gamma\rangle} d\mu_g(\gamma),$$

with $g = \hat{\mu}_g$. By lemmas 3.1 and 3.2, and some calculations we have that the latter expressing is equal to:

$$\sqrt{\frac{\Omega}{2\pi i \sin(\Omega t)}}\int_{\mathbb{R}} e^{\frac{i}{2\hbar}\left(x\Omega\cot(\Omega t)x + y\Omega\cot(\Omega t)y - 2x\Omega\sin(\Omega t)^{-1}y\right)}\psi_0(y)dy, \quad (3.15)$$

that is the solution of the Schrödinger equation (3.12) with initial datum $\psi_0 \in \mathcal{S}(\mathbb{R})$. □

Remark 3.1. By diagonalizing the symmetric matrix Ω, Eq. (3.15) can be generalized to arbitrary dimension d and becomes

$$\sqrt{\det\Big(\frac{\Omega}{2\pi i\sin(\Omega t)}\Big)}\int_{\mathbb{R}^d} e^{\frac{i}{2\hbar}\left(x\Omega\cot(\Omega t)x + y\Omega\cot(\Omega t)y - 2x\Omega\sin(\Omega t)^{-1}y\right)}\psi_0(y)dy.$$

We can now state the main result of the present section. The proof is taken by [114] and is based on a technique present in [263].

Theorem 3.2. *Let* $\psi_0, V \in \mathcal{F}(\mathbb{R}^d)$ *and* $t \neq \left(n+\frac{1}{2}\right)\frac{\pi}{\Omega_j}$, $n \in \mathbb{N}$. *Then the solution of the Schrödinger equation*

$$\begin{cases} i\hbar\frac{\partial}{\partial t}\psi(t,x) = -\frac{\hbar^2}{2}\Delta\psi(t,x) + \frac{1}{2}x\Omega^2 x\psi(t,x) + V(x)\psi(t,x) \\ \psi(0,x) = \psi_0(x) \end{cases}$$

is given by the infinite dimensional oscillatory integral on the Cameron-Martin space

$$\widetilde{\int_{\mathcal{H}_t}} e^{\frac{i}{2\hbar}\int_0^t \dot\gamma(s)^2 ds} e^{-\frac{i}{2\hbar}\int_0^t (\gamma(s)+x)\Omega^2(\gamma(s)+x)ds} e^{-\frac{i}{\hbar}\int_0^t V(\gamma(s)+x)ds}\psi_0(\gamma(0)+x)d\gamma$$

$$= \widetilde{\int_{\mathcal{H}_t}} e^{\frac{i}{2\hbar}\langle\gamma,(I-L)\gamma\rangle} e^{-\frac{i}{2\hbar}x\Omega^2 xt} e^{-\frac{i}{\hbar}\int_0^t \gamma(s)\Omega^2 x ds} e^{-\frac{i}{\hbar}\int_0^t V(\gamma(s)+x)ds}\psi_0(\gamma(0)+x)d\gamma. \tag{3.16}$$

Proof. By lemma 3.3 and by repeating the reasoning in the proofs of theorem 3.1, the function on $\mathcal{H}_t$

$$\gamma \mapsto e^{-\frac{i}{2\hbar}\int_0^t (\gamma(s)+x)\Omega^2(\gamma(s)+x)ds} e^{-\frac{i}{\hbar}\int_0^t V(\gamma(s)+x)ds}\psi_0(\gamma(0)+x), \quad \gamma \in \mathcal{H}_t$$

is Fresnel integrable.

For $u \in [0,t]$ let $\mu_u(V,x)$, $\nu_u^t(V,x)$, $\lambda_u^t(x)$ be the measures on $\mathcal{H}_t$, whose Fourier transforms when evaluated at $\gamma \in \mathcal{H}_t$ are respectively $V(x+\gamma(u))$, $e^{-\frac{i}{\hbar}\int_u^t V(x+\gamma(s))ds}$, and $e^{-\frac{i}{\hbar}\int_u^t x\Omega^2\gamma(s)ds}$. We shall use the following notation

$$\mu_u \equiv \mu_u(V,x),$$

$$\nu_u^t \equiv \nu_u^t(V,x), \tag{3.17}$$

$$\lambda_u^t \equiv \lambda_u^t(x). \tag{3.18}$$

Moreover, for $\psi_0 \in \mathcal{F}(\mathbb{R}^d)$, we shall denote with μ_{ψ_0} the measure on $\mathcal{H}_t$ whose Fourier transform is the function $\gamma \mapsto \psi_0(\gamma(0)+x)$.

If $\{\mu_u : a \leq u \leq b\}$ is a family in $\mathcal{M}(\mathcal{H}_t)$, we shall let $\int_a^b \mu_u du$ denote the measure on $\mathcal{H}_t$ defined by:

$$f \mapsto \int_a^b \int_{\mathcal{H}_t} f(\gamma)\mu_u(d\gamma)du, \qquad f \in C_0(\mathcal{H}_t)$$

whenever it exists. Since for any continuous path γ we have

$$\exp\Big(-\frac{i}{\hbar}\int_0^t V(\gamma(s)+x)ds\Big)$$

$$= 1 - \frac{i}{\hbar}\int_0^t V(\gamma(u)+x)\exp\Big(-\frac{i}{\hbar}\int_u^t V(\gamma(s)+x)ds\Big)du,$$

we get

$$\nu_0^t = \delta_0 - \frac{i}{\hbar}\int_0^t (\mu_u * \nu_u^t)du, \tag{3.19}$$

where δ_0 is the Dirac measure at $0 \in \mathcal{H}_t$.

We set for $t>0$ and $x\in\mathbb{R}^d$

$$U(t)\psi_0(x)=\widetilde{\int_{\mathcal{H}_t}} e^{\frac{i}{2\hbar}\int_0^t|\dot\gamma(s)|^2ds-\frac{i}{2\hbar}\int_0^t(\gamma(s)+x)\Omega^2(\gamma(s)+x)ds}$$
$$e^{-\frac{i}{\hbar}\int_0^t V(\gamma(s)+x)ds}\psi_0(\gamma(0)+x)d\gamma, \quad (3.20)$$

and

$$U_0(t)\psi_0(x)=\widetilde{\int_{\mathcal{H}_t}} e^{\frac{i}{2\hbar}\int_0^t|\dot\gamma(s)|^2ds-\frac{i}{2\hbar}\int_0^t(\gamma(s)+x)\Omega^2(\gamma(s)+x)ds}$$
$$\psi_0(\gamma(0)+x)d\gamma. \quad (3.21)$$

By Parseval-type equality (theorem 2.5), we have

$$U(t)\psi_0(x)=e^{-i\frac{t}{2\hbar}x\Omega^2x}(\det(I-L))^{-1/2}\int_{\mathcal{H}_t}e^{-\frac{i\hbar}{2}\langle\gamma,(I-L)^{-1}\gamma\rangle}\lambda_0^t*\nu_0^t*\mu_{\psi_0}(d\gamma).$$

By applying Eq. (3.19) we obtain:

$$\begin{aligned}U(t)\psi_0(x)&=C(t)\int_{\mathcal{H}_t}e^{-\frac{i\hbar}{2}\langle\gamma,(I-L)^{-1}\gamma\rangle}\lambda_0^t*\mu_{\psi_0}(d\gamma)\\&\quad-\frac{i}{\hbar}C(t)\int_0^t\int_{\mathcal{H}_t}e^{-\frac{i\hbar}{2}\langle\gamma,(I-L)^{-1}\gamma\rangle}\lambda_0^t*\mu_u*\nu_u^t*\mu_{\psi_0}(d\gamma)du,\end{aligned}$$

where $C(t)=e^{-i\frac{t}{2\hbar}x\Omega^2x}(\det(I-L))^{-1/2}$. By applying the Parseval-type equality (2.33) in the other direction we get

$$U(t)\psi_0(x)=U_0(t)\psi_0(x)-\frac{i}{\hbar}e^{-i\frac{t}{2\hbar}x\Omega^2x}$$
$$\int_0^t\widetilde{\int_{\mathcal{H}_t}}e^{\frac{i}{2\hbar}\int_0^t|\dot\gamma(s)|^2ds-\frac{i}{2\hbar}\int_0^t\gamma(s)\Omega^2\gamma(s)ds}e^{-\frac{i}{\hbar}\int_0^t x\Omega^2\gamma(s)ds}$$
$$V(\gamma(u)+x)e^{-\frac{i}{\hbar}\int_u^t V(\gamma(s)+x)ds}\psi_0(\gamma(0)+x)d\gamma du\,. \quad (3.22)$$

Denoting by $\mathcal{H}_{r,s}$ the Cameron-Martin space of paths $\gamma:[r,s]\to\mathbb{R}^d$, we have $\mathcal{H}_t\equiv\mathcal{H}_{0,t}=\mathcal{H}_{0,u}\oplus\mathcal{H}_{u,t}$, indeed each $\gamma\in\mathcal{H}_t$ can uniquely be associated to a couple (γ_1,γ_2), with $\gamma_1\in\mathcal{H}_{0,u}$ and $\gamma_2\in\mathcal{H}_{u,t}$, $\gamma(s)=\gamma_2(s)$ for $s\in[u,t]$ and $\gamma(s)=\gamma_1(s)+\gamma_2(u)$ for $s\in[0,u)$. By means of these notations and by Fubini's theorem for oscillatory integrals (see theorem 2.7), Eq. (3.22) can be written in the following form:

$$U(t)\psi_0(x) = U_0(t)\psi_0(x) - \frac{i}{\hbar}\int_0^t \widetilde{\int_{\mathcal{H}_{u,t}}} e^{\frac{i}{2\hbar}\int_u^t |\dot\gamma_2(s)|^2 ds}$$
$$e^{-\frac{i}{2\hbar}\int_u^t (\gamma_2(s)+x)\Omega^2(\gamma_2(s)+x)ds} e^{-\frac{i}{\hbar}\int_u^t V(\gamma_2(s)+x)ds} V(\gamma_2(u)+x)$$
$$\Big(\widetilde{\int_{\mathcal{H}_{0,u}}} e^{\frac{i}{2\hbar}\int_0^u |\dot\gamma_1(s)|^2 ds - \frac{i}{2\hbar}\int_0^u (\gamma_1(s)+\gamma_2(u)+x)\Omega^2(\gamma_1(s)+\gamma_2(u)+x)ds}$$
$$\psi_0(\gamma_1(0)+\gamma_2(u)+x)d\gamma_1\Big)d\gamma_2 du,$$

and by Eqs. (3.20) and (3.21) the latter is equal to

$$U(t)\psi_0(x) = U_0(t)\psi_0(x) - \frac{i}{\hbar}\int_0^t U(t-u)VU_0(u)\psi_0(x)du. \tag{3.23}$$

For $\psi \in \mathcal{S}(\mathbb{R})$, by theorem 3.1, the function $U_0(t)\psi_0(x)$ is the vector $e^{-\frac{i}{\hbar}H_0 t}\psi_0 \in L^2(\mathbb{R})$ evaluated at the point x. By Eq. (3.23), the operator $U(t)$ on $L^2(\mathbb{R})$ is obtained as the solution of an integral equation, whose iterative solution is the convergent Dyson series [246] for $e^{-\frac{i}{\hbar}Ht}$ (see also [114, 17]). □

It is interesting to remark that, by choosing in the sequential definition of the infinite dimensional oscillatory integral

$$\widetilde{\int_{\mathcal{H}_t}} e^{\frac{i}{2\hbar}\int_0^t |\dot\gamma(s)|^2 ds - \frac{i}{\hbar}\int_0^t V(\gamma(s)+x)ds}\psi_0(\gamma(0)+x)d\gamma$$
$$= \lim_{n\to\infty} \frac{\int^o_{P_n\mathcal{H}_t} e^{\frac{i}{2\hbar}\int_0^t |\dot\gamma_n(s)|^2 ds - \frac{i}{\hbar}\int_0^t V(\gamma_n(s)+x)ds}\psi_0(\gamma_n(0)+x)d\gamma_n}{\int^o_{P_n\mathcal{H}_t} e^{\frac{i}{2\hbar}\int_0^t |\dot\gamma_n(s)|^2 ds}d\gamma_n}, \tag{3.24}$$

(with $\gamma_n := P_n\gamma$) a suitable sequence of finite dimensional projection operators $\{P_n\}_{n\in\mathbb{N}}$, it is possible to recover in the right hand side of Eq. (3.24), an analogous of the Feynman formula (1.9) derived by the Trotter product formula, i.e.

$$e^{-\frac{i}{\hbar}Ht}\psi_0(x) = \lim_{n\to\infty}\Big(\frac{2\pi i\hbar t}{mn}\Big)^{-\frac{nd}{2}}\int_{\mathbb{R}^{nd}} e^{\frac{i}{\hbar}\sum_{j=1}^n\left(\frac{m}{2}\frac{(x_j-x_{j-1})^2}{(t/n)^2}-V(x_j)\right)\frac{t}{n}}$$
$$\psi_0(x_0)dx_0\dots dx_{n-1}. \tag{3.25}$$

Let us consider indeed the *polygonal path approximation* [283]. For any $n \in \mathbb{N}$ and $\gamma \in \mathcal{H}_t$, let $P_n\gamma$ be the piecewise linear polygonal approximation of γ, that is:

$$P_n\gamma(s) = \gamma_j + \Big(s - \frac{jt}{n}\Big)(\gamma_{j+1}-\gamma_j)\frac{n}{t}, \qquad \frac{jt}{n} \le s \le \frac{(j+1)t}{n},$$

where $j = 0, \ldots, n-1$, $\gamma_j := \gamma(jt/n)$. Clearly $P_n^2 = P_n$, moreover for any $\gamma, \eta \in \mathcal{H}_t$

$$\langle \eta, P_n \gamma \rangle = \sum_{j=0}^{n-1} (\eta_{j+1} - \eta_j)(\gamma_{j+1} - \gamma_j)\frac{n}{t} = \langle P_n \eta, \gamma \rangle,$$

and we can conclude that the operators P_n are orthogonal projections. Moreover the following holds:

Lemma 3.4. *The sequence of operators $\{P_n\}_{n\in\mathbb{N}}$ converges to the identity $I : \mathcal{H}_t \to \mathcal{H}_t$ in the strong operator topology.*

Proof. [283] Let us denote with $U \subset \mathcal{H}_t$ the set

$$U := \{\gamma \in \mathcal{H}_t, \mid \lim_{n\to\infty} \|P_n\gamma - \gamma\| = 0\},$$

and let us prove that $U = \mathcal{H}_t$ by showing that it is a closed subspace of $\mathcal{H}_t$ containing its basis functions.

U is a subspace of $\mathcal{H}_t$ because P_n is linear, indeed given $\eta, \gamma \in U$, $\alpha, \beta \in \mathbb{R}$

$$\begin{aligned}\|P_n(\alpha\eta + \beta\gamma) - (\alpha\eta + \beta\gamma)\| &= \|\alpha(P_n\eta - \eta) + \beta(P_n\gamma - \gamma)\| \\ &\leq |\alpha|\|P_n\eta - \eta\| + |\beta|\|P_n\gamma - \gamma\| \to 0, \qquad n \to \infty\,.\end{aligned}$$

U is closed, indeed given a sequence $\{\gamma_n\} \subset U$, with $\|\gamma_n - \gamma\| \to 0$, the limit vector γ belongs to U. Indeed

$$\|P_n\gamma - \gamma\| \leq \|P_n(\gamma - \gamma_m)\| + \|\gamma - \gamma_m\| + \|P_n\gamma_m - \gamma_m\| \tag{3.26}$$

$$\leq 2\|\gamma - \gamma_m\| + \|P_n\gamma_m - \gamma_m\|\,. \tag{3.27}$$

Given $\epsilon > 0$, $\exists N_\epsilon > 0$ such that $\|\gamma - \gamma_m\| < \epsilon/4$ for $m = N_\epsilon$. There exists also $N(m,\epsilon)$, such that $\|P_n\gamma_m - \gamma_m\| < \epsilon/2$ for $n > N(m,\epsilon)$, and consequently $\|P_n\gamma - \gamma\| < \epsilon$.

The inclusion of a basis of $\mathcal{H}_t$ in U can be proved in the following way [280].

Given $\gamma \in \mathcal{H}_t$, let $\alpha_0, \alpha_n, \beta_n$, with $\sum_{n=1}^\infty \alpha_n^2 + \beta_n^2 < \infty$, be the Fourier coefficients of $\dot\gamma \in L^2([0,t])$ and let $\dot\gamma_N$, γ_N be the corresponding partial sums, i.e.:

$$\dot\gamma_N(s) = \alpha_0 + \sum_{n=1}^N \alpha_n \cos\frac{2\pi ns}{t} + \sum_{n=1}^N \beta_n \sin\frac{2\pi ns}{t}, \quad s \in [0,t],$$

$$\gamma_N(s) = \alpha_0(s-t) + \sum_{n=1}^N \frac{\alpha_n t}{2\pi n}\sin\frac{2\pi ns}{t} + \sum_{n=1}^N \frac{\beta_n t}{2\pi n}\left(1 - \cos\frac{2\pi ns}{t}\right), \quad s \in [0,t].$$

One has

$$\|\gamma_N - \gamma\| = \|\dot\gamma_N - \dot\gamma\|_{L^2([0,t])} \to 0, \qquad \text{as } N \to \infty,$$

and it is quite simple to see that $\gamma_N \in U$ for each N. □

3.2 Time dependent potentials

The results of the previous section can be generalized to the case the potential in the Schrödinger equation depends explicitly on the time variable t [29].

Let us consider first of all a linearly forced harmonic oscillator. Let us assume that the quantum mechanical Hamiltonian is given on the smooth vectors $\psi \in C_0(\mathbb{R}^d)$ by

$$H\psi(x) = -\frac{\hbar^2}{2}\Delta\psi(x) + V(t,x)\psi(x), \quad V(t,x) = \frac{1}{2}x\Omega^2 x + f(t)\cdot x, \quad x \in \mathbb{R}^d, \tag{3.28}$$

where Ω is a positive symmetric constant $d \times d$ matrix with eigenvalues Ω_j, $j = 1 \dots d$, and $f : I \subset \mathbb{R} \to \mathbb{R}^d$ is a continuous function (I being a closed interval).

This potential is particularly interesting from a physical point of view as it is used in simple models for a large class of processes, as the vibration-relaxation of a diatomic molecule in gas kinetics and the interaction of a particle with the field oscillators in quantum electrodynamics. Feynman calculated heuristically the Green function for the Schrödinger equation associated to (3.28) in his famous paper on the path integral formulation of quantum mechanics [122]. Our aim is to give meaning to the Feynman path integral representation of the solution of Schrödinger equation

$$i\hbar\frac{d}{dt}\psi = H\psi,$$

with H given by Eq. (3.28):

$$\psi(t,x) = \int_{\gamma(t)=x} e^{\frac{i}{2\hbar}\int_0^t |\dot\gamma(s)|^2 ds - \frac{i}{2\hbar}\int_0^t \gamma(s)\Omega^2\gamma(s)ds - \frac{i}{\hbar}\int_0^t f(s)\cdot\gamma(s)ds}\psi_0(\gamma(0))d\gamma \tag{3.29}$$

in terms of a well defined infinite dimensional oscillatory integral on the Cameron-Martin space $\mathcal{H}_t$. We recall that a similar result has been obtained in the case $d = 1$ by means of the white-noise approach [118].

Let $L : \mathcal{H}_t \to \mathcal{H}_t$ be the symmetric operator on $\mathcal{H}_t$ given by Eq. (3.5). By lemmas 3.1 and 3.2, if

$$t \neq \left(n + \frac{1}{2}\right)\frac{\pi}{\Omega_j}, \qquad n \in \mathbb{N},\ j = 1, \dots, d,$$

then L is trace class and $(I - L)$ is invertible and its inverse is given by Eq. (3.6).

Let w, v be the vectors in $\mathcal{H}_t$ defined by:

$$w(s) \equiv \frac{\Omega^2 x}{2\hbar}(s^2 - t^2), \tag{3.30}$$

$$v(s) \equiv \frac{1}{\hbar}\int_t^s \int_0^{s'} f(s'')ds''ds' \qquad s \in [0,t]. \tag{3.31}$$

With these notations heuristic expression (3.29) can be written in terms of a well defined infinite dimensional oscillatory integrals, i.e.

$$\psi(t,x) = \int_{\gamma(t)=0} e^{\frac{i}{2\hbar}\int_0^t |\dot\gamma(s)|^2 ds - \frac{i}{2\hbar}\int_0^t (\gamma(s)+x)\Omega^2(\gamma(s)+x)ds}$$

$$e^{-\frac{i}{\hbar}\int_0^t f(s)\cdot(\gamma(s)+x)ds}\psi_0(\gamma(0)+x)d\gamma = e^{-i\frac{t}{2\hbar}x\Omega^2 x}e^{-i\frac{x}{\hbar}\cdot\int_0^t f(s)ds}$$

$$\widetilde{\int_{\mathcal{H}_t}} e^{\frac{i}{2\hbar}\langle\gamma,(I-L)\gamma\rangle}e^{i\langle v,\gamma\rangle}e^{i\langle w,\gamma\rangle}\psi_0(\gamma(0)+x)d\gamma. \tag{3.32}$$

Under the assumption that $\psi_0 \in \mathcal{F}(\mathbb{R}^d)$, by lemma 3.3 the functional on $\mathcal{H}_t$ given by

$$\gamma \mapsto \psi_0(\gamma(0)+x), \qquad \gamma \in \mathcal{H}_t$$

belongs to $\mathcal{F}(\mathcal{H}_t)$. Indeed if $\psi_0(x) = \int_{\mathbb{R}^d} e^{ik\cdot x}d\mu_0(k)$, then

$$\psi_0(\gamma(0)+x) = \int_{\mathcal{H}_t} e^{i\langle\eta,\gamma\rangle}\mu_{\psi_0}(d\eta),$$

where

$$\int_{\mathcal{H}_t} f(\gamma)d\mu_{\psi_0}(\gamma) = \int_{\mathbb{R}^d} e^{ik\cdot x}f(k\gamma_0)d\mu_0(k) \tag{3.33}$$

and, for any $k \in \mathbb{R}^d$, $k\gamma_0$ is the element in $\mathcal{H}_t$ such that $\langle k\gamma_0, \gamma\rangle = k\cdot\gamma(0)$, that is

$$k\gamma_0(s) = k(t-s), \quad s \in [0,t].$$

In this case the functional on $\mathcal{H}_t$ defined by

$$\gamma \mapsto e^{i\langle v,\gamma\rangle}e^{i\langle w,\gamma\rangle}\psi_0(\gamma(0)+x), \qquad \gamma \in \mathcal{H}_t,$$

belongs to $\mathcal{F}(\mathcal{H}_t)$ and the infinite dimensional oscillatory integral on the right hand side of Eq. (3.32) on $\mathcal{H}_t$ can be explicitly computed by means of the Parseval-type equality (theorem 2.5):

$$\widetilde{\int_{\mathcal{H}_t}} e^{\frac{i}{2\hbar}\langle\gamma,(I-L)\gamma\rangle}e^{i\langle v,\gamma\rangle}e^{i\langle w,\gamma\rangle}\psi_0(\gamma(0)+x)d\gamma$$

$$= (\det(I-L))^{-1/2}\int_{\mathcal{H}_t} e^{-\frac{i\hbar}{2}\langle\gamma,(I-L)^{-1}\gamma\rangle}\delta_v * \delta_w * \mu_{\psi_0}(d\gamma)$$

and we have

$$\psi(t,x) = \frac{e^{-i\frac{t}{2\hbar}x\Omega^2 x}e^{-i\frac{x}{\hbar}\cdot\int_0^t f(s)ds}}{\sqrt{\det(\cos(\Omega t))}} \int_{\mathbb{R}^d} e^{ik\cdot x} e^{-\frac{i\hbar}{2}\langle(v+w+k\gamma_0),(I-L)^{-1}(v+w+k\gamma_0)\rangle} d\mu_0(k).$$

If $\psi_0 \in \mathcal{S}(\mathbb{R}^d)$, we can proceed further and compute explicitly the Green function $G(0,t,x,y)$ for the Schrödinger equation:

$$\psi(t,x) = \int_{\mathbb{R}^d} G(0,t,x,y)\psi_0(y)dy,$$

where

$$G(0,t,x,y) = (2\pi i\hbar)^{-d/2}\sqrt{\det\Big(\frac{\Omega}{\sin(\Omega t)}\Big)} e^{\frac{i\Omega\sin(\Omega t)^{-1}}{2\hbar}(x\cos(\Omega t)x + y\cos(\Omega t)y - 2xy)}$$
$$e^{-\frac{i}{\hbar}x\sin(\Omega t)^{-1}\int_0^t \sin(\Omega s)f(s)ds - \frac{i}{\hbar}y(\int_0^t \cos(\Omega s)f(s)ds - \cos(\Omega t)\sin(\Omega t)^{-1}\int_0^t \sin(\Omega s)f(s)ds)}$$
$$e^{\frac{i}{\hbar}\Omega^{-1}(\frac{1}{2}\cos(\Omega t)\sin(\Omega t)^{-1}(\int_0^t \sin(\Omega s)f(s)ds)^2 - \int_0^t \sin(\Omega s)f(s)ds\int_0^t \cos(\Omega s)f(s)ds)}$$
$$e^{\frac{i}{\hbar}\Omega^{-1}\int_0^t \cos(\Omega s)f(s)\int_s^t \sin(\Omega s')f(s')ds'ds}. \quad (3.34)$$

One can easily verify by a direct computation that (3.34) is the Green's function for the Schrödinger equation with the time dependent Hamiltonian (3.28).

Remark 3.2. Our result can be obtained even if the initial assumption on the continuity of the function $f : [0,t] \to \mathbb{R}^d$ is weakened, in fact it is sufficient to assume that the function v in (3.30) belongs to $\mathcal{H}_t$, that is $\int_0^t |\int_0^s f(s')ds'|^2 ds < \infty$.

Let us consider now the Schrödinger equation with an harmonic-oscillator Hamiltonian with a time-dependent frequency:

$$H\psi(x) = -\frac{\hbar^2}{2}\Delta\psi(x) + \frac{1}{2}x\Omega^2(t)x\psi(x), \qquad x \in \mathbb{R}^d, \psi \in C_0^2(\mathbb{R}^d), \quad (3.35)$$

where $\Omega : [0,t] \to L(\mathbb{R}^d,\mathbb{R}^d)$ is a continuous map from the time interval $[0,t]$ to the space of symmetric positive $d \times d$ matrices.

This problem has been analyzed by several authors (see for instance [239, 202] and references therein) as an approximated description for the vibration of complex physical systems, as well as an exact model for some physical phenomena, as the motion of an ion in a Paul trap, quantum mechanical description of highly cooled ions, the emergence of non classical optical states of light owing to a time-dependent dielectric constant, or even in cosmology for the study of a three-dimensional isotropic harmonic oscillator in a spatially flat universe such that $g_{ij} = R(t)\delta_{ij}$, with $R(t)$ being the scale factor at time t and $i, j = 1, 2, 3$ the indexes relative to the spatial coordinates.

If $d = 1$ it is possible to solve the Schrödinger equation with Hamiltonian (3.35) (and also the corresponding classical equation of motion) by adopting a suitable transformation of the time and space variables which allows to map the solution of the time-independent harmonic oscillator to the solution of the time-dependent one (see [247, 251, 250] and references therein). Let us consider the classical equation of motion for the time-dependent harmonic oscillator (3.35)

$$\ddot{u}(s) + \Omega^2(s)u(s) = 0. \tag{3.36}$$

Let u_1 and u_2 be two independent solutions of (3.36) such that $u_1(0) = \dot{u}_2(0) = 0$ and $u_2(0) = \dot{u}_1(0) = 1$. Then it easy to prove that the Wronskian $w(u_1, u_2) = u_1\dot{u}_2 - \dot{u}_1 u_2$ is the constant function $w = 1$. Let us define the function $\xi := u_1^2 + u_2^2$. It is possible to prove that $\xi(s) > 0$ $\forall s$ and it satisfies the following differential equation:

$$2\xi\ddot{\xi} - \dot{\xi}^2 + 4\xi^2 - 4 = 0.$$

Moreover the function $\eta : [0, \infty] \to \mathbb{R}$, given by

$$\eta(s) = \int_0^s \xi(\tau)^{-1} d\tau$$

is well defined and strictly increasing. It is possible to verify that

$$u(s) = \xi(s)^{1/2}(A\cos(\eta(s)) + B\sin(\eta(s))) \tag{3.37}$$

is the general solution of the classical equation of motion (3.36). In other words by rescaling the time variable $s \mapsto \eta(s)$ and the space variable $x \mapsto \xi^{-1/2}x$ it is possible to map the solution of the equation of motion for the time-independent harmonic oscillator $\ddot{u}(s) + u(s) = 0$ into the solution of (3.36). By another point of view, it is possible to find (see, for instance, [194] for more details) a general canonical transformation $(x, p, t) \mapsto (X, P, \tau)$, given by

$$\begin{cases} X = \xi(t)^{-1/2}x \\ \frac{d\tau(t)}{dt} = \xi(t)^{-1} \\ P = \frac{dX}{d\tau} = (\xi^{1/2}\dot{x} - \frac{1}{2}\xi^{-1/2}\dot{\xi}x). \end{cases} \tag{3.38}$$

The Hamiltonian in the new variables is time independent

$$H(X, P, \tau) = \frac{1}{2}(P^2 + X^2).$$

The generating function of the transformation $(x,p,t) \mapsto (X,P,\tau)$ is given by $F(x,P,t) = \xi(t)^{-1/2}xP + \frac{\xi(t)^{-1}\dot{\xi}}{4}x^2$ and the transformation is given more explicitly as

$$\begin{cases} p = \frac{\partial}{\partial x}F(x,P,t) \\ X = \frac{\partial}{\partial P}F(x,P,t) \\ H(X,P;\tau)\dot{\tau} = H(x,p;t) + \frac{\partial}{\partial t}F(x,P,t). \end{cases} \tag{3.39}$$

A similar result holds also in the quantum case. In fact by considering the Schrödinger equations for the time-independent and time-dependent harmonic oscillator respectively,

$$(i\hbar\frac{\partial}{\partial t} + \frac{\hbar^2}{2}\Delta - \frac{1}{2}x^2)\phi(t,x) = 0, \tag{3.40}$$

$$(i\hbar\frac{\partial}{\partial t} + \frac{\hbar^2}{2}\Delta - \frac{1}{2}\Omega^2(t)x^2)\psi(t,x) = 0, \tag{3.41}$$

where $\phi(t,x)$ and $\psi(t,x)$ are continuously differentiable with respect to t and twice continuously differentiable with respect to x, it is possible to prove the following [247]:

Theorem 3.3. *Let $\phi(t,x)$ be a solution of* (3.40). *Then*

$$\psi(t,x) = \xi(t)^{-1/4}\exp[i\dot{\xi}(t)x^2/4\hbar\xi(t)]\phi(\eta(t),\xi(t)^{-1/2}x)$$

is a solution of (3.41).

In an analogous way, by denoting with $K_{TI}(t,0;x,y)$ and $K_{TD}(t,0;x,y)$ the Green functions for the Schrödinger equations (3.40) and (3.41) respectively, it is possible to prove that the following holds:

$$K_{TD}(t,0;x,y) = \xi(t)^{-1/4}\exp[i\dot{\xi}(t)x^2/4\hbar\xi(t)]K_{TI}(\eta(t),0;\xi(t)^{-1/2}x,y). \tag{3.42}$$

It is interesting to note that the "correction term"

$$\xi(t)^{-1/4}\exp[i\dot{\xi}(t)x^2/4\hbar\xi(t)]$$

in theorem 3.3 can be interpreted in terms of the classical canonical transformation (3.39) (see [194] for more details).

We can give a rigorous mathematical meaning to the Feynman path integral representation of the solution of equation (3.41) by means of an well defined infinite dimensional oscillatory integral on the Cameron-Martin space $\mathcal{H}_t$ and prove formula (3.42). A similar result has been obtained in the framework of the white-noise approach [150].

Let us consider the following linear operator $L : \mathcal{H}_t \to \mathcal{H}_t$:

$$(L\gamma)(s) = -\int_t^s \int_0^r \Omega^2(u)\gamma(u)dudr, \qquad \gamma \in \mathcal{H}_t.$$

One can easily verify that L is self-adjoint and positive, as for any $\gamma_1, \gamma_2 \in \mathcal{H}_t$ one has:

$$\langle \gamma_1, L\gamma_2 \rangle = \int_0^t \gamma_1(s)\Omega^2(s)\gamma_2(s)ds.$$

By using formula (3.37), it is possible to prove that, if $t \neq \eta^{-1}(\pi/2+n\pi)$, $n \in \mathbb{N}$, the operator $I - L$ is invertible and its inverse is given by:

$$\begin{aligned}(I-L)^{-1}\gamma(s) = \Big[-\frac{\sin(\eta(t))}{\cos(\eta(t))}\Big(\int_0^t \xi(s')^{1/2}\cos(\eta(s'))\ddot{\gamma}(s')ds' \\ + \dot{\gamma}(0)\Big) - \int_t^s \xi(s')^{1/2}\sin(\eta(s'))\ddot{\gamma}(s')ds'\Big]\xi(s)^{1/2}\cos(\eta(s)) \\ + \Big[\int_0^t \xi(s')^{1/2}\cos(\eta(s'))\ddot{\gamma}(s')ds' + \dot{\gamma}(0) \\ + \int_t^s \xi(s')^{1/2}\cos(\eta(s'))\ddot{\gamma}(s')ds'\Big]\xi(s)^{1/2}\sin(\eta(s)), \quad s \in [0,t]. \qquad (3.43)\end{aligned}$$

The Fredholm determinant of the operator $I - L$ can be computed by exploiting the general relation between infinite dimensional determinants of the form

$$\det(I + \epsilon L), \qquad \epsilon \in \mathbb{C}, L \text{ is of trace class},$$

and finite dimensional determinants associated with the solution of a certain Sturm-Liouville problem [5, 250]. According to [5], by using the fact that $v(s) = L\gamma(s)$ is the unique solution of the problem:

$$\begin{cases} \ddot{v}(s) = -\Omega^2(s)\gamma(s), & s \in (0,t), \\ \dot{v}(0) = 0, \quad v(t) = 0 \end{cases} \qquad (3.44)$$

and by using the ellipticity of the problem (3.44), it is possible to prove that the range of L is contained in $H^3((0,t);\mathbb{R})$, the Sobolev space of functions belonging to $L^2((0,t);\mathbb{R})$, whose derivatives up to order 3 belong also to $L^2((0,t);\mathbb{R})$, hence L is a trace class operator. Moreover by considering the solution K_ϵ of the initial value problem

$$\begin{cases} \ddot{K}_\epsilon(s) + \epsilon\Omega^2(s)K_\epsilon(s) = 0, \\ \dot{K}_\epsilon(0) = 0, \quad K_\epsilon(0) = 1 \end{cases} \qquad (3.45)$$

one has

$$K_\epsilon(t) = \det(I - \epsilon L).$$

By substituting $\epsilon = 1$ in (3.45) and by using formula (3.37) for the general solution of the differential equation (3.36) one has:

$$\det(I-L) = \xi(t)^{1/2}\cos(\eta(t)). \tag{3.46}$$

Let us consider now the vectors $\gamma_0, w \in \mathcal{H}_t$ given by

$$\gamma_0(s) = t - s,$$
$$w(s) = \frac{x}{\hbar}\int_t^s \int_0^u \Omega^2(r)\,dr\,du, \qquad s \in [0,t]. \tag{3.47}$$

With the notations introduced so far and by assuming that the initial vector ψ_0 belongs to $\mathcal{F}(\mathbb{R})$, so that $\psi_0 = \hat{\mu}_0$, the heuristic Feynman path integral representation for the solution of the Schrödinger equation with the time-dependent Hamiltonian (3.35)

$$\psi(t,x) = \text{“}\int_{\{\gamma|\gamma(t)=0\}} e^{\frac{i}{2\hbar}\int_0^t \dot\gamma^2(s)ds - \frac{i}{2\hbar}\int_0^t \Omega^2(s)(\gamma(s)+x)^2 ds}\psi_0(\gamma(0)+x)D\gamma\text{”}$$

can be rigorously realized as the infinite dimensional oscillatory integral on the Cameron-Martin space $\mathcal{H}_t$:

$$\psi(t,x) = e^{-\frac{ix^2}{2\hbar}\int_0^t \Omega^2(s)ds} I_t, \qquad I_t = \widetilde{\int_{\mathcal{H}_t}} e^{\frac{i}{2\hbar}\langle\gamma,(I-L)\gamma\rangle} e^{i\langle w,\gamma\rangle}\hat{\mu}_{\psi_0}(\gamma)d\gamma,$$

where μ_{ψ_0} is given by formula (3.33). By Parseval-type equality I_t can be explicitly computed and one has:

$$I_t = \det(I-L)^{-1/2}\int_{\mathcal{H}_t} e^{\frac{i}{2\hbar}\langle\gamma,(I-L)^{-1}\gamma\rangle}\delta_w * \mu_{\psi_0}(d\gamma). \tag{3.48}$$

By assuming $\psi_0 \in \mathcal{S}(\mathbb{R})$, we can proceed further and compute explicitly the Green function of the problem, that is

$$\psi(t,x) = \int_{\mathbb{R}} K_{TD}(t,0;x,y)\psi_0(y)dy.$$

By substituting in (3.48) formulae (3.43) and (3.46), and performing a simple calculation we obtain:

$$K_{TD}(t,0;x,y) = \xi(t)^{-1/4} e^{\frac{ix^2}{4\hbar}\xi(t)^{-1}\dot\xi(t)} \frac{e^{\frac{i}{2\hbar}\left(\frac{\cos(\eta(t))}{\sin(\eta(t))}\right)(\xi(t)^{-1}x^2+y^2) - \frac{2\xi(t)^{-1/2}xy}{\sin(\eta(t))}}}{(2\pi i\hbar\sin(\eta(t)))^{1/2}},$$

and by recalling the well known formula for the Green function $K_{TI}(t,0;x,y)$ of the Schrödinger equation with a time-independent harmonic oscillator Hamiltonian (see, e.g., [263]):

$$K_{TI}(t,0;x,y) = \frac{e^{\frac{i}{2\hbar}\left(\frac{\cos(t)}{\sin(t)}(x^2+y^2) - \frac{2xy}{\sin(t)}\right)}}{(2\pi i\hbar\sin(t))^{1/2}},$$

one can verify directly formula (3.42).

Remark 3.3. The case where $d > 1$ is more complicated. In fact neither a transformation formula analogous to (3.38) exists in general, nor a formula analogous to (3.42) relating the Green function of the Schrödinger equation with a time-dependent harmonic oscillator potential with the Green function of the Schrödinger equation with a time-independent harmonic oscillator potential (see for instance [251, 250] for some partial results in this direction).

Let us finally consider a quantum mechanical Hamiltonian of the following form:

$$H\psi(x) = H_0\psi(x) + V(t,x)\psi(x), \qquad \psi \in C_0^2(\mathbb{R}^d), \tag{3.49}$$

where H_0 is of the type (3.28) or (3.35) and $V : [0,t] \times \mathbb{R}^d \to \mathbb{R}$ satisfies the following assumptions:

(1) for each $s \in [0,t]$, the application $V(s,\cdot) : \mathbb{R}^d \to \mathbb{R}$ belongs to $\mathcal{F}(\mathbb{R}^d)$, i.e. $V(s,x) = \int_{\mathbb{R}^d} e^{ikx}\mu_s(dk)$, $\mu_s \in \mathcal{M}(\mathbb{R}^d)$;
(2) the application $s \in [0,t] \mapsto \mu_s \in \mathcal{M}(\mathbb{R}^d)$ is continuous in the norm $\|\cdot\|$ of the Banach space $\mathcal{M}(\mathbb{R}^d)$.

We remark that condition (1) implies that for each $s \in [0,t]$, the function $V(s,\cdot) : \mathbb{R}^d \to \mathbb{R}$ is bounded. Moreover by condition (2) one can easily verify that the application $s \in [0,t] \mapsto V(s,\cdot) \in C(\mathbb{R}^d) \cap L^\infty(\mathbb{R}^d)$ in continuous in the sup-norm.

Under these assumptions it is possible to prove that the application on $\mathcal{H}_t$, given by

$$\gamma \mapsto \int_0^t V(s,\gamma(s)+x)ds, \qquad \gamma \in \mathcal{H}_t$$

belongs to $\mathcal{F}(\mathcal{H}_t)$. More precisely it is the Fourier transform of the complex bounded variation measure μ_v on the Cameron-Martin space $\mathcal{H}_t$ defined by:

$$\mu_v(I) = \int_0^t \int_{\mathbb{R}^d} e^{ikx}\chi_I(k\gamma_s)\mu_s(dk)ds, \qquad I \in \mathcal{B}(\mathcal{H}_t),$$

where χ_I is the characteristic function of the Borel set $I \subset \mathcal{H}_t$ and, for any $k \in \mathbb{R}^d$, $k\gamma_s$ is the element in $\mathcal{H}_t$ given by

$$\begin{aligned} k\gamma_s(s') &= k(t-s), \qquad s' \leq s \\ k\gamma_s(s') &= k(t-s'), \qquad s' > s. \end{aligned} \tag{3.50}$$

As a consequence also the application

$$\gamma \mapsto e^{-\frac{i}{\hbar}\int_0^t V(s,\gamma(s)+x)ds}, \qquad \gamma \in \mathcal{H}_t,$$

belongs to $\mathcal{F}(\mathcal{H}_t)$ (let us denote by ν_v the bounded variation measure on $\mathcal{H}_t$ associated to it) and the infinite dimensional oscillatory integral associated with the Cameron-Martin space $\mathcal{H}_t$ giving the rigorous mathematical realization of the Feynman path integral representation of the solution of the Schrödinger equation with time dependent Hamiltonian (3.49) is well defined. In the following we give some details of the case where the "free Hamiltonian" H_0 is given by (3.28), but the same reasoning can be repeated in the case where H_0 is given by (3.35). One has:

$$\psi(t,x) = \int_{\gamma(t)=0} e^{\frac{i}{2\hbar}\int_0^t |\dot\gamma(s)|^2 ds - \frac{i}{2\hbar}\int_0^t (\gamma(s)+x)\Omega^2(\gamma(s)+x)ds} e^{-\frac{i}{\hbar}\int_0^t f(s)\cdot(\gamma(s)+x)ds}$$
$$e^{-\frac{i}{\hbar}\int_0^t V(s,\gamma(s)+x)ds}\psi_0(\gamma(0)+x)d\gamma = e^{-i\frac{t}{2\hbar}x\Omega^2 x} e^{-i\frac{x}{\hbar}\cdot\int_0^t f(s)ds} I_t, \quad (3.51)$$

where

$$I_t \equiv \widetilde{\int_{\mathcal{H}_t}} e^{\frac{i}{2\hbar}\langle\gamma,(I-L)\gamma\rangle} e^{i\langle v,\gamma\rangle} e^{i\langle w,\gamma\rangle} e^{-\frac{i}{\hbar}\int_0^t V(s,\gamma(s)+x)ds}\psi_0(\gamma(0)+x)d\gamma \quad (3.52)$$

is well defined and can be explicitly computed using the Parseval type equality (theorem 2.5):

$$I_t = (\det(I-L))^{-1/2}\int_{\mathcal{H}_t} e^{-\frac{i\hbar}{2}\langle\gamma,(I-L)^{-1}\gamma\rangle}\delta_v * \delta_w * \nu_v * \mu_{\psi_0}(d\gamma). \quad (3.53)$$

The detailed proof that the right hand side of (3.51) is the solution of the Schrödinger equation with Hamiltonian (3.49) is completely similar to the proof of theorem 3.2 and we refer to [29] for more details.

3.3 Phase space Feynman path integrals

Let us recall that Feynman's original aim was to give a Lagrangian formulation of quantum mechanics. On the other hand an Hamiltonian formulation could be preferable from many points of view. For instance the discussion of the approach from quantum mechanics to classical mechanics, i.e the study of the behavior of physical quantities taking into account that $\hbar$ is small, is more natural in an Hamiltonian setting (see for instance [7, 221, 220, 223] and section 4.4 for a discussion of this behavior). In other words the "phase space" rather then the "configuration space" is the natural framework of classical mechanics.

As a consequence one is tempted to propose a "phase space Feynman path integral" representation for the solution of the Schrödinger equation, that is an heuristic formula of the following form:

$$\text{“}\psi(t,x) = \text{const} \int_{q(t)=x} e^{\frac{i}{\hbar}S(q,p)}\psi_0(q(0))dqdp\text{”}. \tag{3.54}$$

Here the integral is meant on the space of paths $q(s), p(s)$, $s \in [0,t]$, in the phase space of the system, where $q(s)_{s\in[0,t]}$ is the path in configuration space and $p(s)_{s\in[0,t]}$ is the path in momentum space, and S is the action functional in the Hamiltonian formulation:

$$S(q,p) = \int_0^t (\dot{q}(s)p(s) - H(q(s),p(s)))ds,$$

(H being the classical Hamiltonian of the system).

An approach of phase space Feynman path integrals via analytic continuation of "phase space Wiener integrals" has been presented by I. Daubechies and J. Klauder [90, 91, 88, 89]. Analytic continuation was also used in other "path space" approaches, see [235, 180, 74] and references therein.

Analogously to Feynman Lagrangian formula (1.6), a formal derivation of its Hamiltonian version (3.54) can be given by means of Lie-Trotter product formula [279, 78, 79]. Let us first recall here an abstract version of it, which will be also used in the definition of the quantum dynamics.

Lemma 3.5. *Let A and B be self-adjoint operators in a Hilbert space $\mathcal{H}$ and let $A+B$ be essentially self-adjoint on $D(A)\cap D(B)$. Then*

$$s - \lim_{n\to\infty} (e^{itA/n}e^{itB/n})^n = e^{i(A+B)t}, \qquad t \in \mathbb{R}. \tag{3.55}$$

Here $s-\lim$ is the strong operator limit[1]. For a proof and a discussion of this lemma see e.g. [79, 245].

Let us now define the quantum Hamiltonian operator H on $L^2(\mathbb{R}^d)$ associated to a classical potential V depending both on position and on momentum in the following way

$$H = -\frac{\hbar^2}{2m}\Delta_x + V_1(x) + V_2(p).$$

[1]A sequence $(A_n)_{n\in\mathbb{N}}$ of linear operators $A_n : D \subseteq \mathcal{H} \to \mathcal{H}$ with a common domain D in a Hilbert space $(\mathcal{H}, \|\cdot\|)$ converges strongly to an operator A is for each $\psi \in D$, one has $\lim_{n\to\infty} \|A_n\psi - A\psi\| = 0$.

The operator V_1 is defined as a self-adjoint operator on $L^2(\mathbb{R}^d)$, with its natural domain as a multiplication operator. V_2 is the operator in $L^2(\mathbb{R}^d)$ with domain

$$D(V_2(p)) = \{\psi \in L^2(\mathbb{R}^d) \mid \alpha \to V_2(\alpha)\hat{\psi}(\alpha) \in L^2(\mathbb{R}^d)\},$$

where $\hat{\psi}$ is the Fourier transform of ψ. It coincides with the operator defined by functional calculus as $V_2(p)$, with p denoting the self-adjoint operator in $\mathcal{H}$ given by $p := -i\hbar\nabla_x$, with its natural definition domain. The operator V is then the sum, as a self-adjoint operator in $L^2(\mathbb{R}^d)$, of the self-adjoint operators V_1 and V_2. We assume that the functions V_1 and V_2 are such that the corresponding operators have a common dense domain of essentially self-adjointness D. This is the case, e.g., when $V_1 \in L^2(\mathbb{R}^d) + L^\infty(\mathbb{R}^d)$, V_2 is bounded measurable, and $D = C_0^\infty(\mathbb{R}^d)$ or $D = \mathcal{S}(\mathbb{R}^d)$.

In order to define the quantum dynamics by applying lemma 3.5, let us assume that V_1 and V_2 are such that the operators $-\frac{\hbar^2}{2m}\Delta + V_2$ and $-\frac{\hbar^2}{2m}\Delta + V_1 + V_2$ are essentially self-adjoint on D. We denote by H the closure of the latter operator. H (which we also write simply as $-\frac{\hbar^2}{2m}\Delta + V_1 + V_2$), is then the quantum Hamiltonian.

Let $U(t)_{t\in\mathbb{R}}$ be the one-parameter group of unitary operators on $L^2(\mathbb{R}^d)$ generated by the self-adjoint operator $H/\hbar$:

$$U(t) = e^{-\frac{it}{\hbar}H} = e^{-\frac{it(p^2/2m+V)}{\hbar}}.$$

Given an initial vector $\psi_0 \in L^2(\mathbb{R}^d)$, the solution of the Cauchy problem

$$\begin{cases} \frac{d}{dt}\psi = -\frac{i}{\hbar}H\psi \\ \psi(0,x) = \psi_0(x) \end{cases} \tag{3.56}$$

is given by $\psi(t) = e^{-\frac{it(p^2/2m+V)}{\hbar}}\psi_0$.

By lemma 3.5 we have

$$e^{-\frac{it(p^2/2m+V)}{\hbar}} = s - \lim_{n\to\infty}\left(e^{-\frac{i\epsilon(p^2/2m+V_2)}{\hbar}}e^{-\frac{i\epsilon(V_1)}{\hbar}}\right)^n, \qquad \epsilon \equiv \frac{t}{n}$$

$$\psi(t) = e^{-\frac{it(p^2/2m+V)}{\hbar}}\psi_0 = \lim_{n\to\infty}\left(e^{-\frac{i\epsilon(p^2/2m+V_2)}{\hbar}}e^{-\frac{i\epsilon V_1}{\hbar}}\right)^n\psi_0,$$

(see e.g. [78, 79, 246] for related uses of the Lie-Trotter formula).

Let us consider a smooth vector $\psi_0 \in C_0^\infty(\mathbb{R}^d)$. By shifting from the position representation to the momentum representation and vice versa and by assuming that V_1 and V_2 are continuous, we can write in the strong $L^2(\mathbb{R}^d)$-sense, for all $t > 0$:

$$\psi(t,x) = \lim_{n\to\infty}\int_{\mathbb{R}^d} e^{-\frac{i\epsilon(p_{n-1}^2/2m+V_2(p_{n-1}))}{\hbar}}.$$

$$\left(e^{-\frac{i\epsilon V_1}{\hbar}}\widetilde{\left(e^{-\frac{i\epsilon(p^2/2m+V_2)}{\hbar}}e^{-\frac{i\epsilon(V_1)}{\hbar}}\right)^{n-1}\psi_0}\right)(p_{n-1})\frac{e^{i\frac{xp_{n-1}}{\hbar}}}{(2\pi\hbar)^{d/2}}dp_{n-1}$$

$$= \lim_{n\to\infty}\int_{\mathbb{R}^{2d}} e^{-\frac{i\epsilon(p_{n-1}^2/2m+V_2(p_{n-1}))}{\hbar}} e^{-\frac{i\epsilon V_1(x_{n-1})}{\hbar}}.$$

$$\left(\left(e^{-\frac{i\epsilon(p^2/2m+V_2)}{\hbar}} e^{-\frac{i\epsilon(V_1)}{\hbar}}\right)^{n-1}\psi_0\right)(x_{n-1})\frac{e^{i\frac{xp_{n-1}}{\hbar}}}{(2\pi\hbar)^{d/2}}\frac{e^{-i\frac{x_{n-1}p_{n-1}}{\hbar}}}{(2\pi\hbar)^{d/2}}dp_{n-1}dx_{n-1}$$

$$= \lim_{n\to\infty}\left(\frac{1}{\sqrt{2\pi\hbar}}\right)^{2nd}.$$

$$\int_{\mathbb{R}^{2nd}} e^{-\frac{i\epsilon}{\hbar}\sum_{j=0}^{n-1}\left(\frac{p_j^2}{2m}+V_1(x_j)+V_2(p_j)-p_j\frac{(x_{j+1}-x_j)}{\epsilon}\right)}\psi_0(x_0)\prod_{j=0}^{n-1}dp_jdx_j, \tag{3.57}$$

where[2] $x_n \equiv x$.

Now the latter expression suggests the following heuristic formula for the limit

$$\psi(t,x) = \text{const}\int_{q(t)=x} e^{\frac{i}{\hbar}\left(\int_0^t p(s)\dot{q}(s)-H(q(s),p(s))ds\right)}\psi_0(q(0))dqdp, \tag{3.58}$$

which can be seen phase space-Hamiltonian version of Feynman's path integral (1.6). The aim of the present section is the rigorous mathematical realization of the heuristic formula (3.58) in terms of a well defined infinite dimensional oscillatory integral and to prove that, under suitable assumptions on the initial datum ψ_0 and on the classical potential V, it gives a representation of the solution of the Schrödinger equation.

Let us consider first of all Eq. (3.58) in the particular case of the free particle, namely when the Hamiltonian is just the kinetic energy: $H = p^2/2m$. In this case we have heuristically

$$\psi(t,x) = \text{const}\int_{q(t)=x} e^{\frac{i}{\hbar}\int_0^t(p(s)\dot{q}(s)-p(s)^2/2m)ds}\psi_0(q(0))dqdp \tag{3.59}$$

From now on we will assume for notational simplicity that $m = 1$, but the whole discussion can be generalized to arbitrary m.

Following [284, 285], let us introduce the Hilbert space $\mathcal{H}_t \times \mathcal{L}_t$, namely the space of paths in the d–dimensional phase space $(q(s),p(s))_{s\in[0,t]}$, such that the path $(q(s))_{s\in[0,t]}$ belongs to the Cameron Martin space $\mathcal{H}_t$, while

[2] The integrals in Eq. (3.57) are to be understood as limits as $\Lambda \uparrow \mathbb{R}^d$, $n \to \infty$ in the $L^2(\mathbb{R}^{2nd})$ sense of the corresponding integrals over Λ^{2nd}, with Λ bounded (see [235]). Formula (3.57) holds first as a strong L^2-limit, but then (possibly by subsequences) also for Lebesgue almost everywhere in $\mathbb{R}^d$. It also follows from this that Eq. (3.57) gives the solution to the Cauchy problem (3.56).

the path in the momentum space $(p(s))_{s\in[0,t]}$ belongs to $\mathcal{L}_t = L^2([0,t],\mathbb{R}^d)$. $\mathcal{H}_t \times \mathcal{L}_t$ is an Hilbert space with the natural inner product

$$\langle q,p;Q,P\rangle = \int_0^t \dot{q}(s)\dot{Q}(s)ds + \int_0^t p(s)P(s)ds.$$

Let us introduce the following bilinear form:

$$\begin{aligned}&[q,p;Q,P]\\ &= \int_0^t \dot{q}(s)P(s)ds + \int_0^t p(s)\dot{Q}(s)ds - \int_0^t p(s)P(s)ds = \langle q,p;A(Q,P)\rangle,\end{aligned}$$

where A is the following operator in $\mathcal{H}_t \times \mathcal{L}_t$,:

$$A(Q,P)(s) = \Big(\int_t^s P(u)du, \dot{Q}(s) - P(s)\Big). \tag{3.60}$$

A is densely defined, e.g. on $C^1([0,t];\mathbb{R}^d) \times C^1([0,t];\mathbb{R}^d)$.

Moreover A is invertible with inverse given by

$$A^{-1}(Q,P)(s) = \Big(\int_t^s P(u)du + Q(s), \dot{Q}(s)\Big) \tag{3.61}$$

(on the range of A).

Let $\psi_0 \in \mathcal{F}(\mathbb{R}^d)$, it is easy to see that the functional on $\mathcal{H}_t \times \mathcal{L}_t$ given by

$$(q,p) \mapsto \psi_0(q(0)+x), \qquad (q,p) \in \mathcal{H}_t \times \mathcal{L}_t,$$

belongs to $\mathcal{F}(\mathcal{H}_t \times \mathcal{L}_t)$:

$$\psi_0(q(0)+x) = \int_{\mathcal{H}_t\times\mathcal{L}_t} e^{i\langle q,p;Q,P\rangle} d\mu_0(Q,P).$$

Now expression (3.58) can be realized rigorously as the normalized oscillatory integral with respect to the operator A (in the sense of definition 2.5) on the Hilbert space $\mathcal{H}_t \times \mathcal{L}_t$:

$$\widetilde{\int_{\mathcal{H}_t\times\mathcal{L}_t}^{A}} e^{\frac{i}{2\hbar}\langle q,p;A(q,p)\rangle}\psi_0(q(0)+x)dqdp, \tag{3.62}$$

and by theorem 2.8 the normalized oscillatory integral is well defined and can be computed by means of Parseval type equality (2.41):

$$\widetilde{\int_{\mathcal{H}_t\times\mathcal{L}_t}^{A}} e^{\frac{i}{2\hbar}\langle q,p;A(q,p)\rangle}\psi_0(q(0)+x)dqdp = \int_{\mathcal{H}_t\times\mathcal{L}_t} e^{-\frac{i\hbar}{2\hbar}\langle q,p;A^{-1}(q,p)\rangle} d\mu_0(q,p).$$

By choosing a suitable sequence of finite dimensional projection operators $\{P_n\}$ on $\mathcal{H}_t \times \mathcal{L}_t$, it is possible to recover in the definition of the normalized oscillatory integral the expression (3.57), obtained by means of the Lie-Trotter product formula.

Let us consider a sequence of partitions π_n of the interval $[0,t]$ into n subintervals of amplitude $\epsilon \equiv t/n$:

$$t_0 = 0, t_1 = \epsilon, \ldots, t_i = i\epsilon, \ldots, t_n = n\epsilon = t.$$

To each π_n we associate a projector $P_n : \mathcal{H}_t \times \mathcal{L}_t \to \mathcal{H}_t \times \mathcal{L}_t$ onto a finite dimensional subspace of $\mathcal{H}_t \times \mathcal{L}_t$, namely the subspace of polygonal paths. In other words each projector P_n acts on a phase space path $(q,p) \in \mathcal{H}_t \times \mathcal{L}_t$ in the following way:

$P_n(q,p)(s) =$

$$\Big(\textstyle\sum_{i=1}^n \chi_{[t_{i-1},t_i]}(s)\Big(q(t_{i-1}) + \frac{(q(t_i)-q(t_{i-1})}{t_i-t_{i-1}}(s-t_{i-1})\Big), \sum_{i=1}^n \chi_{[t_{i-1},t_i]}(s)p_i\Big),$$

where

$$p_i = \frac{\int_{t_{i-1}}^{t_i} p(s)ds}{t_i - t_{i-1}} = \frac{1}{\epsilon}\int_{t_{i-1}}^{t_i} p(s)ds.$$

Analogously to lemma 3.4, it is possible to prove the following

Lemma 3.6. *For each $n \in \mathbb{N}$, P_n is a projector in $\mathcal{H}_t \times \mathcal{L}_t$. Moreover for $n \to \infty$, $P_n \to I$ as a strong operator limit.*

Proof.

- P_n is symmetric, indeed for all $(Q,P) \in \mathcal{H}_t \times \mathcal{L}_t$ and all $(q,p) \in \mathcal{H}_t \times \mathcal{L}_t$

$$\begin{aligned}\langle Q,P;P_n(q,p)\rangle &= \int_0^t \dot{Q}(s)\sum_{i=1}^n \chi_{[t_{i-1},t_i]}(s)\frac{(q(t_i)-q(t_{i-1})}{t_i-t_{i-1}}ds \\ &+ \int_0^t P(s)\sum_{i=1}^n \chi_{[t_{i-1},t_i]}(s)p_i ds = \sum_{i=1}^n \frac{(q(t_i)-q(t_{i-1})(Q(t_i)-Q(t_{i-1})}{t_i-t_{i-1}} \\ &+ \sum_{i=1}^n \frac{\int_{t_{i-1}}^{t_i} p(s)ds\int_{t_{i-1}}^{t_i} P(s)ds}{t_i-t_{i-1}} = \langle P_n(Q,P);q,p\rangle \end{aligned} \quad (3.63)$$

- $P_n^2 = P_n$, indeed

$P_n^2(q,p)(s)$

$$= \Big(\textstyle\sum_{i=1}^n \chi_{[t_{i-1},t_i]}(s)\Big(q(t_{i-1}) + \frac{(q(t_i)-q(t_{i-1})}{t_i-t_{i-1}}(s-t_{i-1})\Big), \sum_{i=1}^n \chi_{[t_{i-1},t_i]}(s)p_i\Big)$$

$= P_n(q,p)(s)$

- $\forall (q,p) \in \mathcal{H}_t \times \mathcal{L}_t$, $\|P_n(q,p) - (q,p)\| \to 0$ as $n \to \infty$:
  Let us consider the subset $\mathcal{K} \subseteq \mathcal{H}_t \times \mathcal{L}_t$,
  $$\mathcal{K} = \{(q,p) \in \mathcal{H}_t \times \mathcal{L}_t \,:\, \|P_n(q,p) - (q,p)\| \to 0\,,\, n \to \infty\}.$$
  It is enough to prove that the closure of $\mathcal{K}$ is $\mathcal{H}_t \times \mathcal{L}_t$. To prove this, it is sufficient to show that $\mathcal{K}$ is a closed subspace of $\mathcal{H}_t \times \mathcal{L}_t$ and contains a dense subset of $\mathcal{H}_t \times \mathcal{L}_t$. This follows from the density of the piecewise linear paths in $\mathcal{H}_t$ (see lemma 3.4 and [283]) and the density of the piecewise constant paths in $\mathcal{L}_t$. □

By means of these results and under the assumption that $\psi_0 \in \mathcal{F}(\mathbb{R}^d)$, it is possible to prove that the infinite dimensional oscillatory integral (3.62) coincides with the limit (3.57), which can be taken for Lebesgue almost every $x \in \mathbb{R}^d$.

Theorem 3.4. *Let the function $(q,p) \to \psi_0(x + q(0))$, $\psi_0 \in L^2(\mathbb{R}^d)$, be Fresnel integrable[3] with respect to A (with A defined by (3.60)). Then the phase space Feynman path integral, namely the limit*

$$\lim_{n\to\infty} \frac{\sqrt{\det(P_n A P_n)}}{(2\pi i\hbar)^{nd}} \int_{P_n(\mathcal{H}_t \times \mathcal{L}_t)} e^{\frac{i}{2\hbar}\langle P_n(q,p), A P_n(q,p)\rangle} \psi_0(x + q(0)) dP_n(q,p) \tag{3.64}$$

coincides with the limit (3.57), namely with the solution of the Schrödinger equation with a free Hamiltonian.

Proof. The result follows by direct computation, by showing that the two limits (3.57) and (3.64) coincide. Indeed (3.64) is a pointwise limit by hypothesis. On the other hand (3.57) is a limit in the L_2 sense, hence, passing if necessary to a subsequence, it is also a pointwise limit. □

Remark 3.4. The latter result is equivalent to the "traditional" formulation of the Feynman path integral in the configuration space. Indeed it can be obtained by means of Fubini theorem and an integration with respect to the momentum variables:

$$\lim_{n\to\infty} \left(\frac{1}{\sqrt{2\pi\hbar}}\right)^{2nd} \int_{\mathbb{R}^{2nd}} e^{-\frac{i\epsilon}{\hbar}\sum_{j=0}^{n-1}\left(\frac{p_j^2}{2m} - p_j\frac{(x_{j+1}-x_j)}{\epsilon}\right)} \psi_0(x_0) \prod_{j=0}^{n-1} dp_j dx_j$$

$$= \lim_{n\to\infty} \left(\frac{1}{\sqrt{2\pi i\hbar}}\right)^{nd} \int_{\mathbb{R}^{nd}} e^{-\frac{i\epsilon}{\hbar}\sum_{j=0}^{n-1} m\frac{(x_{j+1}-x_j)^2}{2\epsilon^2}} \psi_0(x_0) \prod_{j=0}^{n-1} dx_j\,. \tag{3.65}$$

[3]This condition is satisfied if for instance $\psi_0 \in \mathcal{F}(\mathbb{R}^d)$.

The latter expression yields the Feynman functional on the configuration space, i.e. heuristically

$$\text{const}\int e^{\int_0^t \mathcal{L}(q(s),\dot{q}(s))ds}dq,$$

($\mathcal{L}$ being the classical Lagrangian density).

Even if the integration with respect to the momentum variables might seem to be superfluous, it is very useful when we introduce a potential depending explicitly on the momentum variables, as the following theorem shows.

Theorem 3.5. *Let us consider a semibounded potential V depending explicitly on the momentum: $V = V(p)$ and the corresponding quantum mechanical Hamiltonian $H = -\frac{\hbar^2}{2}\Delta + V(p)$. Let us assume H is an essentially self-adjoint operator on $L_2(\mathbb{R}^d)$. Let the functional on the Hilbert space $\mathcal{H}_t \times \mathcal{L}_t$, given by*

$$(q,p) \mapsto e^{-\frac{i}{\hbar}\int_0^t V(p(s))ds}\psi_0(x+q(0)), \qquad (q,p) \in \mathcal{H}_t \times \mathcal{L}_t$$

be Fresnel integrable with respect to the operator A, with A defined by (3.60). Then the solution to the Schrödinger equation

$$\begin{cases} \frac{d}{dt}\psi = -\frac{i}{\hbar}H\psi \\ \psi(0,x) = \psi_0(x), \qquad \psi_0 \in \mathcal{S}(\mathbb{R}^d) \end{cases} \tag{3.66}$$

is given by the phase space path integral

$$\lim_{n\to\infty}(2\pi i\hbar)^{-nd}(\det(P_nAP_n))^{1/2}\int_{P_n(\mathcal{H}_t\times\mathcal{L}_t)} e^{\frac{i}{2\hbar}\langle P_n(q,p),AP_n(q,p)\rangle} e^{-\frac{i}{\hbar}\int_0^t V(P_n(p(s)))ds}\psi_0(x+q(0))dP_n(q,p)\,.$$

Proof. We can proceed in a completely analogous way as in the proof of theorem 3.4, therefore we shall omit the details. □

We can now handle the case of a classical potential V depending both on position Q and on momentum P of the form $V = V(Q,P) = V_1(Q)+V_2(P)$ (the general case presents problems due to the non commutativity of the quantized expression of Q and P). For a different approach with more general Hamiltonians see [265]).

Let us consider an initial wave function $\psi_0 \in \mathcal{F}(\mathbb{R}^d)$ and let us assume that $V_1 \in \mathcal{F}(\mathbb{R}^d)$ and the function on $\mathcal{L}_t$ given by

$$p(s)_{s\in[0,t]} \to e^{-\frac{i}{\hbar}\int_0^t V_2(p(s))ds}$$

belongs to $\mathcal{F}(\mathcal{L}_t)$. Then it is easy to see that the functional $f : \mathcal{H}_t \times \mathcal{L}_t \to \mathbb{C}$

$$f(q,p) = \psi_0(x + q(0))e^{-\frac{i}{\hbar}\int_0^t V(q(s)+x,p(s))ds},$$

belongs to $\mathcal{F}(\mathcal{H}_t \times \mathcal{L}_t)$, i.e. $f = \hat{\mu}_f$, $\mu_f \in \mathcal{M}(\mathcal{H}_t \times \mathcal{L}_t)$. Indeed $f(q,p)$ is the product of two functions: the first, say f_1, depends only on the first variable q, while the second f_2 depends only on the variable p, more precisely

$$f_1(q) = \psi_0(x + q(0))e^{-\frac{i}{\hbar}\int_0^t V_1(q(s)+x)ds}, \qquad f_2(p) = e^{-\frac{i}{\hbar}\int_0^t V_2(p(s))ds}.$$

Under the given hypothesis on V_1 and ψ_0, f_1 belongs to $\mathcal{F}(\mathcal{H}_t)$ (see lemma 3.3). For f_2 one must pay more attention: indeed the same proof given for f_1 does not work, as f_2 is defined on a different Hilbert space and we have to require explicitly that $e^{-\frac{i}{\hbar}\int_0^t V_2(p(s))ds} \in \mathcal{F}(\mathcal{L}_t)$. This holds for instance if V_2 is linear, i.e. $V_2(p) = a \cdot p$, $a \in \mathbb{R}^d$, or if the function $p(s)_{s\in[0,t]} \to \int_0^t V_2(p(s))ds \in \mathcal{F}(\mathcal{L}_t)$.
Now if $f_1 = \hat{\mu}_{f_1} \in \mathcal{F}(\mathcal{H}_t)$, f_1 can be extended to a function, denoted again by f_1, in $\mathcal{F}(\mathcal{H}_t \times \mathcal{L}_t)$: it is the Fourier transform of the product measure on $\mathcal{H}_t \times \mathcal{L}_t$ of $\mu_{f_1}(dq)$ and $\delta_0(dp)$. The same holds for $f_2 = \hat{\mu}_{f_2}$: $f_2 = (\widehat{\delta_0(dq)\mu_{f_2}}(dp))$.

Finally, as $\mathcal{F}(\mathcal{H}_t \times \mathcal{L}_t)$ is a Banach algebra, the product of two elements $f_1 f_2$ is again an element of $\mathcal{F}(\mathcal{H}_t \times \mathcal{L}_t)$: more precisely it is the Fourier transform of the convolution of the two measures in $\mathcal{M}(\mathcal{H}_t \times \mathcal{L}_t)$ corresponding to f_1 and f_2 respectively.

By applying theorem 2.8, the phase space Feynman path integral of the function f is well defined and can be computed in terms of a Parseval type equality:

$$\widetilde{\int_{\mathcal{H}_t\times\mathcal{L}_t}^A} e^{\frac{i}{2\hbar}\langle(q,p),A(q,p)\rangle} f(q,p)dqdp = \int_{\mathcal{H}_t\times\mathcal{L}_t} e^{-\frac{i\hbar}{2}\langle(q,p),A^{-1}(q,p)\rangle} d\mu_f(q,p).$$

The next theorem shows that the above oscillatory integral

$$\widetilde{\int_{\mathcal{H}_t\times\mathcal{L}_t}^A} e^{\frac{i}{2\hbar}\langle(q,p),A(q,p)\rangle} e^{-\frac{i}{\hbar}\int_0^t V_1(q(s)+x)ds} e^{-\frac{i}{\hbar}\int_0^t V_2(p(s))ds} \psi_0(x + q(0))dqdp \tag{3.67}$$

gives the solution to the Schrödinger equation (3.56).

Theorem 3.6. *Let us consider the following Hamiltonian*

$$H(Q;P) = \frac{P^2}{2} + V_1(Q) + V_2(P)$$

in $L^2(\mathbb{R}^d)$ and the corresponding Schrödinger equation

$$\begin{cases} \frac{d}{dt}\psi = -\frac{i}{\hbar}H\psi \\ \psi(0,x) = \psi_0(x), \quad x \in \mathbb{R}^d. \end{cases}$$

Let us suppose that $V_1, \psi_0 \in \mathcal{F}(\mathbb{R}^d)$ and $\int_0^t V_2(p(s))ds \in \mathcal{F}(\mathcal{L}_t)$. Then the solution to the Cauchy problem (3.56) is given by the phase space Feynman path integral (3.67).

Proof. As in the proof of theorem 3.2, we follow the technique proposed by Elworthy and Truman in [114].

For $0 \leq u \leq t$ let $\mu_u(V_1, x) \equiv \mu_u$, $\nu_u^t(V_1, x) \equiv \nu_u^t$, $\eta_u^t(V_2) \equiv \eta_u^t$ and $\mu_0(\psi)$ be the measures on $\mathcal{H}_t \times \mathcal{L}_t$, whose Fourier transforms when evaluated at $(q,p) \in \mathcal{H}_t \times \mathcal{L}_t$ are $V_1(x + q(u))$, $\exp\Big(- \frac{i}{\hbar} \int_u^t V_1(x + q(s))ds\Big)$, $\exp\Big(- \frac{i}{\hbar} \int_u^t V_2(p(s))ds\Big)$ and $\psi_0(q(0) + x)$.
We set

$$U(t)\psi_0(x) = \widetilde{\int_{\mathcal{H}_t \times \mathcal{L}_t}^{A}} e^{\frac{i}{2\hbar}\langle q,p;A(q,p)\rangle} e^{-\frac{i}{\hbar}\int_0^t (V_1(q(s)+x)+V_2(p(s)))ds} \psi_0(q(0)+x)dqdp$$

and

$$U_0(t)\psi_0(x) = \widetilde{\int_{\mathcal{H}_t \times \mathcal{L}_t}^{A}} e^{\frac{i}{2\hbar}\langle q,p;A(q,p)\rangle} e^{-\frac{i}{\hbar}\int_0^t V_2(p(s)))ds} \psi_0(q(0) + x)dqdp.$$

By theorem 2.8 we have:

$$U(t)\psi_0(x) = \int_{\mathcal{H}_t \times \mathcal{L}_t} e^{\frac{-i\hbar}{2}\langle q,p;A^{-1}(q,p)\rangle} (\eta_0^t * \nu_0^t * \mu_0(\psi))(dqdp). \qquad (3.68)$$

Now, if $\{\mu_u : a \leq u \leq t\}$ is a family in $\mathcal{M}(\mathcal{H}_t \times \mathcal{L}_t)$, we shall let $\int_a^b \mu_u du$ denote the measure on $\mathcal{H}_t \times \mathcal{L}_t$ given by :

$$f \to \int_a^b \int_{\mathcal{H}_t \times \mathcal{L}_t} f(q,p)d\mu_u(q,p)du, \qquad f \in C_0(\mathcal{H}_t \times \mathcal{L}_t),$$

whenever it exists.
Since for any continuous path q we have

$$\exp\Big(-\frac{i}{\hbar}\int_0^t V_1(q(s)+x)ds\Big) = 1 - \frac{i}{\hbar}\int_0^t V_1(q(u)+x)\exp\Big(-\frac{i}{\hbar}\int_u^t V_1(q(s)+x)ds\Big)du$$

the following relation holds

$$\nu_0^t = \delta_0 - \frac{i}{\hbar}\int_0^t (\mu_u * \nu_u^t)du \qquad (3.69)$$

where δ_0 is the Dirac measure at $0 \in \mathcal{H}_t \times \mathcal{L}_t$.

Applying this relation to (3.68) we obtain:

$$U(t)\psi_0(x) = \int_{\mathcal{H}_t\times\mathcal{L}_t} e^{\frac{-i\hbar}{2}\langle q,p;A^{-1}(q,p)\rangle}(\eta_0^t * \mu_0(\psi))(dqdp)$$

$$-\frac{i}{\hbar}\int_0^t\int_{\mathcal{H}_t\times\mathcal{L}_t} e^{\frac{-i\hbar}{2}\langle q,p;A^{-1}(q,p)\rangle}(\eta_0^t * \mu_u(V_1,x) * \nu_u^t * \mu_0(\psi))(dqdp)du$$

$$= U_0(t)\psi_0(x) - \frac{i}{\hbar}\int_0^t \widetilde{\int_{\mathcal{H}_t\times\mathcal{L}_t}^A} e^{\frac{i}{2\hbar}\langle q,p;A(q,p)\rangle} e^{-\frac{i}{\hbar}\int_u^t V_1(q(s)+x)ds}$$

$$e^{-\frac{i}{\hbar}\int_0^t V_2(p(s)))ds}V_1(q(u)+x)\psi_0(q(0)+x)dqdpdu.$$

The conclusion follows by repeating the same procedure, as in the proof of theorem 3.2. □

3.4 Magnetic field

Let us consider a charged quantum particle (with unitary charge and mass) moving in the plane and submitted to the influence of a constant magnetic field orthogonal to it. By choosing a system of Cartesian coordinates in $\mathbb{R}^3$ such that the magnetic field $\vec{B}$ is directed along the x_3–axis, i.e. $\vec{B} = (0,0,a)$, the Schrödinger equation describing the time evolution of the wave function $\psi \in L^2(\mathbb{R}^2)$ is

$$\begin{cases} i\hbar\frac{\partial}{\partial t}\psi(t,x) = \frac{1}{2}\Big(\big(i\hbar\frac{\partial}{\partial x_1}+ax_2\big)^2 + \big(i\hbar\frac{\partial}{\partial x_2}-ax_1\big)^2\Big)\psi(t,x) + V_0(x)\psi(t,x) \\ \psi(0,x) = \psi_0(x), \qquad x=(x_1,x_2)\in\mathbb{R}^2,\ t>0, \end{cases} \tag{3.70}$$

where $a \in \mathbb{R}$ and V_0 is a scalar potential. Let us assume that the scalar potential V_0 is an element of $\mathcal{F}(\mathbb{R}^2)$. In this case it is possible to prove (see for instance [263]) that the Cauchy problem (3.70) is well posed in the sense that for any $\psi_0 \in L^2(\mathbb{R}^2)$ there exists a unique continuous function $\psi : \mathbb{R} \to L^2(\mathbb{R}^2)$, such that $\psi(0) = \psi_0$ and the Schrödinger equation (3.70) is satisfied in a mild form.

In [7] a Feynman path integral representation for the fundamental solution $G(t,x,y)$ of Eq. (3.70) is rigorously defined in terms of infinite dimensional oscillatory integrals on a suitable Hilbert space. Indeed let us consider the space

$$Y_t = \{\gamma \in H^1([0,t],\mathbb{R}^2) \,:\, \gamma(0)=\gamma(t)=0\},$$

endowed with the norm

$$\|\gamma\|^2 := \int_0^t \big(|\dot\gamma_1(s) - a\gamma_2(s)|^2 + |\dot\gamma_2(s) + a\gamma_1(s)|^2\big)ds.$$

It is not difficult to show that $(Y_t, \| \ \|)$ is an Hilbert space.

The action functional $S : Y_t \to \mathbb{R}$ is given by

$$S(\gamma) = \frac{1}{2}\|\gamma\|^2 - \frac{1}{2}a^2\langle\gamma, L\gamma\rangle + V(\gamma), \tag{3.71}$$

where $L : Y_t \to Y_t$ is the linear self-adjoint operator defined by

$$\langle\gamma, L\gamma\rangle = \int_0^t |\gamma(s)|^2 ds,$$

and $V : Y_t \to \mathbb{R}$ is given by

$$V(\gamma) = \int_0^t V_0(\gamma(s))ds.$$

By reasoning as in the proof of lemma 3.2, it is possible to prove that L is a trace class operator, its spectrum has the form

$$\sigma(L) = \{0, \lambda_j = \left(\frac{t}{\pi j}\right)^2, j = 1, 2, \ldots\},$$

where each λ_j is of multiplicity 2, and consequently

$$\det(I - a^2 L) = \left(\frac{\sin(at)}{at}\right)^2.$$

Let $\beta_{t,x,y}$ be a solution of the following boundary value problem:

$$\begin{cases} -\ddot{\beta}(s) + 2aC\dot{\beta}(s) = 0, & 0 < s < t \\ \beta(0) = y \quad \beta(t) = x \end{cases} \tag{3.72}$$

where $C = \begin{pmatrix} 0 & 1 \\ -1 & 0 \end{pmatrix}$. With these notations it is possible to prove that the heuristic Feynman path integral representation (1.5) for the fundamental solution of the Schrödinger equation (3.70)

$$G(t, x, y) = \int_{\gamma(0)=y,\gamma(t)=x} e^{\frac{i}{2\hbar}S(\gamma)} D\gamma$$

can be rigorously mathematically realized as an infinite dimensional oscillatory integral on the Hilbert space $(Y_t, \| \ \|)$. More precisely the following holds:

Theorem 3.7. [7] *If $V_0 \in \mathcal{F}(\mathbb{R}^2)$ and if $\sin(at) \neq 0$, the fundamental solution of the Schrödinger equation (3.70) is given by the infinite dimensional oscillatory integral*

$$G(t, x, y) = (2\pi i t)^{-1} e^{\frac{i}{\hbar}S(\beta_{t,x,y})} \widetilde{\int_{Y_t}} e^{\frac{i}{2\hbar}\left(\|\gamma\|^2 ds - a^2\langle\gamma, L\gamma\rangle\right)} e^{-\frac{i}{\hbar}V(\gamma+\beta_{t,x,y})} d\gamma.$$

Analogous results have been obtained in [9] in terms of class-2 normalized integrals (see definition 2.6).

3.5 Quartic potential

The examples we have seen so far have shown that infinite dimensional oscillatory integrals are a flexible tool and provide a rigorous mathematical realization for a large class of Feynman path integrals representations. However all the examples have a common problem, that is the restriction on the class of classical potentials V which can be handled. Indeed, in order to apply theorem 2.5 (or theorems 2.8 and 2.9), we have to assume that the potential V describing the dynamics of the quantum particle is of the type "harmonic oscillator plus a bounded perturbation which is the Fourier transform of a measure". This situation is rather unsatisfactory from a physical point of view, as this class of potentials does not include several interesting unbounded ones. There are some extensions of the theory to unbounded potentials which are Laplace transforms of measures, such as those with exponential growth at infinity [10, 210, 19], but even this class does not include the functions with a generic polynomial growth at infinity. In fact the problem for such polynomial potentials is not simple, as it has been proved [293] that in one dimension, if the potential is time independent and super-quadratic in the sense that

$$V(x) \geq C(1+|x|)^{2+\epsilon}, \qquad x \to \infty,$$

where $C > 0$ and $\epsilon > 0$, then, as a function of (t, x, y), the fundamental solution $G(t, 0, x, y)$ of the time dependent Schrödinger equation is nowhere C^1. In other words, we do not have even the hope to define in terms of a functional integral an expression of the form

$$G(0, x; t, y) = \int e^{\frac{i}{\hbar}S_t(\gamma)} D\gamma, ,$$

depending from the variables (t, x, y), that can be differentiated in order to prove that it is a fundamental solution for the Schrödinger equation.

Even if the problem is not only technical, but rather fundamental, some interesting results can be obtained in the case where the potential V has a quartic polynomial growth at infinity, by means of the results of section 2.5.

Let us consider the Schrödinger equation in $L^2(\mathbb{R}^d)$

$$\begin{cases} i\hbar\frac{\partial}{\partial t}\psi(t,x) = -\frac{\hbar^2}{2}\Delta\psi(t,x) + V(x)\psi(t,x) \\ \psi(0,x) = \psi_0(x) \end{cases} \tag{3.73}$$

where the potential V is of the following form:

$$V(x) = \frac{1}{2}x\Omega^2 x + \lambda|x|^4, \qquad x \in \mathbb{R}^d, \tag{3.74}$$

where Ω^2 is a positive symmetric $d \times d$ matrix, and $\lambda \in \mathbb{R}$ is a real constant. More generally, we can also consider a polynomial potential of the following form

$$V(x) = \frac{1}{2} x\Omega^2 x + \lambda C(x, x, x, x), \qquad x \in \mathbb{R}^d, \tag{3.75}$$

where C is a completely symmetric positive fourth order covariant tensor on $\mathbb{R}^d$. In the following we shall focus on expression (3.74) because of notational simplicity, but all the reasoning can be repeated in the general case of Eq. (3.75).

In the case where λ is positive, the quantum mechanical Hamiltonian $H : D(H) \subset L^2(\mathbb{R}^d) \to L^2(\mathbb{R}^d)$ given on vector $\psi \in C_0^\infty(\mathbb{R}^d)$ by

$$H\psi(x) = -\frac{\Delta}{2}\psi(x) + V(x)\psi(x) \tag{3.76}$$

is essentially self-adjoint [246] and determines uniquely the quantum dynamics. In the case where λ is strictly negative, i.e. when the potential V represents a quartic (double well) polynomial potentials unbounded from below, the quantum Hamiltonian is not essentially self-adjoint as one can deduce by a limit point argument (see [246], theorem X.9) and the quantum evolution is not uniquely determined.

In the following we shall present the results of [27, 25, 224] and show that the infinite dimensional oscillatory integrals with polynomial phase function studied in section 2.5 can provide a mathematical realization for the Feynman path integral representation for the weak solution of the Schrödinger equation (3.73) with potential V given by Eq. (3.74), i.e. the matrix elements $\langle \phi, e^{-\frac{i}{\hbar}Ht}\psi_0 \rangle$, $\phi, \psi_0 \in L^2(\mathbb{R}^d)$:

$$\int_{\mathbb{R}^d} \bar{\phi}(x) \int_{\gamma(t)=x} e^{\frac{i}{2\hbar}\int_0^t \dot{\gamma}^2(s)ds - \frac{i}{\hbar}\int_0^t \gamma(s)\Omega^2\gamma(s)ds - \frac{i\lambda}{\hbar}\int_0^t |\gamma(s)|^4 ds} \psi_0(\gamma(0)) d\gamma dx. \tag{3.77}$$

Let us consider the Cameron-Martin space[4] $\mathcal{H}_t$, that is the Hilbert space of absolutely continuous paths $\gamma : [0, t] \to \mathbb{R}^d$, with $\gamma(0) = 0$ and inner product

$$\langle \gamma_1, \gamma_2 \rangle = \int_0^t \dot{\gamma}_1(s)\dot{\gamma}_2(s)ds.$$

The cylindrical Gaussian measure on $\mathcal{H}_t$ with covariance operator the identity extends to a σ-additive measure on the Wiener space $C_t = \{\omega \in$

[4]With an abuse of notation we call here Cameron-Martin space the space of paths γ belonging to the Sobolev space $H^{1,2}_{(t)}([0, t], \mathbb{R}^d)$ such that $\gamma(0) = 0$, while in the previous sections with the same name we denoted the space of paths $\gamma \in H^{1,2}_{(t)}([0, t], \mathbb{R}^d)$ such that $\gamma(t) = 0$.

$C([0,t];\mathbb{R}^d) \mid \gamma(0)=0\}$: the Wiener measure W. $(i, \mathcal{H}_t, C_t)$ is an abstract Wiener space (see the appendix for the definition and the main properties of abstract Wiener spaces).

Let us consider moreover the Hilbert space $\mathcal{H} = \mathbb{R}^d \times \mathcal{H}_t$ and the Banach space $\mathcal{B} = \mathbb{R}^d \times C_t$ endowed with the product measure $N(dx) \times W(d\omega)$, N being the Gaussian measure on $\mathbb{R}^d$ with covariance equal to the $d \times d$ identity matrix. $(i, \mathcal{H}, \mathcal{B})$ is an abstract Wiener space.

Let us consider the operator $B : \mathcal{H} \to \mathcal{H}$ given by:

$$(x,\gamma) \mapsto (y,\eta) = B(x,\gamma), \qquad (x,\gamma) \in \mathbb{R}^d \times \mathcal{H}_t \tag{3.78}$$

$$\begin{aligned} y &= t\Omega^2 x + \Omega^2 \int_0^t \gamma(s)ds, \\ \eta(s) &= \Omega^2 x\Big(ts - \frac{s^2}{2}\Big) - \int_0^s \int_t^u \Omega^2 \gamma(r) dr du, \qquad s \in [0,t]. \end{aligned} \tag{3.79}$$

It is easy to verify that $B : \mathcal{H} \to \mathcal{H}$ is positive and symmetric, indeed

$$\langle (y,\eta), B(x,\gamma)\rangle = \int_0^t (\eta(s)+y)\Omega^2(\gamma(s)+x)ds, \qquad (y,\eta),(x,\gamma) \in \mathcal{H}.$$

Let us introduce the fourth order tensor operator $A : \mathcal{H} \times \mathcal{H} \times \mathcal{H} \times \mathcal{H} \to \mathbb{R}$ given by:

$$\begin{aligned} &A((x_1,\gamma_1),(x_2,\gamma_2),(x_3,\gamma_3),(x_4,\gamma_4)) \\ &\quad = \int_0^t (\gamma_1(s)+x_1)(\gamma_2(s)+x_2)(\gamma_3(s)+x_3)(\gamma_4(s)+x_4)ds, \end{aligned} \tag{3.80}$$

and the homogeneous fourth order polynomial function $V_4 : \mathcal{H} \to \mathbb{R}$ given by

$$V_4(x,\gamma) = \lambda A((x,\gamma),(x,\gamma),(x,\gamma),(x,\gamma)) = \lambda \int_0^t |\gamma(s)+x|^4 ds.$$

Given two vectors $\phi, \psi_0 \in L^2(\mathbb{R}^d) \cap \mathcal{F}(\mathbb{R}^d)$, let us consider the function $f : \mathcal{H} \to \mathbb{C}$ given by

$$f(x,\gamma) = (2\pi i\hbar)^{d/2} e^{-\frac{i}{2\hbar}|x|^2} \bar{\phi}(x)\psi_0(\gamma(t)+x), \qquad (x,\gamma) \in \mathbb{R}^d \times \mathcal{H}_t. \tag{3.81}$$

By means of these notations, and by imposing suitable conditions on the vectors ϕ, ψ_0 as well as to the time variable t, expression (3.77) can be realized as the following infinite dimensional oscillatory integral on $\mathcal{H}$:

$$\widetilde{\int_{\mathcal{H}}} e^{\frac{i}{2\hbar}(|x|^2+|\gamma|^2)} e^{-\frac{i}{2\hbar}\langle (x,\gamma), B(x,\gamma)\rangle} e^{-\frac{i}{\hbar}V_4(x,\gamma)} f(x,\gamma) dx d\gamma\,. \tag{3.82}$$

In the following we will denote by Ω_i, $i = 1, \ldots, d$, the eigenvalues of the matrix Ω.

Lemma 3.7. *The operator $B : \mathcal{H}_t \to \mathcal{H}_t$ given by Eq. (3.78) is trace class. Moreover, if for each $i = 1, \ldots, d$ the following inequalities are satisfied*

$$\Omega_i t < \frac{\pi}{2}, \qquad 1 - \Omega_i \tan(\Omega_i t) > 0. \tag{3.83}$$

then the operator $(I - B)$ is strictly positive.

Proof. Let us study the spectrum of the self-adjoint operator B on $\mathcal{H}$ given by (3.78). In order to avoid the use of too many indexes we will assume $d = 1$, but our reasoning remains valid also in the case $d > 1$. A positive real number c_l and a vector $(x_l, \gamma_l) \in \mathcal{H}$ are respectively an eigenvalue and an eigenvector of B if and only if:

$$\begin{cases} t\Omega^2 x_l + \Omega^2 \int_0^t \gamma_l(s)ds = c_l x_l \\ \Omega^2 x_l(ts - \frac{s^2}{2}) - \int_0^s \int_t \Omega^2 \gamma_l(r) dr du = c_l \gamma_l(s) \,. \end{cases}$$

By differentiating twice, the vector $(x_l, \gamma_l) \in \mathcal{H}$ is a solution of the following system:

$$\begin{cases} t\Omega^2 x_l + \Omega^2 \int_0^t \gamma_l(s)ds = c_l x_l \\ c_l \ddot{\gamma}_l(s) + \Omega^2 \gamma_l(s) = -\Omega^2 x_l \\ \gamma_l(0) = 0 \\ \dot{\gamma}_l(t) = 0 \,. \end{cases}$$

By a direct calculation one can verify that the latter system indeed admits a (unique) solution if and only if c_l satisfies the following equation

$$\frac{\Omega}{\sqrt{c_l}} \tan \frac{\Omega t}{\sqrt{c_l}} = 1 \,.$$

A graphical representation of the position of the solutions shows that the operator B is trace class. Moreover if the conditions (3.83) are fulfilled the maximum eigenvalue of B is strictly less than 1, so that $(I - B)$ is positive definite. □

Lemma 3.8. *Let $\phi, \psi_0 \in L^2(\mathbb{R}^d) \cap \mathcal{F}(\mathbb{R}^d)$. Let μ_0 be the complex bounded variation measure on $\mathbb{R}^d$ such that $\hat{\mu}_0 = \psi_0$. Let μ_ϕ be the complex bounded variation measure on $\mathbb{R}^d$ such that $\hat{\mu}_\phi(x) = (2\pi i\hbar)^{d/2} e^{-\frac{i}{2\hbar}|x|^2} \bar{\phi}(x)$.*
Let us assume that t satisfies inequalities (3.83) *and that the measures μ_0, μ_ϕ satisfy the following assumption:*

$$\int_{\mathbb{R}^d} \int_{\mathbb{R}^d} e^{\frac{\hbar}{4}(y - \cos(\Omega t)^{-1}x)(1 - \Omega \tan(\Omega t))^{-1}(y - \cos(\Omega t)^{-1}x)}$$
$$e^{\frac{\hbar}{4} x \Omega^{-1} \tan(\Omega t) x} |\mu_0|(dx) |\mu_\phi|(dy) < \infty. \tag{3.84}$$

Then the function $f : \mathcal{H} \to \mathbb{C}$, given by (3.81) *is the Fourier transform of a bounded variation measure μ_f on $\mathcal{H}$ satisfying*

$$\int_{\mathcal{H}} e^{\frac{\hbar}{4}\langle (y,\eta),(I-B)^{-1}(y,\eta)\rangle} |\mu_f|(dy d\eta) < \infty, \tag{3.85}$$

(B being given by (3.78)*).*

Proof. By the assumptions on ϕ, one can easily verify that the function $(2\pi i\hbar)^{d/2} e^{-\frac{i}{2\hbar}|x|^2} \bar{\phi}(x)$ is the Fourier transform of the bounded variation measure on $\mathbb{R}^d \times \mathcal{H}_t$, which is the product measure $\mu_\phi(dx) \times \delta_0(d\gamma)$, where $\delta_0(d\gamma)$ is the measure on $\mathcal{H}_t$ concentrated on the vector $0 \in \mathcal{H}_t$. Analogously the function $(x, \gamma) \mapsto \psi_0(\gamma(t) + x)$ is the Fourier transform of the bounded variation measure μ_ψ on $\mathbb{R}^d \times \mathcal{H}_t$ given by:

$$\int_{\mathbb{R}^d \times \mathcal{H}_t} f(x,\gamma)\mu_\psi(dx d\gamma) = \int_{\mathbb{R}^d \times \mathcal{H}_t} f(x, x\gamma)\delta_{\gamma_t}(d\gamma)\mu_0(dx),$$

where γ_t is the vector in $\mathcal{H}_t$ given by

$$\gamma_t(s) = s, \qquad s \in [0, t].$$

As $\mathcal{F}(\mathbb{R}^d \times \mathcal{H}_t)$ is a Banach algebra, the product

$$f(x,\gamma) := (2\pi i\hbar)^{d/2} e^{-\frac{i}{2\hbar}|x|^2} \bar{\phi}(x)\psi(\gamma(t) + x)$$

still belongs to $\mathcal{F}(\mathbb{R}^d \times \mathcal{H}_t)$, in fact it is the Fourier transform of the convolution $\mu_f \equiv (\mu_\phi \times \delta_0) * \mu_\psi$.

Let us now prove that the measure μ_f satisfies assumptions (2.68) of theorem 2.11, that is (3.85), if μ_0 and μ_ϕ satisfy (3.84).

By theorem A.5 in appendix, we have

$$e^{\frac{\hbar}{4}\langle (y,\eta),(I-B)^{-1}(y,\eta)\rangle} = \sqrt{\det(I-B)} F(y/\sqrt{2}, \eta/\sqrt{2}),$$

where the function $F : \mathcal{H} \to \mathbb{C}$ is given by the Gaussian integral

$$F(y,\eta) = \int_{\mathbb{R}^d \times C_t} e^{\sqrt{\hbar} xy + \sqrt{\hbar} n(\eta)(\omega)} e^{\frac{1}{2}\langle (x,\omega), B(x,\omega)\rangle} N(dx) W(d\omega).$$

By a direct computation and by Fubini theorem, F is equal to

$$F(y,\eta) = (2\pi)^{-\frac{d}{2}} \int_{\mathbb{R}^d} e^{\sqrt{\hbar} xy} e^{-x\frac{(I-t\Omega^2)}{2}x} \Big(\int_{C_t} e^{\sqrt{\hbar} n(\eta)(\omega)} e^{x\int_0^t \Omega^2 \omega(s) ds}$$
$$e^{\frac{1}{2}\int_0^t \omega(s)\Omega^2\omega(s) ds} W(d\omega)\Big) dx$$

$$= (2\pi)^{-\frac{d}{2}} \int_{\mathbb{R}^d} e^{\sqrt{\hbar} xy} e^{-x\frac{(I-t\Omega^2)}{2}x} \Big(\int_{C_t} e^{\sqrt{\hbar} n(\eta)(\omega)} e^{n(v_x)(\omega)} e^{\frac{1}{2}\langle \omega, L\omega\rangle} W(d\omega)\Big) dx, \tag{3.86}$$

where $L : \mathcal{H}_t \to \mathcal{H}_t$ is the operator given by

$$L\gamma(s) = -\int_0^s \int_t^{s'} \Omega^2 \gamma(s'')ds''ds'$$

and $v_x \in \mathcal{H}_t$ is the vector given by

$$v_x(s) = \Omega^2 x\Big(ts - \frac{s^2}{2}\Big).$$

One can easily verify that L is symmetric and trace class. Indeed by denoting by α^2, γ_α respectively the eigenvalues and the eigenvectors of the operator L, we have

$$\alpha^2 \ddot{\gamma}_\alpha(s) + \Omega^2 \gamma_\alpha(s) = 0, \qquad s \in [0, t],$$

with the conditions

$$\gamma_\alpha(0) = 0, \quad \dot{\gamma}_\alpha(t) = 0.$$

Without loss of generality we can assume Ω^2 is diagonal with eigenvalues Ω_i^2, $i = 1, \dots d$. The components $\gamma_{\alpha,i}$, $i = 1, \dots d$, of the eigenvector γ_α corresponding to the eigenvalue α^2 are equal to

$$\gamma_{\alpha,i}(s) = A_i \sin \frac{\Omega_i s}{\alpha}, \qquad s \in [0, t].$$

By imposing the condition $\dot{\gamma}(t) = 0$, we have

$$\Omega_i t/\alpha = \pi/2 + n_i \pi, \qquad n_i \in \mathbb{Z}.$$

The possible α^2 are of the form

$$\alpha^2 = \frac{\Omega_i^2 t^2}{\left(n_i + \frac{1}{2}\right)^2 \pi^2}, \qquad n_i \in \mathbb{Z}.$$

It follows that the operator $I - L$ is positive definite if and only if $\Omega_i t < \pi/2$ for all $i = 1, \dots d$. Moreover the Fredholm determinant of L can easily be computed by means of the equality $\cos x = \prod \left(1 - \frac{x^2}{\pi^2 (n+1/2)^2}\right)$ and it is equal to $\det \cos \Omega t$.

By theorem A.5 in appendix, it possible to verify that the function $G : \mathbb{R}^d \to \mathbb{R}$ given by

$$G(x) = \int_{C_t} e^{\sqrt{\hbar} n(\eta)(\omega) + n(v_x)(\omega)} e^{\frac{1}{2}\langle \omega, L\omega \rangle} W(d\omega)$$

is equal to

$$\frac{1}{\sqrt{\det \cos \Omega t}} e^{\frac{1}{2}\langle \sqrt{\hbar}\eta + v_x, (I-L)^{-1}(\sqrt{\hbar}\eta + v_x)\rangle},$$

where $(I-L)^{-1}\gamma$, for $\gamma \in \mathcal{H}_t$ sufficiently regular, is given by

$$(I-L)^{-1}\gamma(s) = \Omega^{-1}\int_0^s \sin[\Omega(s-s')]\ddot{\gamma}(s')ds' + \frac{\sin(\Omega s)}{\Omega\cos(\Omega t)}\Big[\dot{\gamma}(t) - \int_0^t \cos[\Omega(t-s')]\ddot{\gamma}(s')ds'\Big].$$

Moreover by direct computation we see that

$$G(x) = \frac{1}{\sqrt{\det\cos\Omega t}} e^{\frac{1}{2}\langle\sqrt{\hbar}\eta,(I-L)^{-1}\sqrt{\hbar}\eta\rangle} e^{\frac{1}{2}x(-t\Omega^2+\Omega\tan\Omega t)x} e^{\langle v_x,(I-L)^{-1}\sqrt{\hbar}\eta\rangle}.$$

By inserting this into (3.86), we have

$$F(y,\eta) = \frac{(2\pi)^{-d/2}}{\sqrt{\det\cos\Omega t}} e^{\frac{1}{2}\langle\sqrt{\hbar}\eta,(I-L)^{-1}\sqrt{\hbar}\eta\rangle} \int_{\mathbb{R}^d} e^{\sqrt{\hbar}xy} e^{-\frac{1}{2}x(I-\Omega\tan\Omega t)x} e^{\langle v_x,(I-L)^{-1}\sqrt{\hbar}\eta\rangle}dx. \quad (3.87)$$

By imposing Eq. (3.85), that is the condition

$$\int_{\mathcal{H}} F(y/\sqrt{2},\eta/\sqrt{2})d|\mu_f|(y,\eta) < \infty,$$

we get

$$\int_{\mathcal{H}\times\mathcal{H}} F((x+y)/\sqrt{2},(\gamma+\eta)/\sqrt{2})d|(\mu_\phi\times\delta_0)|(y,\eta)d|\mu_\psi|(x,\gamma)$$
$$= \int_{\mathbb{R}^d\times\mathbb{R}^d} F((x+y)/\sqrt{2},(x\gamma_t)/\sqrt{2})d|\mu_\phi|(y)d|\mu_0|(x) < \infty. \quad (3.88)$$

By substituting into Eq. (3.87) we have

$$F((x+y)/\sqrt{2},(x\gamma_t)/\sqrt{2}) = \frac{(2\pi)^{-d/2}}{\sqrt{\det\cos\Omega t}} e^{\frac{\hbar}{4}\langle x\gamma_t,(I-L)^{-1}x\gamma_t\rangle} \int_{\mathbb{R}^d} e^{\sqrt{\frac{\hbar}{2}}z(x+y)} e^{-\frac{1}{2}z(I-\Omega\tan\Omega t)z} e^{\sqrt{\frac{\hbar}{2}}\langle v_z,(I-L)^{-1}x\gamma_t\rangle}dz$$
$$= \frac{(2\pi)^{-d/2}}{\sqrt{\det\cos\Omega t}} e^{\frac{\hbar}{4}x\frac{\sin(\Omega t)}{\Omega\cos(\Omega t)}x} \int_{\mathbb{R}^d} e^{\sqrt{\frac{\hbar}{2}}z(y-\cos(\Omega t)^{-1}x)} e^{-\frac{1}{2}z(I-\Omega\tan\Omega t)z}dz$$
$$= \frac{(2\pi)^{-d/2}}{\sqrt{\det(\cos\Omega t-\Omega\sin\Omega t)}} \cdot e^{\frac{\hbar}{4}x\frac{\sin(\Omega t)}{\Omega\cos(\Omega t)}x} e^{\frac{\hbar}{4}(y-\cos(\Omega t)^{-1}x)(I-\Omega\tan\Omega t)^{-1}(y-\cos(\Omega t)^{-1}x)}.$$

By substituting in Eq. (3.88), we obtain condition (3.84). □

Theorem 3.8. *Let us assume that $\lambda < 0$ and that the time variable t satisfies conditions (3.83). Let $\phi, \psi_0 \in L^2(\mathbb{R}^d) \cap \mathcal{F}(\mathbb{R}^d)$, with $\hat{\mu}_0 = \psi_0$ and $\hat{\mu}_\phi(x) = (2\pi i\hbar)^{d/2} e^{-\frac{i}{2\hbar}|x|^2} \bar{\phi}(x)$. Assume in addition that the measures $\mu_0, \mu_\phi \in \mathcal{M}(\mathbb{R}^d)$ satisfy the assumption (3.84).*

Then the function $f : \mathcal{H} \to \mathbb{C}$, given by (3.81) is the Fourier transform of a bounded variation measure μ_f on $\mathcal{H}$ satisfying (3.85) and the infinite dimensional oscillatory integral (3.82) is well defined and is given by:

$$\int_{\mathbb{R}^d \times H_t} \Big(\int_{\mathbb{R}^d \times C_t} e^{ie^{i\pi/4}(x\cdot y + \sqrt{\hbar} n(\gamma)(\omega))} e^{\frac{1}{2\hbar}\int_0^t (\sqrt{\hbar}\omega(s)+x)\Omega^2(\sqrt{\hbar}\omega(s)+x)ds}$$

$$e^{-i\frac{\lambda}{\hbar}\int_0^t |\sqrt{\hbar}\omega(s)+x|^4 ds} W(d\omega) \frac{e^{-\frac{|x|^2}{2\hbar}}}{(2\pi\hbar)^{d/2}} dx \Big) \mu_f(dyd\gamma). \quad (3.89)$$

This is also equal to

$$(i)^{d/2} \int_{\mathbb{R}^d \times C_t} e^{\frac{1}{2\hbar}\int_0^t (\sqrt{\hbar}\omega(s)+x)\Omega^2(\sqrt{\hbar}\omega(s)+x)ds}$$

$$e^{-i\frac{\lambda}{\hbar}\int_0^t |\sqrt{\hbar}\omega(s)+x|^4 ds} \bar{\phi}(e^{i\pi/4}x)\psi_0(e^{i\pi/4}\sqrt{\hbar}\omega(t) + e^{i\pi/4}x) W(d\omega)dx. \quad (3.90)$$

Proof. By lemma 3.7, the operator $B : \mathcal{H} \to \mathcal{H}$ given by (3.78) is bounded symmetric and trace class. Moreover if assumptions (3.83) are satisfied, $I - B$ is positive definite.

By lemma 3.8 the function $f : \mathcal{H} \to \mathbb{C}$, given by (3.81) is the Fourier transform of a bounded variation measure μ_f on $\mathcal{H}$ satisfying assumptions (2.68) of theorem 2.11, that is Eq. (3.85).

A direct computation shows that the function $V_4 : \mathcal{H} \to \mathbb{R}$,

$$V_4(x,\gamma) = \int_0^t |\gamma(s) + x|^4 ds, \qquad (x,\gamma) \in \mathcal{H},$$

is continuous in the norm of the Banach space B and extends to a function $\bar{V}_4$ on it.

By applying theorems 2.11 and 2.12, the conclusion follows. □

Remark 3.5. The class of states $\phi, \psi_0 \in L^2(\mathbb{R}^d) \cap \mathcal{F}(\mathbb{R}^d)$ satisfying assumption (3.84) is sufficiently rich. Indeed both ϕ and ψ_0 can be chosen in two dense subsets of the Hilbert space $L^2(\mathbb{R}^d)$.

More precisely one can take for instance $\psi_0, \phi \in \mathcal{S}(\mathbb{R}^d)$ of the form

$$|\hat{\psi}_0(x)| = P(x) e^{-\frac{\alpha\hbar}{2}x^2}, \qquad |\hat{\phi}(x)| = Q(x) e^{-\frac{\beta\hbar}{2}x^2}, \qquad x \in \mathbb{R}^d,$$

and with $\alpha, \beta \in \mathbb{R}^+$ and with P, Q arbitrary polynomials. Moreover α and β have to satisfy the following conditions, for all $i = 1, \dots, d$:

$$\begin{cases} \beta - (1-\Omega_i \tan \Omega_i t)^{-1} > 0 \\ \alpha - \frac{(1-\Omega_i \tan \Omega_i t)^{-1}}{\cos^2 \Omega_i t} + \frac{\sin \Omega_i t}{\Omega \cos \Omega_i t} > 0 \\ (\beta - (1-\Omega_i \tan \Omega_i t)^{-1})\Big(\alpha - \frac{(1-\Omega_i \tan \Omega_i t)^{-1}}{\cos^2 \Omega_i t} + \frac{\sin \Omega_i t}{\Omega \cos \Omega_i t}\Big) \\ -\Big(\frac{(1-\Omega_i \tan \Omega_i t)^{-1}}{\cos^2 \Omega_i t}\Big)^2 > 0\,. \end{cases} \tag{3.91}$$

By the density of the finite linear combinations of Hermite functions in $L^2(\mathbb{R}^d)$ (see for instance [245, 246]), it is easy to see that the vectors of the above form are dense in $L^2(\mathbb{R}^d)$.

Remark 3.6. In [28, 31] the above result has been generalized to the case the quartic potential is explicitly time-dependent.

Let us consider the (Wiener Gaussian) integrals (3.89) and (3.90). According to theorem 3.8, for $\lambda < 0$ they are equal to the infinite dimensional oscillatory integral (3.82).

By the considerations in remark 2.9 the absolutely convergent integrals (3.89) and (3.90) are analytic functions of the complex variable λ if $Im(\lambda) > 0$, continuous in $Im(\lambda) = 0$. The following theorem shows that they solve the Schrödinger equation (3.73) with potential given by Eq. (3.74) in a weak sense (see also [36] for further comments).

Theorem 3.9. *Let $\lambda \in \mathbb{R}$. Let $t \in \mathbb{R}^+$ satisfying conditions* (3.83). *Let $\phi, \psi_0 \in L^2(\mathbb{R}^d) \cap \mathcal{F}(\mathbb{R}^d)$ satisfying the assumption* (3.84).
Then the Gaussian integral $I_t(\phi, \psi_0)$, given by

$$I_t(\phi,\psi_0) := (i)^{d/2} \int_{\mathbb{R}^d \times C_t} e^{\frac{1}{2\hbar}\int_0^t (\sqrt{\hbar}\omega(s)+x)\Omega^2(\sqrt{\hbar}\omega(s)+x)ds}$$
$$e^{-i\frac{\lambda}{\hbar}\int_0^t |\sqrt{\hbar}\omega(s)+x|^4 ds}\bar{\phi}(e^{i\pi/4}x)\psi_0(e^{i\pi/4}\sqrt{\hbar}\omega(t)+e^{i\pi/4}x)W(d\omega)dx,$$

is a quadratic form in the variables (ϕ, ψ_0) and it satisfies the Schrödinger equation (3.73) *in the following weak sense, namely:*

$$I_0(\phi,\psi) = \langle \phi, \psi \rangle, \tag{3.92}$$

$$i\hbar \frac{d}{dt} I_t(\phi,\psi_0) = I_t(\phi, H\psi_0) = I_t(H\phi,\psi_0), \tag{3.93}$$

(H being given on the smooth vector $\phi \in C_0^2(\mathbb{R}^d)$ by (3.76)*).*

Remark 3.7. The result of theorem 3.9 does not depend on the sign of the coupling constant λ.

Proof. Equation (3.93) follows by an application of Ito's formula (see for instance [195]). Equation (3.92), that is

$$(i)^{d/2}\int_{\mathbb{R}^d\times C_t}\bar{\phi}(e^{i\pi/4}x)\psi_0(e^{i\pi/4}x)W(d\omega)dx$$

$$= (i)^{d/2}\int_{\mathbb{R}^d}\bar{\phi}(e^{i\pi/4}x)\psi_0(e^{i\pi/4}x)dx \tag{3.94}$$

$$= \int_{\mathbb{R}^d}\bar{\phi}(x)\psi_0(x)dx \tag{3.95}$$

follows by the analyticity of the functions ϕ, ψ_0 and by a rotation of the integration contour in the complex plane. □

The particular case where the coupling constant λ is negative is rather interesting. Indeed in this case the potential (3.74) is unbounded from below, and the quantum Hamiltonian

$$H = -\frac{\Delta}{2m} + V$$

is not essentially self-adjoint as one can deduce by a limit point argument (see [246], theorem X.9). In this case the quantum evolution is not uniquely determined. Nelson [235] was the first mathematician proposing Feynman path integrals as a tool defining the quantum dynamics in the case of not essentially self-adjoint Hamiltonians. In [235], by means of a generalized Trotter product formula and an analytic continuation technique, a strongly continuous contraction semigroup

$$U(t) : L^2(\mathbb{R}^d) \to L^2(\mathbb{R}^d), \qquad t \geq 0$$

is constructed and, given a $\psi \in L^2(\mathbb{R}^d)$, the vector $\psi(t) \equiv U(t)\psi$ satisfies the Schrödinger equation in a distributional way, i.e. for any $\phi \in L^2(\mathbb{R}^d)$ sufficiently regular, one has

$$i\hbar\frac{d}{dt}\langle\phi, \psi(t)\rangle = \langle H\phi, \psi(t)\rangle.$$

Even if the starting point of Nelson's derivation is a Wiener integral representation of the solution of an heat equation with imaginary potential, the evolution operators $U(t)$ are defined in an abstract way by means of a limiting procedure and, in general, a path integral representation for its matrix elements $\langle\phi, U(t)\psi\rangle$ cannot be defined (even for very regular vectors $\phi, \psi \in L^2(\mathbb{R}^d)$). A technical problem of Nelson's result, directly connected with the method of the proof (i.e. the application of the Fatou-Privaloff theorem) is a restriction to the allowed values of the mass parameter m, which cannot belong to a certain set N of Lebesgue measure 0.

In the particular case of quartic potential however, following [224], we can show that both problems of Nelson's paper can be overcome. Indeed we can prove that the infinite dimensional oscillatory integral with polynomial phase function (3.82) studied in theorem 3.8 provides a Feynman path integral representation for the matrix elements $\langle \phi, e^{-\frac{i}{\hbar}Ht}\psi\rangle$ of the evolution operator defined by means of Nelson's method.

This result provides a link between two different approaches to the mathematical definition of Feynman path integral (the analytic continuation approach and the infinite dimensional oscillatory integral approach).

Theorem 3.10. *Let $\lambda \in \mathbb{R}$, $\lambda < 0$. Let $t \in \mathbb{R}^+$ satisfying conditions* (3.83). *Let $\phi, \psi_0 \in L^2(\mathbb{R}^d) \cap \mathcal{F}(\mathbb{R}^d)$ satisfying the assumption* (3.84). *Then the infinite dimensional oscillatory integral* (3.82) *is equal to the Gaussian integral $I_t(\phi, \psi_0)$, given by Eq. (3.90), and the latter is equal to the inner product $\langle \phi, U(t)\psi_0\rangle$, with $U(t)$, $t \geq 0$, strongly continuous contraction semigroup and*

$$i\hbar\frac{d}{dt}\langle \phi, U(t)\psi_0\rangle = \langle H\phi, U(t)\psi_0\rangle,$$

(H being given on the smooth vector $\phi \in C_0^2(\mathbb{R}^d)$ by (3.76)).

Proof. Let us consider the heat equation with complex potential

$$\begin{cases} -\hbar\frac{\partial}{\partial t}\psi(t,x) = -\frac{\hbar^2}{2m}\Delta\psi(t,x) + iV(x)\psi(t,x) \\ \psi(0,x) = \psi_0(x) \end{cases} \tag{3.96}$$

with V given by (3.74). For any $\psi_0 \in L_2(\mathbb{R}^d)$, the integral

$$U_m(t)\psi_0(x) \equiv \int_{C_t} e^{i\frac{\lambda}{\hbar}\int_0^t|\sqrt{\hbar/m}\,\omega(s)+x|^4 ds}$$
$$e^{-\frac{i}{2\hbar}\int_0^t(\sqrt{\hbar/m}\,\omega(s)+x)\Omega^2(\sqrt{\hbar}\,\omega(s)+x)ds}\psi_0(\sqrt{\hbar/m}\,\omega(t)+x)W(d\omega) \tag{3.97}$$

is convergent and defines a contraction operator $U_m(t)$, as

$$|U_m(t)\psi_0(x)| \leq \int_{C_t}|\psi_0(\sqrt{\hbar/m}\omega(t)+x)|W(d\omega)$$
$$\leq (2\pi t\hbar/m)^{-d/2}\int_{\mathbb{R}^d} e^{-\frac{m}{2t\hbar}|x-y|^2}\psi_0(y)dy = K_m(t)|\psi_0|(x), \tag{3.98}$$

where $K_m(t)$ is the heat semigroup $K_m(t) = e^{\frac{t\hbar}{2m}\Delta}$. By writing the cylindrical approximations of the Wiener integral (3.97) one has

$$U_m(t)\psi_0(x) = \lim_{n\to\infty}(K_m(t/n)M_V(t/n))^n\psi_0(x), \tag{3.99}$$

where $M_V(t)$ is the group given by the multiplication operator $M_V(t) = e^{-it\hbar V}$. By mimicking Nelson's argument [235] one can see that the limit (3.99) can be taken in L^2-sense, it defines a strongly continuous contraction semigroup $U_m(t)$ and for any $\psi_0 \in C_0^2(\mathbb{R}^d)$ the generator A_m is given by

$$A_m\psi_0 = \lim_{t\to 0}\frac{1}{t}(U_m(t)\psi_0 - \psi_0) = \Big(\frac{\hbar}{2m}\Delta - \frac{i}{\hbar}V\Big)\psi_0. \tag{3.100}$$

As Δ is a negative operator, for any $t \geq 0$ $K_m(t)$ is an holomorphic operator-valued function of m in the half plane $Re(m) > 0$. It follows that for any $\psi_0 \in L^2$ and for any $n \in \mathbb{N}$, the expression

$$F_n(m) := \big(K_m(t/n)M_V(t/n)\big)^n\psi_0$$

defines an L^2-valued function holomorphic in the half plane $Re(m) > 0$ and continuous on $Re(m) \geq 0$. Since the sequence of functions $\{F_n\}_{n\in\mathbb{N}}$ is uniformly bounded on $Re(m) \geq 0$ by $\|\psi_0\|$ and converges for $m > 0$, by Vitali's theorem it converges on the whole domain $Re(m) > 0$ and the limit

$$\lim_{n\to\infty}(K_m(t/n)M_V(t/n))^n\psi_0 \equiv U_m(t)\psi_0$$

defines an holomorphic L^2-valued function $U_m(t)\psi_0$ on $Re(m) > 0$. By analytic continuation, one can prove that $U_m(t)$ is a strongly continuous contraction semigroup whose generator is given on vectors $\psi_0 \in C_0^2(\mathbb{R}^d)$ by Eq. (3.100).

The purely quantum mechanical - Schrödinger case is obtained for m in (3.96) purely imaginary, i.e. $m = -i$. Let us consider ϕ, ψ_0 satisfying the assumptions of the theorem. For $m > 0$, the inner product $\langle\phi, U_m(t)\psi\rangle$ is given by

$$\int_{\mathbb{R}^d\times C_t} e^{-\frac{i}{2\hbar}\int_0^t(\sqrt{\hbar/m}\,\omega(s)+x)\Omega^2(\sqrt{\hbar/m}\,\omega(s)+x)ds}$$
$$e^{i\frac{\lambda}{\hbar}\int_0^t|\sqrt{\hbar/m}\,\omega(s)+x|^4ds}\psi_0(\sqrt{\hbar/m}\,\omega(t)+x)\bar{\phi}(x)W(d\omega)dx\,. \tag{3.101}$$

By a change of variable $x \mapsto x/\sqrt{m}$ the latter becomes

$$\langle\phi, U_m(t)\psi_0\rangle = m^{-d/2}\int_{\mathbb{R}^d\times C_t} e^{-\frac{i}{2\hbar m}\int_0^t(\sqrt{\hbar}\,\omega(s)+x)\Omega^2(\sqrt{\hbar}\,\omega(s)+x)ds}$$
$$e^{i\frac{\lambda}{m^2\hbar}\int_0^t|\sqrt{\hbar}\,\omega(s)+x|^4ds}\psi_0(\sqrt{\hbar/m}\,\omega(t)+x/\sqrt{m})\bar{\phi}(x/\sqrt{m})W(d\omega)dx\,. \tag{3.102}$$

By assumptions (3.83) and (3.84), the right hand side of (3.102) is an holomorphic function of m in the domain $\{Re(m) > 0\}\cap\{Im(m) < 0\}$ and continuous on the boundary. On the other hand, by previous considerations, the matrix element $\langle\phi, U_m(t)\psi_0\rangle$ is an holomorphic function of m in the domain $\{Re(m) > 0\}$ and coincides with the functional integral (3.102) on the half line $m > 0$. By the uniqueness of analytic continuation, both sides of (3.102) coincide on the domain $\{Re(m) > 0\} \cap \{Im(m) < 0\}$. In particular there exists the limit

$$\lim_{m \to -i} \langle\phi, U_m(t)\psi_0\rangle$$

and, by bounded convergence theorem, it is equal to (3.90). □

Remark 3.8. The results of theorems 3.8, 3.9 and 3.10 hold also in the case where the potential V is of the form:

$$V(x) = \lambda|x|^4 - \frac{1}{2}x\Omega^2 x, \qquad x \in \mathbb{R}^d$$

with $\lambda \in \mathbb{R}$ and Ω^2 a positive symmetric $d \times d$ matrix. In this case conditions (3.83) can be dropped and condition (3.84) has to be replaced by:

$$\int_{\mathbb{R}^d}\int_{\mathbb{R}^d} e^{\frac{\hbar}{4}(y-\cosh(\Omega t)^{-1}x)(1+\Omega\tanh(\Omega t))^{-1}(y-\cosh(\Omega t)^{-1}x)} e^{-\frac{\hbar}{4}x\Omega^{-1}\tanh(\Omega t)x}|\mu_0|(dx)|\mu_\phi|(dy) < \infty.$$

Chapter 4

The Stationary Phase Method and the Semiclassical Limit of Quantum Mechanics

4.1 Asymptotic expansions

In the study of several problems, such as the evaluation of integrals or the solution of differential equations, it is very difficult to find an exact analytical solution, given in terms of known functions. On the other hand, in the applications, exact solutions are not always of practical use, both from a computational and an analytical point of view and it is preferable the knowledge of an approximated solution when a parameter or some variable of the problem can be considered either large or small.

The asymptotic analysis studies the techniques allowing to obtain approximated solutions of a large class of problems, when a parameter approaches a value in which the solution is not analytic. It was introduced at the end of the nineteenth century by Poincaré, who gave a precise definition of an *asymptotic expansion*, and was further developed in the twentieth century in connection with the study of equations arising in mathematical physics, in particular in fluid mechanics.

We recall here the definition an the main properties of asymptotic sequence and expansions, that will be used in the next sections. For a detailed treatment see for instance [49, 115, 119, 160, 237, 244].

Let $V \subset \mathbb{C}$ a domain[1] in the complex plane $\mathbb{C}$ (or, more generally, on the Riemann surface $\tilde{\mathbb{C}}$ of the logarithm) such that $0 \in \partial V$ and

$$z \in V \Rightarrow \forall t \in (0,1], \quad tz \in V.$$

Let us denote $\hat{V} := V \cup \{0\}$. Both V and $\hat{V}$ will be called *angular neighborhoods* of zero.

A set $U \subset \mathbb{C}$, which is the closure of an angular neighborhood of zero, will be called *closed angular neighborhood.*

[1]With the word *domain* we mean a connected open nonempty set.

Definition 4.1. Let V be an angular neighborhood of zero. An asymptotic sequence of functions $(\phi_i)_{i\in\mathbb{N}}$ for $z \to 0$ in V is a sequence of functions $\phi_i : \hat{V} \to \mathbb{C}$, which do not vanish in V and such that for every $i \in \mathbb{N}$:

$$\lim_{z\to 0} \frac{\phi_{i+1}}{\phi_i}(z) = 0.$$

In the following we shall focus on the asymptotic sequence (in any angular neighborhood of zero) $\phi_n(z) = z^{n/k}$, $n \in \mathbb{N}$, for fixed $k > 0$ and denote by $\mathbb{C}[z^{1/k}]$ the space of formal power series with complex coefficients

$$\hat{f}(z) = \sum_{n=0}^{\infty} a_n z^{n/k}, \quad \{a_n\} \subset \mathbb{C},\ k > 0. \tag{4.1}$$

Definition 4.2. A formal power series $\hat{f}$ is called a ($z^{1/k}$–) asymptotic expansion for a function $f : V \to \mathbb{C}$ as $z \to 0$ in an angular neighborhood V if for each closed angular neighborhood U with $U \subsetneq V$ and any $N \in \mathbb{N}$, there exists a number $C(N) > 0$ such that

$$\forall z \in U \ : \quad |f(z) - \sum_{n=0}^{N-1} a_n z^{n/k}| \leq C(N)|z^{N/k}|. \tag{4.2}$$

In this case we write $f \sim \hat{f}$, $z \to 0$ in V.

For more details see [160].

Remark 4.1. It is important to recall that the domain V in definition 4.2 plays a crucial role, indeed the existence of an expansion depends strongly on V.

Remark 4.2. An asymptotic expansions is not necessarily convergent (and usually this is the case!). Indeed condition (4.2) means that for fixed N the function f is approximated by the sum $\sum_{n=0}^{N} a_n z^{n/k}$ for z sufficiently small, while if the formal power series (4.1) is convergent at $z = 0$ in some domain V to a function f then the following holds:

$$\forall z \in V \ : \quad \lim_{N\to\infty} |f(z) - \sum_{n=0}^{N} a_n z^{n/k}| = 0, \tag{4.3}$$

which means that for fixed $z \in V$ the value of the function f is approximated by the sum $\sum_{n=0}^{N} a_n z^{n/k}$ for N sufficiently large.

It is easy to see that if a function f admits an ($z^{1/k}-$) asymptotic expansion in a given domain V, i.e. $f \sim \hat{f} = \sum_{n=0}^{\infty} a_n z^{n/k}$ for $z \to 0$ in V, then the coefficients a_n, $n \in \mathbb{N}$, are uniquely determined by

$$a_0 = \lim_{z\to 0} f(z)$$

$$a_1 = \lim_{z\to 0} \frac{f(z) - a_0}{z^{1/k}} \tag{4.4}$$

$$\ldots \tag{4.5}$$

$$a_n = \lim_{z\to 0} \frac{f(z) - \sum_0^{n-1} a_j z^{j/k}}{z^{n/k}}. \tag{4.6}$$

On the other hand different functions can have the same asymptotic expansion, for instance the function $f(z) = 0$ and $g(z) = e^{-1/z}$ have both a zero asymptotic expansion in the domain $\{z \in \mathbb{C},\ Re(z) > 0\}$. In other words, if an asymptotic expansion is not convergent (and this is often the case) it does not characterize uniquely a function f asymptotically equivalent to it.

In order to associate in a unique way to formal power series $\sum a_n z^{n/k}$ a function f which is asymptotically equivalent to it, one can apply, under suitable assumptions, a powerful summation tool: Borel summability (see for instance [216, 217, 160, 244, 267, 249]). It works as follows:

(1) transform the given power series $\hat{f}$ into another convergent power series $\hat{B}$;
(2) compute the analytic function B which has $\hat{B}$ as a convergent power series expansion;
(3) apply an integral transform mapping the analytic function B to an analytic function f;
(4) the function f (the so-called "sum of $\hat{f}$") obtained in this way has the power series $\hat{f}$ we started as asymptotic expansion.

For the applicability of Borel summability method it is necessary to impose suitable growth conditions on the coefficients a_n [244].

Definition 4.3. Given $s > 0$, a formal power series $\hat{f}(z) = \sum_{n=0}^{\infty} a_n z^{n/k} \in \mathbb{C}[z^{1/k}]$ belongs to the s-Gevrey class $\mathbb{C}[z^{1/k}]_s$ if there exist two constants $C, M > 0$, such that

$$|a_n| \leq CM^n (\Gamma(1 + n/k))^s, \qquad \forall n \in \mathbb{N}, \tag{4.7}$$

where Γ is the Euler Γ function.

Remark 4.3. By Stirling formula conditions (4.7) can be replaced by

$$|a_n| \leq CM^n \left(\frac{n}{k}\right)^{ns/k}, \qquad \forall n \in \mathbb{N}, \tag{4.8}$$

or by

$$|a_n| \leq CM^n\Gamma(1 + sn/k), \quad \forall n \in \mathbb{N}. \tag{4.9}$$

The Gevrey classes are connected via the following transform acting on formal series:

Definition 4.4. The map $B_{p,k} : \mathbb{C}[z^{1/k}]_s \to \mathbb{C}[z^{p/k}]_{s-p}$ defined by

$$B_{p,k}[\sum_{n=0}^{\infty} a_n z^{n/k}](t) := \sum_{n=0}^{\infty} \frac{a_n}{\Gamma(1 + np/k)} t^{np/k} \tag{4.10}$$

is called the (formal) $(p, k)-$Borel transform.

It is important to note that the $(s, k)-$Borel transform maps $\mathbb{C}[z^{1/k}]_s$ to convergent series.
We can now define the concept of *$\mu-$Borel summability*:

Definition 4.5. Let $k, s > 0$, $\mu = 1/s$. A formal power series $\hat{f}(z) = \sum_{n=0}^{\infty} a_n z^{n/k}$ is called $\mu-$Borel summable to the sum f if f is an holomorphic function on V for some angular neighborhood of zero V, $f \sim \hat{f}$ as $z \to 0$ in V and the following procedure is possible:

(1) The $(s, k)-$Borel transform $B_{s,k}[\hat{f}](t)$ has nonzero radius of convergence and thus converges in a neighborhood of zero to some function $B(\cdot)$.
(2) This holomorphic function B admits an analytic continuation (denoted again by the symbol $B(\cdot)$) onto some open neighborhood of $\mathbb{R}^+$.
(3) The Borel-Laplace transform $\mathcal{L}(B)$ of B gives a representation of f on a subset of V: $f(z) = \frac{1}{z^\mu}\mathcal{L}(B)(\frac{1}{z^\mu})$, i.e.[2]:

$$f(z) = \frac{1}{z^\mu} \int_0^{\infty} B(t) e^{-t/z^\mu} dt\,. \tag{4.11}$$

If $\mu = 1$, then $\hat{f}$ is simply said *Borel summable*.

In other words if an asymptotic series is Borel summable to a function f, it characterizes uniquely f, even if it is not convergent.
The following criterion for Borel summability is due to F. Nevanlinna [236] and is a improvement of a result by Watson [287], see also [267, 249] for a modern proof and discussion of Nevanlinna's result:

[2] The integral in (4.11) is the conventional Laplace transform in the variable $w = 1/z^\mu$. Note that if the integral converges for some $z_0 \neq 0$ them it converges for all $z \in \mathbb{C}$ such that $Re(1/z^\mu) > Re(1/z_0^\mu)$.

Theorem 4.1. *Let $k > 0$, $R \in (0, +\infty]$ and define $D_R := \{z \in \mathbb{C} : Re(1/z) > 1/R\}$ if $R \neq \infty$ and $D_R := \{z \in \mathbb{C} : Re(z) > 0\}$ else.*
Let f be an holomorphic function on D_R admitting an asymptotic expansion with respect to the asymptotic sequence $z^{n/k}$ in the domain D_R

$$f(z) \sim \sum_{n=0}^{\infty} a_n z^{n/k} =: \hat{f}.$$

Let us assume that $\exists A > 0, \rho > 0$, such that $\forall \epsilon > 0, z \in \{Re(1/z) \geq \epsilon + 1/R\}$, $\hat{\rho} > \rho, n \in \mathbb{N}$, the following holds:

$$|f(z) - \sum_{i=0}^{n-1} a_i z^{i/k}| \leq A\Gamma(1 + n/k)\hat{\rho}^n |z|^{n/k}. \tag{4.12}$$

Then the asymptotic power series $\hat{f}$ is Borel summable to the function f.

Remark 4.4. By Stirling formula condition (4.12) can be replaced by

$$|f(z) - \sum_{i=0}^{n-1} a_i z^{i/k}| \leq A\hat{\rho}^n \left(\frac{n}{k}\right)^{n/k} |z|^{n/k}. \tag{4.13}$$

Proof. The present proof is taken from [267], and it is a generalization of Hardy's proof of Watson's theorem [160], see also [236]. Without loss of generality, we can restrict ourselves to consider the case where $k = 1$. The general case can be handled in a completely similar way.

By estimate (4.12), it is possible to see that the integrals

$$b_m(t) = a_m + \frac{1}{2\pi i} \oint_{Rez^{-1}=r^{-1}} e^{t/z} z^{-(m+1)} (f(z) - \sum_{j=0}^{m} a_j z^j) dz, \tag{4.14}$$

are absolutely convergent for $t \geq 0$ and independent of r for $0 < r < R$. Moreover b_0 is a C^∞ function whose mth derivative is b_m and the following estimate holds:

$$|b_m(t)| \leq K_1 \rho^{m+1} (m+1)! e^{t/R},$$

with K_1 independent of t and m. By performing a contour integral, one finds

$$b_0(t) = \sum_{j=0}^{N-1} \frac{1}{j!} a_j t^j + \frac{1}{2\pi i} \oint_{Rez^{-1}=r^{-1}} e^{t/z} z^{-1} (f(z) - \sum_{j=0}^{N-1} a_j z^j) dz. \tag{4.15}$$

By Eq. (4.12) and by taking $r = t/N$ (with $N > t/R$), the second term in the sum at the right hand side of Eq. (4.15) goes to 0 as $N \to \infty$ and it follows that

$$b_0(t) = B(t) = \sum_{j=0}^{\infty} \frac{1}{j!} a_j t^j. \tag{4.16}$$

By condition (4.12) the series on the right hand side of Eq. (4.16) converges in the circle $|t| < 1/\rho$. Moreover it is possible to see that each series

$$B_{t_0}(t) = \sum_{m=0}^{\infty} \frac{1}{m!} b_m(t_0)(t - t_0)^m, \qquad t_0 \geq 0,$$

converges in the circle $|t - t_0| < 1/\rho$ and satisfies there the bound

$$|B_{t_0}(t)| \leq K_1 e^{t_0/R} \frac{1}{1 - \rho(t - t_0)}.$$

It is not difficult to prove that $B_{t_0}(t) = B_{t_1}(t)$ when both functions are defined and that the union of these functions defines an analytic function $B(t)$ defined in the region $S_\rho \subset \mathbb{C}$ given by

$$S_\rho := \{z \in \mathbb{C}, : dist(z, \mathbb{R}^+) < 1/\rho\},$$

and satisfying there the bound

$$|B(t)| \leq K e^{|t|/R}.$$

By inserting (4.14) with $m = 0$ into the expression for the Laplace transform and by interchanging the order of integration, we obtain:

$$\frac{1}{z}\int_0^\infty e^{-t/z} B(t) dt = \frac{1}{z}\int_0^\infty e^{-t/z} a_0 dt +$$
$$+ \frac{1}{z}\int_0^\infty e^{-t/z} \frac{1}{2\pi i} \oint_{Re z^{-1} = r^{-1}} e^{t/z'} \frac{f(z') - a_0}{z'} dz' dt = f(z).$$ □

4.2 The stationary phase method. Finite dimensional case

The present and the following sections concern the study of the asymptotic behavior as $\hbar \downarrow 0$ of functions of the variable $\hbar \in \mathbb{R}^+$ defined in terms of oscillatory integrals of the form

$$I(\hbar) := \widetilde{\int_{\mathcal{H}}} e^{\frac{i}{\hbar}\Phi(x)} f(x) dx, \qquad \hbar \in \mathbb{R}^+, \tag{4.17}$$

with $\Phi : \mathcal{H} \to \mathbb{R}$ and $f : \mathcal{H} \to \mathbb{C}$.

In the finite dimensional case, i.e. when we can identify the Hilbert space $\mathcal{H}$ with $\mathbb{R}^n$, the fundamental tool for the study of the asymptotics of the integral (4.17) is the *stationary phase method* [115]. It was originally developed by Stokes [269] and Kelvin [201] in the 19th century. The physical relevance of this topic is connected with the fact that integrals of the form (4.17) play a fundamental role in the description of wave phenomena.

More recent investigations can be found in the work of Maslov [221], in connection with the study of the semiclassical limit of quantum mechanics, and in the work by Hörmander [164, 165], in connection with the theory of Fourier integral operators and the study of partial differential equations. When the phase function Φ has some degenerate critical points, the theory of unfoldings of singularities plays a crucial role in the description of the asymptotics of the integral (4.17) (see for instance Arnold's and Duistermaat's work [41, 110]) and brings the stationary phase method to an high level of mathematical rigour and elegance.

Despite the technical difficulties in the rigorous study of the asymptotic behavior of oscillatory integrals, the mathematical and physical ideas behind the stationary phase method are rather simple and intuitive. Indeed, when $\hbar$ can be considered very small, the function

$$x \mapsto e^{\frac{i}{\hbar}\Phi(x)}, \qquad x \in \mathbb{R}^n,$$

oscillates very fast, in such a way that the contributions to the integral coming from the positive and the negative parts of the oscillations annul each other. The key point is the fact that the only regions of $\mathbb{R}^n$ giving a non vanishing contribution to the value of the integral (4.17) are the neighborhoods of the stationary points of the phase function Φ, i.e. the points x_c satisfying the equation

$$\nabla\Phi(x) = 0.$$

As an illustrative example, let us consider the integral $I(\hbar) := \int_{-\pi/4}^{\pi/4} cos(x^2/\hbar)dx$. As a comparison between figure 4.1(a) and figure 4.1(b) shows, when the frequency $\hbar^{-1}$ increases, the region giving a significant contribution to the integral is a neighborhood of the stationary point $x = 0$.

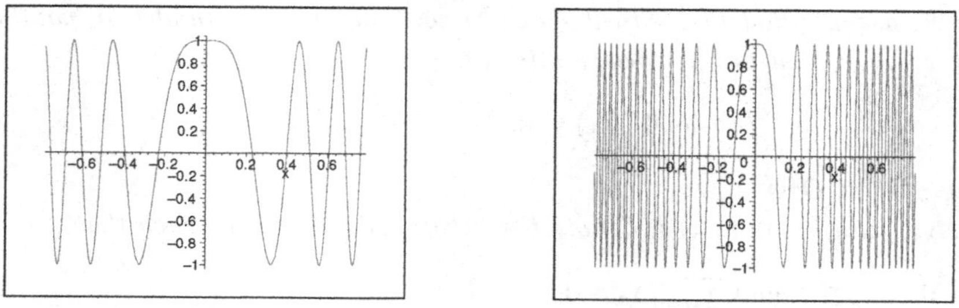

(a) (b)

Fig. 4.1 The function $\cos(x^2/\hbar)$ for different values of the parameter $\hbar$. (a) $\hbar = 1/30$. (b) $\hbar = 1/150$.

These reasoning can be made completely rigorous under rather general assumptions. Following [7], we describe here in some detail the study of the asymptotics of finite dimensional oscillatory integrals of the form

$$I(\hbar) := \widetilde{\int_{\mathbb{R}^n}} e^{\frac{i}{2\hbar}\langle x,Tx\rangle} e^{-\frac{i}{\hbar}V(x)} g(x)dx, \tag{4.18}$$

where $T = (I - L) : \mathbb{R}^n \to \mathbb{R}^n$ is a self-adjoint bijection of $\mathbb{R}^n$, $V : \mathbb{R}^n \to \mathbb{R}$ and $g : \mathbb{R}^n \to \mathbb{C}$ satisfy suitable assumptions. These results will be generalized in the next section to the infinite dimensional case.

Let us consider the phase function $\Phi : \mathbb{R}^n \to \mathbb{R}$,

$$\Phi(x) = \frac{1}{2}\langle x, Tx\rangle - V(x), \qquad x \in \mathbb{R}^n,$$

and the set of critical points of Φ:

$$\mathcal{C}(\Phi) := \{x \in \mathbb{R}^n \, . \, \Phi'(x) = 0\}.$$

Theorem 4.2. *Let $V : \mathbb{R}^n \to \mathbb{R}$ be a C^∞ function with bounded first derivatives and all higher order derivatives of at most linear growth, i.e.*

$$|V'(x)| \leq M, \qquad D^\alpha V(x) \leq m(1 + |x|), \qquad x \in \mathbb{R}^n,$$

for suitable constants m, M, where $V'(x) \equiv \nabla V(x)$ and

$$D^\alpha = D_1^{\alpha_1} \dots D_n^{\alpha_n} = \frac{\partial^{\alpha_1}}{\partial x_1^{\alpha_1}} \cdots \frac{\partial^{\alpha_n}}{\partial x_n^{\alpha_n}}, \qquad |\alpha| = \alpha_1 + \cdots + \alpha_n.$$

Let $g : \mathbb{R}^n \to \mathbb{C}$ a C^∞ function such that for some $p \geq 0$

$$|D^\alpha g(x)| \leq C_\alpha (1 + |x|^2)^{p/2}, \qquad \forall \alpha \in \mathbb{N}^n, \forall x \in \mathbb{R}^n. \tag{4.19}$$

Let us assume that the critical set $\mathcal{C}(\Phi)$ contains a finite number of points $\{c_1, \dots, c_s\}$, that are not degenerate, i.e.

$$\det \Phi''(c_i) \neq 0, \qquad i = 1, \dots, s,$$

where $\Phi'' \equiv D^2\Phi$.
Let χ_i, with $i = 0, 1, \dots s$, denote the $C^\infty(\mathbb{R}^n, \mathbb{R})$ functions such that

(1) $0 \leq \chi_i \leq 1$ and $\sum_{i=0}^s \chi_i = 1$;
(2) $\mathcal{C}(\Phi) \cap supp(\chi_i) = \{c_i\}$ for $i = 1, \dots s$;
(3) $supp(1 - \chi_0) \subset B(0, 3/2r)$ and $\chi_0^{-1}(\{0\}) = \bar{B}(0, r)$;
(4) $\chi_i(x) = 1$ for $x \in B(c_i, r_i)$ and $i = 1, \dots s$, for some $r, r_i > 0$;

where $B(a,r)$ denotes the open ball with center a and radius r and $\bar{B}(a,r)$ its closure.

Then the integral $I(\hbar)$ in Eq. (4.18) is well defined and it is given by:

$$I(\hbar) = \sum_{j=0}^{s} \widetilde{\int_{\mathbb{R}^n}} e^{\frac{i}{2\hbar}\langle x,Tx\rangle} e^{-\frac{i}{\hbar}V(x)} g(x)\chi_j(x)dx$$

$$=: \sum_{j=0}^{s} I_j(\hbar). \tag{4.20}$$

Moreover, by defining $I_j^*(\hbar)$ as

$$I_j(\hbar) = e^{\frac{i}{\hbar}\Phi(c_j)} I_j^*(\hbar), \qquad j = 1, \ldots, s$$

and $I_0^*(\hbar) = I_0(\hbar)$, then each I_j^* is a C^∞ function on $\mathbb{R}$ and in particular

$$I_j^*(0) = (\det(T - D^2V(c_j)))^{-1/2} g(c_j), \qquad j = 1, \ldots, s,$$

$$I_0^{(k)}(0) = 0, \qquad \forall k \geq 0.$$

Proof. We follow [7], see also [164, 165].

Since both the phase function Φ and the function g satisfy the assumptions of theorem 2.1, the oscillatory integral $I(\hbar)$ is well defined.

Let us assume that $s = 1$ (the proof in the general case is completely analogous).

Let $\phi \in \mathcal{S}(\mathbb{R}^n)$ such that $\phi(0) = 1$ and let $\epsilon > 0$. The regularized approximations of the oscillatory integral $I(\hbar)$ are given by:

$$I_\epsilon(\hbar, \phi) = (2\pi i\hbar)^{-n/2} \int_{\mathbb{R}^n} e^{\frac{i}{2\hbar}\langle x,Tx\rangle} e^{-\frac{i}{\hbar}V(x)} g(x)\phi(\epsilon x)dx$$

$$= (2\pi i\hbar)^{-n/2} \int_{\mathbb{R}^n} e^{\frac{i}{2\hbar}\langle x,Tx\rangle} e^{-\frac{i}{\hbar}V(x)} g(x)\chi_1(x)\phi(\epsilon x)dx$$

$$+ (2\pi i\hbar)^{-n/2} \int_{\mathbb{R}^n} e^{\frac{i}{2\hbar}\langle x,Tx\rangle} e^{-\frac{i}{\hbar}V(x)} g(x)(1 - \chi_1(x))\phi(\epsilon x)dx$$

$$= I_\epsilon^1(\hbar, \phi) + I_\epsilon^2(\hbar, \phi). \tag{4.21}$$

The first integral is equal to

$$I_\epsilon^1(\hbar, \phi) = (2\pi i\hbar)^{-n/2} e^{\frac{i}{\hbar}\Phi(c_1)} \int_{\mathbb{R}^n} e^{\frac{i}{\hbar}(\Phi(x)-\Phi(c_1))} g(x)\chi_1(x)\phi(\epsilon x)dx,$$

and, by dominated convergence, it converges as $\epsilon \downarrow 0$ to

$$(2\pi i\hbar)^{-n/2} e^{\frac{i}{\hbar}\Phi(c_1)} \int_{\mathbb{R}^n} e^{\frac{i}{\hbar}(\Phi(x)-\Phi(c_1))} g(x)\chi_1(x)dx,$$

which is a C^∞ function of $\hbar \in \mathbb{R}$ (see [165], theorem 7.7.5). By making in a neighborhood of c_1, i.e. in the support of χ_1, the change of variable

$$\frac{\Phi(x)-\Phi(c_1)}{2\hbar} = \frac{yD^2\Phi(c_1)y}{2}$$

one gets easily

$$I_1^*(0) = (\det D^2\Phi(c_1))^{-1/2}g(c_1) = (\det(T - D^2V(c_1)))^{-1/2}g(c_1).$$

By reasoning as in the proof of theorem 2.1, it is possible to see that also the limit $\lim_{\epsilon\to 0} I_\epsilon^2(\hbar,\phi)$ exists and it is independent of ϕ. Moreover, by denoting this limit with $I^2(\hbar)$, one has

$$|I^2(\hbar)| \leq C_k|\hbar|^k, \quad |\hbar| \leq 1, \forall k \geq 1.$$

Indeed let us consider the C^∞ vector field $a(x) = (a_1(x), \ldots, a_n(x))$ with components given by

$$a_j(x) = \frac{D_j\Phi(x)}{|\nabla\Phi(x)|^2}(1-\chi_1(x)), \qquad j = 1, \ldots, n,$$

and the first order differential operator

$$L_\hbar := (-i\hbar)\sum_{j=1}^n a_j(x)\frac{\partial}{\partial x_j}.$$

It is simple to see that $L_\hbar^+$, the adjoint of $L_\hbar$, is given by

$$L_\hbar^+ := i\hbar\Big(\sum_{j=1}^n a_j(x)\frac{\partial}{\partial x_j} + \operatorname{div} a\Big) \equiv i\hbar A.$$

By the assumptions on the function V, it is simple to verify that for any $\alpha \in \mathbb{N}^n$, there exists $C_\alpha > 0$ such that

$$|D^\alpha a_j(x)| \leq \frac{C_\alpha}{1+|x|}, \qquad x \in \mathbb{R}^d.$$

Moreover, since

$$L_\hbar e^{\frac{i}{\hbar}\Phi(x)} = (1-\chi_1(x))e^{\frac{i}{\hbar}\Phi(x)},$$

by Stokes formula we have:

$$\begin{aligned} I_\epsilon^2(\hbar,\phi) &= (2\pi i\hbar)^{-n/2}\int_{\mathbb{R}^n} e^{\frac{i}{\hbar}\Phi(x)}L_\hbar^+(f_\epsilon)dx \\ &= (2\pi i\hbar)^{-n/2} i\hbar \int_{\mathbb{R}^n} e^{\frac{i}{\hbar}\Phi(x)}A(f_\epsilon)dx \end{aligned} \tag{4.22}$$

where $f_\epsilon(x) = \phi(\epsilon x)g(x)$. By iterating the procedure k times, we get

$$I^2_\epsilon(\hbar, \phi) = (2\pi i\hbar)^{-n/2}(i\hbar)^k \int_{\mathbb{R}^n} e^{\frac{i}{\hbar}\Phi(x)} f_{\epsilon,k}(x)dx,$$

where

$$f_{\epsilon,k}(x) = \sum_{j=0}^{k} \beta^k_j(x) D^j f_\epsilon(x),$$

where $\beta^k_j(x)$ is a system of coefficients with the property that each $\beta^k_j(x)$ is a sum of finite number of terms, and each term is a multiplication of k factors. Each factor is either one of the functions a_j or its derivative, the total number of derivatives in each term being equal to $k - j$.

By taking $k > n + p$, where p is constant in Eq. (4.19), we have that $\beta^k_j(x)D^j g(x) \in L^1(\mathbb{R}^n)$ and, by dominated convergence theorem, the limit

$$I^2(\hbar) := \lim_{\epsilon \to 0} I^2_\epsilon(\hbar, \phi)$$

exists and satisfies the condition

$$|I^2(\hbar)| \leq C_k |\hbar|^k, \quad |\hbar| \leq 1, \forall k \geq 1. \qquad \square$$

Remark 4.5. Under additional assumptions (see for instance [16], Corollaries 2.4 and 2.5) it is also possible to compute all terms of the asymptotic expansion of each integral $I^*_j(\hbar)$, which is given by:

$$I^*_j(\hbar) = (\det T)^{-1/2} \sum_{m=0}^{\infty} \hbar^m \left(\frac{i}{2}\right)^m \sum_{k=0}^{\infty} \left(\frac{1}{2}\right)^k \frac{1}{k!(m+k)!}$$
$$\left[\left(\left(\sum_{l=1}^{k} \nabla_{x_l} + \nabla_y\right) T^{-1} \left(\sum_{l=1}^{k} \nabla_{x_l} + \nabla_y\right)\right)_2^{m+k} V(x_1)\dots V(x_k) g(y) \chi_j(y)\right] \tag{4.23}$$

where the value of [ ] is to be taken at the critical point c_j and $\left(\left(\sum_{l=1}^k \nabla_{x_l} + \nabla_y\right)T^{-1}\left(\sum_{l=1}^k \nabla_{x_l} + \nabla_y\right)\right)_2^{m+k}$ is the sum of all terms in the expansion of $\left(\left(\sum_{l=1}^k \nabla_{x_l} + \nabla_y\right)T^{-1}\left(\sum_{l=1}^k \nabla_{x_l} + \nabla_y\right)\right)^{m+k}$ which are of at least second degree with respect to each ∇_{x_l}, $l = 1, \dots, k$.

In the case where the phase function Φ in the oscillatory integral (4.17) presents some degenerate critical point c, that is

$$\nabla\Phi(c) = 0, \qquad \det \Phi''(c) = 0,$$

the situation is more involved. In this case the asymptotic behavior as $\hbar \downarrow 0$ of

$$I(\hbar) := \int^{o} e^{\frac{i}{\hbar}\Phi(x)} f(x) dx$$

is determined by taking into account the higher derivatives of Φ and the classification of different types of degeneracies [45, 44].

In the case of the generalized Fresnel integral studied in section 2.3, the solution of the problem is simpler and has been described in [30], where the whole asymptotic expansion of an oscillatory integral of the form

$$\int^{o}_{\mathbb{R}^n} e^{\frac{i}{\hbar}\Phi(x)} f(x) dx, \tag{4.24}$$

is studied, where $\Phi : \mathbb{R}^n \to \mathbb{R}$ is an homogeneous polynomial function of even degree $2M$:

$$\Phi(x) = A_{2M}(x, \ldots, x), \qquad x \in \mathbb{R}^d,$$

and A_{2M} is a completely symmetric $2M$th order covariant tensor on $\mathbb{R}^n$ such that $A_{2M}(x, \ldots, x) > 0$ unless $x = 0$.

By the positivity of the phase function it is simple to see that the integral (4.24) is well defined also for $\hbar \in \mathbb{C}$, $Im(\hbar) < 0$, provided that the function f is bounded. For $\hbar \in \mathbb{R}$, sufficient conditions for the definition of the purely oscillatory integral (4.24) are given by theorem 2.4.

Theorem 4.3. *Let $f \in \mathcal{F}(\mathbb{R}^n)$ be the Fourier transform of a bounded variation measure μ_f admitting moments of all orders.*
Let us suppose f satisfies the following conditions, for all $l \in \mathbb{N}$:

(1)

$$\int_{\mathbb{R}^n} |kx|^l e^{-kx} |d\mu_f|(k) \leq F(l) g(|x|) e^{c|x|^{2M-1}}, \qquad \forall x \in \mathbb{R}^n,$$

where $c \in \mathbb{R}$, $F(l)$ is a positive constant depending on l, $g : \mathbb{R}^+ \to \mathbb{R}$ is a positive function with polynomial growth.

(2)

$$\left| \int_{\mathbb{R}^n} (ku)^l e^{ik\rho u \hbar^{1/2M} e^{i\pi/4M}} d\mu_f(k) \right| \leq A c^l C(l, M, n)$$

for all $u \in S_{n-1}$, $\rho \in \mathbb{R}^+$, $\mathrm{Im}(\hbar) \leq 0$ $\hbar \neq 0$, where $A, c, C(l, M, n) \in \mathbb{R}$ (and S_{n-1} is the $(n-1)$-spherical hypersurface of radius 1 and centered at the origin).

Then the oscillatory integral (4.24) admits for $\hbar \in \mathbb{C}$, $\mathrm{Im}(\hbar) \leq 0$, the following asymptotic expansion in powers of $\hbar^{1/2M}$:

$$I(\hbar) = \hbar^{n/2M} \frac{e^{in\pi/4M}}{2M} \sum_{l=0}^{N-1} \frac{(i)^l}{l!} (e^{i\pi/4M})^l \hbar^{l/2M} \Gamma\Big(\frac{l+n}{2M}\Big)$$

$$\int_{\mathbb{R}^n} \int_{S_{n-1}} (ku)^l P(u)^{-\frac{l+n}{2M}} d\Omega_{n-1} d\mu_f(k) + \mathcal{R}_N, \quad (4.25)$$

with $|\mathcal{R}_N| \leq A'|\hbar|^{N/2M}(c')^N \frac{C(N,M,n)}{N!}\Gamma\Big(\frac{n+N}{2M}\Big)$ where $A', c' \in \mathbb{R}$ are suitable constants and $C(N,M,n)$ is the constant in assumption (2). If $C(N,M,n)$ satisfies the following bound:

$$C(N,M,n) \leq N!\Gamma\Big(\frac{n+N}{2M}\Big)^{-1} \quad (4.26)$$

then the series has a positive radius of convergence, while if

$$C(N,M,n) \leq N!\Gamma\Big(1+\frac{N}{2M}\Big)\Gamma\Big(\frac{n+N}{2M}\Big)^{-1} \quad (4.27)$$

then the expansion is Borel summable in the sense of, e.g. [236, 160] and determines $I(\hbar)$ uniquely.

Remark 4.6. The hypothesis of theorem 4.3 are slightly different from those presented in [30].

Proof. Let $\tilde{F}(k) \equiv \int_{\mathbb{R}^n} e^{ikx} e^{iA_{2M}(x,\ldots,x)} dx$, then by theorem 2.4 the oscillatory integral (4.24) is given by:

$$\int_{\mathbb{R}^n} e^{\frac{i}{\hbar}A_{2M}(x,\ldots,x)} f(x) dx = \hbar^{n/2M} \int_{\mathbb{R}^n} \tilde{F}(\hbar^{1/2M} k) \mu_f(dk). \quad (4.28)$$

By lemma 2.1, $\tilde{F}$ is given by

$$\tilde{F}(\hbar^{1/2M} k) = e^{in\pi/4M} \int_{\mathbb{R}^n} e^{i\hbar^{1/2M} kx e^{i\pi/4M}} e^{-A_{2M}(x,\ldots,x)} dx, \qquad k \in \mathbb{R}^n,$$

where, if $\hbar = |\hbar| e^{i\phi}$, $\phi \in [-\pi, 0]$, $\hbar^{1/2M} = |\hbar|^{1/2M} e^{i\phi/2M}$. By representing the latter absolutely convergent integral using polar coordinates in $\mathbb{R}^n$ we get:

$$\tilde{F}(\hbar^{1/2M} k) = e^{in\pi/4M} \int_{S_{n-1}} \int_0^\infty e^{i\hbar^{1/2M} e^{i\pi/4M} \rho k u} e^{-\rho^{2M} A_{2M}(u,\ldots,u)} \rho^{n-1} d\rho d\Omega_{n-1}$$

where $d\Omega_{n-1}$ is the Riemann-Lebesgue measure on the $n-1$-dimensional spherical hypersurface S_{n-1}, $x = \rho u$, $\rho = |x|$, $u \in S_{n-1}$ is a unitary vector.

We can expand the latter integral in a power series of $\hbar^{1/2M}$ and apply Fubini theorem:

$$\tilde{F}(\hbar^{1/2M}k) = e^{in\pi/4M}\int_{S_{n-1}}\int_0^\infty \sum_{l=0}^\infty \frac{(i)^l}{l!}(e^{i\pi/4M})^l \hbar^{l/2M}\rho^l(ku)^l \, e^{-\rho^{2M}A_{2M}(u,\dots,u)}\rho^{n-1}d\rho d\Omega_{n-1}$$

$$= e^{in\pi/4M}\sum_{l=0}^\infty \frac{(i)^l}{l!}(e^{i\pi/4M})^l\hbar^{l/2M}\int_{S_{n-1}}(ku)^l\int_0^\infty \rho^{l+n-1} \, e^{-\rho^{2M}A_{2M}(u,\dots,u)}d\rho d\Omega_{n-1}$$

$$= \frac{e^{in\pi/4M}}{2M}\sum_{l=0}^\infty \frac{(i)^l}{l!}(e^{i\pi/4M})^l\hbar^{l/2M}\Gamma\Big(\frac{l+n}{2M}\Big)\int_{S_{n-1}}(ku)^l P(u)^{-\frac{l+n}{2M}}d\Omega_{n-1} \tag{4.29}$$

where $P(u) \equiv A_{2M}(u,\dots,u)$ is a strictly positive continuous function on the compact set S_{n-1}, so that it admits an absolute minimum denoted by m. This gives

$$\Big|\int_{S_{n-1}}(ku)^l P(u)^{-\frac{l+n}{2M}}d\Omega_{n-1}\Big| \le |k|^l m^{-\frac{l+n}{2M}}\Omega_{n-1}(S_{n-1})$$

$$= |k|^l m^{-\frac{l+n}{2M}} 2\pi^{n/2}\Gamma\Big(\frac{n}{2}\Big)^{-1}. \tag{4.30}$$

The latter inequality and the Stirling formula assure the absolute convergence of the series (4.29). We can now insert this formula into (4.28) and get:

$$\int_{\mathbb{R}^n} e^{\frac{i}{\hbar}A_{2M}(x,\dots,x)}f(x)dx$$

$$= \hbar^{n/2M}\frac{e^{in\pi/4M}}{2M}\sum_{l=0}^{N-1}\frac{(i)^l}{l!}(e^{i\pi/4M})^l\hbar^{l/2M}\Gamma\Big(\frac{l+n}{2M}\Big) \int_{\mathbb{R}^n}\int_{S_{n-1}}(ku)^l P(u)^{-\frac{l+n}{2M}}d\Omega_{n-1}\mu_f(dk) + \mathcal{R}_N. \tag{4.31}$$

Equation (4.31) can also be written in the following form:

$$\int_{\mathbb{R}^n} e^{\frac{i}{\hbar}A_{2M}(x,\dots,x)}f(x)dx$$

$$= \hbar^{n/2M}\frac{e^{iN\pi/4M}}{2M}\sum_{l=0}^{N-1}\frac{1}{l!}(e^{i\pi/4M})^l\hbar^{l/2M}\Gamma\Big(\frac{l+n}{2M}\Big) \int_{S_{n-1}}P(u)^{-\frac{l+n}{2M}}\frac{\partial^l}{\partial u^l}f(0)d\Omega_{n-1} + \mathcal{R}_N \tag{4.32}$$

where $\frac{\partial^l}{\partial u^l} f(0)$ denotes the l_{th} partial derivative of f at 0 in the direction u, and

$$\mathcal{R}_N = \hbar^{n/2M} e^{in\pi/4M} \int_{\mathbb{R}^n} \int_{S_{n-1}} \int_0^\infty \sum_{l=N}^\infty \frac{(i)^l}{l!} (e^{i\pi/4M})^l \hbar^{l/2M} \rho^l (ku)^l$$
$$e^{-\rho^{2M} A_{2M}(u,\ldots,u)} \rho^{n-1} d\rho d\Omega_{n-1} \mu_f(dk). \quad (4.33)$$

In the case where assumptions (1) and (2) are satisfied, we can prove the asymptoticity of the expansion (4.31), indeed

$$\mathcal{R}_N = \hbar^{n/2M} e^{in\pi/4M} \frac{(i)^N}{N-1!} (e^{i\pi/4M})^N \hbar^{N/2M}$$
$$\int_{\mathbb{R}^n} \int_{S_{n-1}} \int_0^\infty \int_0^1 (1-t)^{N-1} e^{iku\rho t \hbar^{1/2M} e^{i\pi/4M}} e^{-\rho^{2M} A_{2M}(u,\ldots,u)}$$
$$(ku)^N \rho^{n+N-1} dt d\rho d\Omega_{n-1} \mu_f(dk). \quad (4.34)$$

By assumptions (1), (2) and Fubini theorem, the latter is bounded by

$$|\mathcal{R}_N| \le \frac{A}{M} \pi^{n/2} \Gamma\left(\frac{n}{2}\right)^{-1} |\hbar|^{(n+N)/2M} c^N m^{-\frac{n+N}{2M}} \frac{C(N,M,n)}{N!} \Gamma\left(\frac{n+N}{2M}\right).$$

If assumption (4.26) is satisfied, then the latter becomes

$$|\mathcal{R}_N| \le \frac{A}{M} \pi^{n/2} \Gamma\left(\frac{n}{2}\right)^{-1} |\hbar|^{(n+N)/2M} c^N m^{-\frac{n+N}{2M}}$$

and the series has a positive radius of convergence, while if assumption (4.27) holds, we get the estimate

$$|\mathcal{R}_N| \le \frac{A}{M} \pi^{n/2} \Gamma\left(\frac{n}{2}\right)^{-1} |\hbar|^{(n+N)/2M} c^N m^{-\frac{n+N}{2M}} \Gamma\left(1 + \frac{N}{2M}\right).$$

This bound, the analyticity of the function $I(\hbar)$ in a sector of the complex plane of amplitude π (i.e. in $\mathrm{Im}(\hbar) < 0$) and Nevanlinna's theorem 4.1 assure the Borel summability of the power series expansion in $\hbar^{1/2M}$ (4.25). □

4.3 The stationary phase method. Infinite dimensional case

The implementation of an infinite dimensional version of the stationary phase method allowing the study of the asymptotic behavior, when $\hbar \to 0$, of infinite dimensional oscillatory integrals of form

$$\widetilde{\int_{\mathcal{H}}} e^{\frac{i}{\hbar}\Phi(x)} f(x) dx$$

is not trivial.

The first results can be found in [16] and were further developed in [7] and in [248]. The authors consider infinite dimensional oscillatory integrals of the form

$$I(\hbar) = \widetilde{\int_{\mathcal{H}}} e^{\frac{i}{2\hbar}\langle x,Tx\rangle} e^{-\frac{i}{\hbar}V(x)} g(x)dx, \tag{4.35}$$

with $V, g \in \mathcal{F}(\mathcal{H})$, $T = I - L$ self-adjoint and invertible operator, $L : \mathcal{H} \to \mathcal{H}$ of trace class. Under additional regularity assumptions on V and g, it is possible to prove that the phase function

$$\Phi(x) = \frac{1}{2}\langle x, (I-L)x\rangle - V(x), \qquad x \in \mathcal{H}, \tag{4.36}$$

has only non degenerate critical points, that the essential part in $I(\hbar)$ is a C^∞ function of $\hbar$ and its asymptotic expansion at $\hbar = 0$ depends only on the derivatives of V and g at these critical points (see for instance [16, 7]).

The following lemma gives sufficient conditions for the existence and uniqueness of a non degenerate stationary point of the phase function (4.36).

Lemma 4.1. [7] *Let $V \in \mathcal{F}(\mathcal{H})$, with*

$$V(x) = \int_{\mathcal{H}} e^{i\langle x,y\rangle} d\mu(y).$$

Let us assume that the measure $|\mu|$ of variation of the measure μ satisfies the following inequality

$$\|T^{-1}\| \int_{\mathcal{H}} \|y\|^2 d\mu(y)| < 1. \tag{4.37}$$

Then there exists a unique $a \in \mathcal{H}$ such that

$$\Phi'(a) = Ta - V'(a) = 0.$$

Moreover $\Phi''(a)$ is invertible.

Proof.
As $V'(x) = i\int_{\mathcal{H}} ye^{i\langle x,y\rangle} d\mu(y)$, from the triangle inequality

$$\begin{aligned}\|T^{-1}V'(x) - T^{-1}V'(y) \leq \|T^{-1}\| \int_{\mathcal{H}} \|z\| |1 - e^{i\langle x-y,z\rangle}| d|\mu|(z) \\ \leq \|T^{-1}\| \|x-y\| \int_{\mathcal{H}} \|z\|^2 \left|\frac{1 - e^{i\langle x-y,z\rangle}}{\|x-y\|\|z\|}\right| d|\mu|(z).\end{aligned} \tag{4.38}$$

As

$$\left|\frac{1-e^{it}}{t}\right| = 2\left|\frac{\sin\frac{t}{2}}{t}\right| \leq 1,$$

we have

$$\|T^{-1}V'(x) - T^{-1}V'(y) \leq \|\|x-y\|\|T^{-1}\| \int_{\mathcal{H}} \|y\|^2 d|\mu|(y)| < \|\|x-y\|,$$

and by the contraction mapping principle, it follows that there exists a unique $a \in \mathcal{H}$ such that $Ta - V'(a) = 0$.

Concerning the study of the operator $\Phi''(a)$, it is simple to see that

$$\Phi''(a) = T - V''(a),$$

and that

$$V''(a)(x)(z) = i^2 \int_{\mathcal{H}} \langle y, z\rangle\langle y, x\rangle e^{i\langle y,a\rangle} d\mu(y), \qquad x, z \in \mathcal{H}.$$

By condition (4.37), it follows that $\|T^{-1}V''(a)\| < 1$, so $T - V''(a)$ is invertible. □

The following theorem gives sufficient conditions for the existence of the asymptotic expansion of the integral (4.35) around a unique stationary point of the phase function (4.36). It also states the Borel summability of the asymptotic expansion. We skip here the proof, which is rather technical and involves a large amount of calculations, but we point out to the interested reader the original paper of J. Rezende [248] for more details.

Theorem 4.4. [248] *Let $V, g \in \mathcal{F}(\mathcal{H})$, with*

$$V(x) = \int_{\mathcal{H}} e^{i\langle x,y\rangle} d\mu(y),$$

$$g(x) = \int_{\mathcal{H}} e^{i\langle x,y\rangle} d\nu(y). \tag{4.39}$$

Let us assume that the measures $\mu, \nu \in \mathcal{M}(\mathcal{H})$ admit finite moments of all orders and there exist constants $L, M, \epsilon > 0$ such that the following inequalities hold:

$$\int_{\mathcal{H}} \|x\|^j d|\mu|(x) \leq L\frac{j!}{\epsilon^j}, \qquad j \in \mathbb{N}, \tag{4.40}$$

$$\int_{\mathcal{H}} \|x\|^j d|\nu|(x) \leq M\frac{j!}{\epsilon^j}, \qquad j \in \mathbb{N}, \tag{4.41}$$

where $|\mu|, |\nu|$ denote measure of variation of the measures μ, ν respectively. Let us assume moreover that the constants $L, \epsilon > 0$ satisfy the following inequality

$$2L\|T^{-1}\|(3 + 2\sqrt{2}) < \epsilon^2 \tag{4.42}$$

($\|T^{-1}\|$ denoting the operator norm of T^{-1}).

Then the following holds:

(1) *There is a unique point* $a \in \mathcal{H}$ *such that* $V'(a) = Ta$; $T^{-1}V''(a)$ *is of trace class and its trace-norm* $\| \|_1$ *satisfies the inequality*

$$\|T^{-1}V''(a)\|_1 < 1.$$

(2) *The infinite dimensional oscillatory integral* $I(\hbar)$ *given by Eq. (4.35) is analytic in* $\mathrm{Im}(\hbar) < 0$ *and the function* I^*, *given by*

$$I^*(\hbar) = I(\hbar)e^{\frac{i}{\hbar}V(a)-\frac{i}{2\hbar}\langle a,Ta\rangle}, \qquad \text{for } \mathrm{Im}(\hbar) \leq 0, \hbar \neq 0,$$

is a continuous function of $\hbar$ *in* $\mathrm{Im}(\hbar) \leq 0$, *with*

$$I^*(0) = \det(T - V''(a))^{-1/2}g(a),$$

where $\det(T - V''(a))$ *is the Fredholm determinant of the operator* $(T - V''(a))$.

(3) $I^*(\hbar)$ *has the following asymptotic expansion and estimate*

$$\Big|I^*(\hbar) - \det T^{-1/2}\sum_{m=0}^{l-1}\hbar^m\Big(-\frac{i}{2}\Big)^m\sum_{n=0}^{\infty}\frac{(-2)^{-n}}{n!(m+n)!}$$

$$\int\cdots\int\Big\{T^{-1/2}\Big(\sum_{j=1}^{n}\alpha_j+\beta\Big)\Big\}^{2m+2n}_{(\alpha,2)}\prod_{j=1}^{n}e^{i\langle a,\alpha_j\rangle}d\mu(\alpha_j)e^{i\langle a,\beta\rangle}d\nu(\beta)\Big|$$

$$= \Big|I^*(\hbar) - \det(T - V''(a))^{-1/2}\sum_{m=0}^{l-1}\hbar^m\Big(-\frac{i}{2}\Big)^m\sum_{n=0}^{2m}\frac{(-2)^{-n}}{n!(m+n)!}$$

$$\int\cdots\int\Big\{(T-V''(a))^{-1/2}\Big(\sum_{j=1}^{n}\alpha_j+\beta\Big)\Big\}^{2m+2n}_{(\alpha,3)}$$

$$\prod_{j=1}^{n}e^{i(a,\alpha_j)}d\mu(\alpha_j)e^{i\langle a,\beta\rangle}d\nu(\beta)\Big|$$

$$\leq \frac{M}{(2-\sqrt{2})\sqrt{\pi}}\Big(\frac{2|\hbar|\|T^{-1}\|}{\epsilon^2(6-4\sqrt{2})}\Big)^l\Big(1-\frac{2L\|T^{-1}\|(3+2\sqrt{2})}{\epsilon^2}\Big)^{-l-1/2}\Big(l-\frac{1}{2}\Big)!,$$

where

$$(x_1 + \cdots + x_n + y)^s_{(x,m)}$$

$$= \frac{s!}{(s-mn)!(m-1)!^n}\int_0^t\cdots\int_0^t[(1-t_1)\dots(1-t_n)]^{m-1}$$

$$(x_1^m,\dots,x_n^m,(t_1x_1+\cdots+t_nx_n+y)^{s-mn})dt_1\dots dt_n,$$

and

$$(x_1,\dots,x_{2n}) = \frac{1}{(2n)!}\sum_{\sigma}(x_{\sigma(1)},x_{\sigma(2)})\cdots(x_{\sigma(2n-1)},x_{\sigma(2n)}),$$

the summation being over all permutations σ *of* $\{1,\dots,2n\}$.

(4) The asymptotic expansion is Borel summable and determines $I^*(\hbar)$ uniquely.

As we have seen in the previous section, in the finite dimensional case, if the phase function Φ has several critical points, the asymptotic expansion of the integral $I(\hbar)$ is just the sum over all the stationary points of the corresponding expansion for each critical point (see theorem 4.2). In the finite dimensional case this result is obtained by writing the function g as a sum of functions, each one having compact support containing a unique stationary point. In the infinite dimensional case the generalization of this technique is not straightforward, as we do not know whether the integral has an asymptotic expansion at all if condition (4.41) in theorem 4.4 is not satisfied. Indeed the condition $g(x) = \int_{\mathcal{H}} e^{i\langle x,y\rangle} d\nu(y)$, with

$$\int_{\mathcal{H}} \|x\|^j d|\nu|(x) \leq M\frac{j!}{\epsilon^j}, \qquad j \in \mathbb{N},$$

implies that there exists a $\lambda \in \mathbb{R}^+$ such that

$$\int_{\mathcal{H}} e^{\lambda\|x\|} d|\nu|(x) < \infty. \tag{4.43}$$

By condition (4.43) it is simple to see that, if $z = x + iy$ with $x, y \in \mathcal{H}$, the function g defined on the complexification of the real Hilbert space $\mathcal{H}$ by

$$g(z) = \int_{\mathcal{H}} e^{i\langle z,\alpha\rangle} d\nu(\alpha)$$

is analytic in $\|y\| < \lambda$, hence the support of g cannot be compact.

In order to overcome this problem, in [16] an alternative technique has been implemented. Indeed by imposing suitable assumption on the function $V \in \mathcal{F}(\mathcal{H})$, it is possible to prove that there exists a decomposition of the Hilbert space $\mathcal{H}$ into a direct sum $\mathcal{H} = \mathcal{H}_1 \oplus \mathcal{H}_2$, with $\mathcal{H}_2$ finite dimensional and the phase Φ when restricted to $\mathcal{H}_1$ presents a unique stationary point. The oscillatory integral is then studied by means of Fubini theorem (see theorem 2.7). We present here the results obtained in [16], where the case $T = I$ is handled. Analogous results can be obtained also for $T = I - L$ self-adjoint and invertible operator, $L : \mathcal{H} \to \mathcal{H}$ of trace class (see [7]).

Lemma 4.2. *Let $V \in \mathcal{F}(\mathcal{H})$, with $V(x) = \int_{\mathcal{H}} e^{i\langle x,y\rangle} d\mu(y)$, such that there exists a $\lambda > 0$ with $\|\mu\| < \lambda^2$ and*

$$\int_{\mathcal{H}} e^{\sqrt{2}\lambda\|y\|} d|\mu|(y) < \infty.$$

Then there exists a decomposition $\mathcal{H} = \mathcal{H}_1 \oplus \mathcal{H}_2$, with finite dimensional $\mathcal{H}_2$, such that, with $V(x) = V(y,z)$ for $x = y \oplus z$, $V(y,z)$ satisfies, as a function of y, the following condition uniformly in z:

$$\frac{1}{\lambda^2}\int_{\mathcal{H}_1} e^{\sqrt{2}\lambda\|\beta\|}d|\mu_z|(\beta) < 1,$$

where

$$V(y,z) = \int_{\mathcal{H}_1} e^{i\langle\beta,y\rangle}d\mu_z(\beta).$$

Moreover the equation

$$d_1V(y,z) = y$$

has a unique solution $y = b(z)$ for all $z \in \mathcal{H}_2$ and the mapping $z \mapsto b(z)$ is real analytic from $\mathcal{H}_2 \to \mathcal{H}_1$.

Proof. Let $\{P_n\}$ be a sequence of orthogonal projections on $\mathcal{H}$ with finite dimensional ranges such that P_n converges strongly to the identity. By Lebesgue dominated convergence theorem

$$\lim_{m\to\infty}\frac{1}{\lambda^2}\int_{\mathcal{H}} e^{\sqrt{2}\lambda\|y-P_ny\|}d|\mu|(y) = \frac{1}{\lambda^2}\int_{\mathcal{H}} d|\mu|(y) < 1,$$

so there exists a finite dimensional projection operator P such that

$$\frac{1}{\lambda^2}\int_{\mathcal{H}} e^{\sqrt{2}\lambda\|y-Py\|}d|\mu|(y) < 1.$$

Let us decompose $\mathcal{H} = \mathcal{H}_1 \oplus \mathcal{H}_2 = (I-P)\mathcal{H} \oplus P\mathcal{H}$ and introduce the notation $x = (y,z)$ for $x = y \oplus z$. The measure $d\mu(\alpha)$ can be considered as a measure $d\mu(\beta,\gamma)$ on the product space $\mathcal{H}_1 \times \mathcal{H}_2$, which is isomorphic as a metric and as a measure space with $\mathcal{H}_1 \oplus \mathcal{H}_2$. We then have

$$V(y,z) = \int_{\mathcal{H}_1\times\mathcal{H}_2} e^{i\langle\beta,y\rangle}e^{i\langle\gamma,z\rangle}d\mu(\beta,\gamma).$$

Let us denote with μ_z the measure on $\mathcal{H}_1$ given by

$$\int_{\mathcal{H}_1} f(\beta)d\mu_z(\beta) = \int_{\mathcal{H}_1\times\mathcal{H}_2} f(\beta)e^{i\langle\gamma,z\rangle}d\mu(\beta,\gamma).$$

Clearly we have

$$V(y,z) = \int_{\mathcal{H}_1} e^{i\langle\beta,y\rangle}d\mu_z(\beta),$$

moreover by the Minkowski inequality

$$\frac{1}{\lambda^2}\int_{\mathcal{H}_1} e^{\sqrt{2}\lambda\|\beta\|}d|\mu_z|(\beta) \le \frac{1}{\lambda^2}\int_{\mathcal{H}} e^{\sqrt{2}\lambda\|\beta\|}d|\mu|(\beta,\gamma) < 1.$$

The last inequality implies that

$$\int_{\mathcal{H}_1} \|\beta\|^2 d|\mu_z|(\beta) < 1$$

and, as in the proof of the first part of lemma 4.1, this assures that for any $z \in \mathcal{H}_2$, the equation

$$d_1 V(y, z) = y$$

has a unique solution $b(z)$:

$$d_1 V(b(z), z) = b(z) \tag{4.44}$$

(where $d_1 V(y, z)$ denotes the derivative of $V(y, z)$ with respect to y.

In order to prove that $b(z)$ is a smooth function from $\mathcal{H}_2$ to $\mathcal{H}_1$, let us take the derivative of Eq. (4.44):

$$d_1^2 V(b(z), z) db(z) + d_1 d_2 V(b(z), z) = db(z). \tag{4.45}$$

As

$$\|d_1^2 V(y, z)\|_1 \leq \int_{\mathcal{H}_1} \|\beta\|^2 d|\mu_z|(\beta) < 1,$$

($\| \quad \|_1$ denoting the trace norm) we have that $I - d_1^2 V(y, z)$ has a uniformly bounded inverse and from Eq. (4.45)

$$db(z) = (I - d_1^2 V(b(z), z))^{-1} d_1 d_2 V(b(z), z).$$

This proves that $b(z)$ is uniformly continuous and bounded in z, so that $z \mapsto b(z)$ is a smooth mapping.
The assumptions on V implies that $d_1 V(y, z)$ is analytic in $|\text{Im}(y)|^2 + |\text{Im}(z)|^2 < 2\lambda^2$ and since z is a regular solution of

$$d_1 V(b(z), z) = b(z)$$

it is possible to conclude that $b(z)$ is real analytic from $\mathcal{H}_2$ to $\mathcal{H}_1$. □

Lemma 4.3. *Let $V \in \mathcal{F}(\mathcal{H})$ satisfy the assumptions of lemma 4.2. Then the equation*

$$V'(x) = x$$

(V' being the Frechet derivative of the function V) has at most a discrete set S of solutions, i.e. S has no limit points in $\mathcal{H}$.

Proof. Under the given assumptions, all the results of lemma 4.2 hold. Let us consider the partial derivative $d_2V(y,z)$ of $V(y,z)$ with respect to z, and the equation

$$d_2V(b(z),z) = z. \tag{4.46}$$

Since $d_2V(y,z)$ is analytic in y and in z, the function $z \mapsto d_2V(b(z),z)$ is analytic on the finite dimensional space $\mathcal{H}_2$. It follows that Eq. (4.46) has at most a discrete set of solutions. □

Theorem 4.5. *Let $\mathcal{H}$ be a real separable Hilbert space, and V and g in $\mathcal{F}(\mathcal{H})$, with*

$$V(x) = \int_{\mathcal{H}} e^{i\langle x,y\rangle} d\mu(y),$$

$$g(x) = \int_{\mathcal{H}} e^{i\langle x,y\rangle} d\nu(y). \tag{4.47}$$

Let us assume V and g are C^∞ functions, i.e. all moments of μ and ν exist. Moreover we assume $\mathcal{H} = \mathcal{H}_1 \oplus \mathcal{H}_2$ where dim $\mathcal{H}_2 < \infty$, and if $d\mu(\beta,\gamma), d\nu(\beta,\gamma)$ are the measures on $\mathcal{H}_1 \times \mathcal{H}_2$ given by μ and ν, then there is a λ such that $\|\mu\| < \lambda^2$ and

$$\int_{\mathcal{H}} e^{\sqrt{2}\lambda|\beta|} d|\mu|(\beta,\gamma) < \infty, \qquad \int_{\mathcal{H}} e^{\sqrt{2}\lambda|\beta|} d|\nu|(\beta,\gamma) < \infty.$$

If the equation $dV(x) = x$ has only a finite number of solutions $x_1, \ldots, x_n$ on the support of the function g, such that none of the operators $I - d^2V(x_i)$, $i = 1, \ldots, n$, has zero as an eigenvalue, then the function

$$I(\hbar) = \widetilde{\int_{\mathcal{H}}} e^{\frac{i}{2\hbar}\|x\|^2} e^{-\frac{i}{\hbar}V(x)} g(x) dx$$

is of the following form

$$I(\hbar) = \sum_{k=1}^{n} e^{\frac{i}{2\hbar}\|x_k\|^2 - V(x_k)} I_k^*(\hbar),$$

where $I_k^*(\hbar)$ $k = 1, \ldots, n$ are C^∞ functions of $\hbar$ such that

$$I_k^*(0) = e^{\frac{i\pi}{2}n_k} |\det(I - d^2V(x_k))|^{-\frac{1}{2}} g(x_k),$$

where n_k is the number of negative eigenvalues of the operator $d^2V(x_k)$ which are larger than 1.
Moreover if $V(x)$ is gentle, that is there exists a constant $\bar\lambda > 0$ with

$$\|\mu\| < \bar\lambda^2 \quad \text{and} \quad \int_{\mathcal{H}} e^{\sqrt{2}\bar\lambda\|\alpha\|} d|\mu|(\alpha) < \infty, \tag{4.48}$$

then the solutions of equation $dV(x) = x$ have no limit points.

Proof. By applying lemma 4.2 to the Hilbert space $\mathcal{H}_1$, we obtain a decomposition $\mathcal{H} = \mathcal{H}'_1 \oplus \mathcal{H}'_2$, with $\mathcal{H}_2 \subseteq \mathcal{H}'_2$ and $\mathcal{H}'_2$ finite dimensional such that with $x = y' \oplus z'$, $V(x) = V(y', z')$ satisfies, as a function of $y' \in \mathcal{H}'_1$, the following condition uniformly in $z' \in \mathcal{H}'_2$:

$$\frac{1}{\bar{\lambda}^2} \int_{\mathcal{H}'_1} e^{\bar{\lambda}\sqrt{2}\|\beta'\|} d|\mu|(\beta') \leq \frac{1}{\bar{\lambda}^2} \int_{\mathcal{H}} e^{\bar{\lambda}\sqrt{2}\|\beta'\|} d|\mu|(\beta', \gamma') < 1. \tag{4.49}$$

So, if necessary by using the decomposition $\mathcal{H} = \mathcal{H}'_1 \oplus \mathcal{H}'_2$ instead of $\mathcal{H} = \mathcal{H}_1 \oplus \mathcal{H}_2$, we may assume that, with the notation of the theorem,

$$\frac{1}{\bar{\lambda}^2} \int_{\mathcal{H}_1} e^{\bar{\lambda}\sqrt{2}\|\beta\|} d|\mu|(\beta) \leq \frac{1}{\bar{\lambda}^2} \int_{\mathcal{H}} e^{\bar{\lambda}\sqrt{2}\|\beta\|} d|\mu|(\beta, \gamma) < 1. \tag{4.50}$$

Condition (4.50) implies that the equation

$$d_1 V(y, z) = y$$

has a unique solution $y = b(z)$ and $z \mapsto b(z)$ is a smooth mapping of $\mathcal{H}_1$ into $\mathcal{H}_2$. By using the Fubini theorem for oscillatory integrals (theorem 2.7), $I(\hbar)$ is equal to

$$I(\hbar) = \widetilde{\int_{\mathcal{H}_2}} e^{\frac{i}{2\hbar}\|z\|^2} e^{\frac{i}{2\hbar} b(z)^2 - \frac{i}{\hbar} V(b(z), z)} I_2(\hbar, z) dz, \tag{4.51}$$

with $I_2(\hbar, z) = e^{-\frac{i}{\hbar}(\frac{b(z)^2}{2} - V(b(z), z))} I_1(\hbar, z)$ and

$$I_1(\hbar, z) = \widetilde{\int_{\mathcal{H}_1}} e^{\frac{i}{2\hbar}\|y\|^2} e^{-\frac{i}{\hbar} V(y,z)} g(y, z) dy. \tag{4.52}$$

It is now possible to prove that the Fresnel integral $I_2(\hbar)$ on the infinite dimensional Hilbert space $\mathcal{H}_1$ is a C^∞ function of $\hbar$ on the real line and it is analytic in $\mathrm{Im}(\hbar) < 0$. Moreover

$$I_2(0) = |1 - d_1^2 V(b(z), z)|^{-1/2} g(b(z), z).$$

The integral $I(\hbar)$ is now given in terms of a finite dimensional oscillatory integral on $\mathcal{H}_2$ and it can be studied by means of the classical method of stationary phase for the asymptotic expansions of finite dimensional oscillatory integrals (see theorem 4.2).

Since the solutions of the equation

$$d_2 V(b(z), z) = z \tag{4.53}$$

form a discrete set, there exists a partition of unity

$$1 = \sum_j \phi_j(z), \qquad z \in \mathcal{H}_2$$

by smooth functions $\phi_j : \mathcal{H}_2 \to [0,1]$ of compact support such that only one solution of Eq. (4.53) is contained in the support of each ϕ_j. One has then to study the asymptotics of integrals of the form

$$I(\hbar) = \widetilde{\int_{\mathcal{H}_2}} e^{\frac{i}{2\hbar}\|z\|^2} e^{\frac{i}{2\hbar}b(z)^2 - \frac{i}{\hbar}V(b(z),z)} \phi(z) I_2(\hbar, z) dz, \qquad (4.54)$$

with ϕ with compact support containing only one solution of Eq. (4.53). As from the assumptions of the theorem it follows that

$$\left| I_2(\hbar, z) - \sum_{m=0}^{N} \frac{\hbar^m}{m!} I_2^{(m)}(0, z) \right| \leq |\hbar|^{n+1} C_N,$$

with C_N independent on z, up to terms of order $|\hbar|^{N+1}$ the integral (4.54) may be written as

$$\sum_{m=0}^{N} \frac{\hbar^m}{m!} \widetilde{\int_{\mathcal{H}_2}} e^{\frac{i}{2\hbar}\|z\|^2} e^{\frac{i}{2\hbar}b(z)^2 - \frac{i}{\hbar}V(b(z),z)} \phi(z) I_2^{(m)}(0, z) dz.$$

It is then enough to study the asymptotic behavior of the integrals

$$\widetilde{\int_{\mathcal{H}_2}} e^{\frac{i}{2\hbar}\|z\|^2} e^{\frac{i}{2\hbar}b(z)^2 - \frac{i}{\hbar}V(b(z),z)} \phi(z) I_2^{(m)}(0, z) dz,$$

which is determined by the solutions of the equation $d\Phi(z) = 0$, with

$$\Phi(z) = \frac{1}{2}\|z\|^2 + \frac{1}{2}b(z)^2 - V(b(z), z), \qquad z \in \mathcal{H}_2.$$

As

$$d\Phi(z) = z + b(z)db(z) - d_1V(b(z), z)db(z) - d_2V(b(z), z),$$

and $b(z) = d_1(b(z), z)$, we have that

$$d\Phi(z) = z - d_2V(b(z), z),$$

so that the critical points are the solutions of Eq. (4.53) and, by construction, the support of ϕ contains only one of them, which will be denoted by c. By denoting $b = b(c)$ and $a = (b, c)$ it is easy to verify that a is a solution of $dV(x) = x$.

By computing $d^2\Phi(c)$, we have:

$$d^2\Phi(z) = 1 - d_2d_1V(b(z), z)db(z) - d_2^2V(b(z), z).$$

Since $b(z) = d_1V(b(z), z)$, by differentiating we get

$$db(z) = d_1^2V(b(z), z)db(z) + d_1d_2V(b(z), z).$$

By lemma 4.2, $I - d_1^2 V(b(z), z)$ has a bounded inverse, so that

$$db(z) = (I - d_1^2 V(b(z), z))^{-1} d_1 d_2 V(b(z), z),$$

hence

$$d^2\Phi(c) = 1 - d_2 d_1 V(a)(I - d_1^2 V(a))^{-1} d_1 d_2 V(a) - d_2^2 V(a).$$

Let us consider a vector $\xi \in \mathcal{H}_2$, such that $d^2\Phi(c)\xi = 0$. By defining $\eta \in \mathcal{H}_1$ as

$$\eta = -(I - d_1^2 V(a))^{-1} d_1 d_2 V(a)\xi,$$

and $\zeta = (\eta, \xi)$, we have that

$$(I - d^2 V(a))\zeta = 0,$$

as this is equivalent to

$$(I - d_1^2 V(a))\eta + d_1 d_2 V(a)\xi = 0,$$

$$d_2 d_1 V(a)\eta + (I - d_2^2 V(a))\xi = 0.$$

If for some $\zeta \in \mathcal{H}$, $(I - d^2 V(a))\zeta = 0$, then $d^2\Phi(c)\xi = 0$, where ξ is the projection of ζ on $\mathcal{H}_2$. So the condition of non degeneracy of $d^2\Phi(c)$ is equivalent to the condition of non degeneracy of $(I - d^2 V(a))$. Moreover

$$\det(d^2\Phi(c)) = \det(I - d_1^2 V(a))^{-1} \det(I - d^2 V(a)).$$

So, if $I - d^2 V(a)$ is not degenerate, the integral (4.54) is a C^∞ function of $\hbar$ and its asymptotic behavior can be studied by means of the classical method of the stationary phase in the (simplest) non degenerate case (theorem 4.2) and obtaining, by summing all contributions given by the stationary points, the final result:

$$I(\hbar) = \sum_{k=1}^{n} e^{\frac{i}{2\hbar}\|x_k\|^2 - V(x_k)} I_k^*(\hbar),$$

$$I_k^*(0) = e^{\frac{i\pi}{2} n_k} |\det(I - d^2 V(x_k))|^{-\frac{1}{2}} g(x_k).$$

□

If some critical point of the phase function is degenerate, the study of the asymptotic behavior of the oscillatory integral $I(\hbar)$ in Eq. (4.35) becomes more complicated. Indeed, as we we know from the case of a finite dimensional Hilbert space $\mathcal{H}$, in this situation it is possible that the integral $I(\hbar)$, divided by the above leading term, will not tend to a limit as $\hbar \to 0$. A possible approach to the study of this situation can be found for instance in the work of Duistermaat [110] and has been generalized in [16] to case where the integration is performed in an infinite dimensional setting. The problem is solved by letting the functions V and g depend on an additional parameter $y \in \mathbb{R}^k$, for suitable k, and to study, instead of

$$I(\hbar, y) = \widetilde{\int_{\mathcal{H}}} e^{\frac{i}{2\hbar}\|x\|^2} e^{-\frac{i}{\hbar}V(x,y)} g(x,y)dx$$

an oscillatory integral of a larger Hilbert space of the following form:

$$I(\hbar, \psi) = (2\pi i\hbar)^{k/2} \int_{\mathbb{R}^k} e^{-\frac{i}{\hbar}\psi(y)} \chi(y) \widetilde{\int_{\mathcal{H}}} e^{\frac{i}{2\hbar}\|x\|^2} e^{-\frac{i}{\hbar}V(x,y)} g(x,y)dxdy, \tag{4.55}$$

where ψ and χ are C^∞ functions and χ has a compact support. It is also required that the applications

$$y \to V(\,\cdot\,, y), \qquad y \in \mathbb{R}^k,$$

$$y \to g(\,\cdot\,, y), \qquad y \in \mathbb{R}^k,$$

are C^∞ functions from $\mathbb{R}^k$ to $\mathcal{F}(\mathcal{H})$ in the strong topology. By the Fubini theorem (2.7) the integral (4.55) can be regarded as an oscillatory integral on $\mathcal{H} \oplus \mathbb{R}^k$, with a new phase function Φ given by

$$\Phi(x, y) = \frac{1}{2}\|x\|^2 - V(x, y) - \psi(y), \qquad (x, y) \in \mathcal{H} \oplus \mathbb{R}^k.$$

The key point is the non degeneracy of the critical point of this new phase function. Indeed, by the Morse theorem, the set of functions $V(x)$ such that $\frac{1}{2}\|x\|^2 - V(x)$ has only non degenerate critical points form a open and dense set in the space of all C^∞ functions and the complement is in a natural sense of codimension 1. In other words, the case of degenerate critical points is unstable in the sense that degenerate critical points will disappear under arbitrary small perturbations (in the sense of the C^∞ topology). We refer to [110] and to [16] for a detailed discussion of this elegant technique, which involves an amount of differential geometry and allows eventually to prove an expression for the leading term in the asymptotic expansion of the integral (4.55) involving only objects and quantities with an intrinsic geometrical meaning.

A different approach to the study of the degeneracies is described in in [7] and [6], where some particular examples are handled. In particular in [7], by applying the Fubini theorem for infinite dimensional oscillatory integrals (theorem 2.7), the authors reduce to the study of the degeneracy on a finite dimensional subspace of the Hilbert space $\mathcal{H}$ and apply the existing theory for finite dimensional oscillatory integrals. In fact they assume that the phase function $\frac{1}{2}(x, Tx) - V(x)$ has the point $x_c = 0$ as a unique stationary point, which is degenerate, i.e.

$$Z := \mathrm{Ker}(T - d^2V)(0) \neq \{0\}.$$

Under suitable assumptions on T and V, they prove that Z is finite dimensional. By taking the subspace $Y = T(Z^\perp)$ and applying the Fubini theorem 2.7 one has

$$\begin{aligned} I(\hbar) &= \widetilde{\int_{\mathcal{H}}} e^{\frac{i}{2\hbar}\langle x, Tx\rangle} e^{-\frac{i}{\hbar}V(x)} g(x)dx \\ &= C_T \widetilde{\int_Z} e^{\frac{i}{2\hbar}\langle z, T_2 z\rangle} \widetilde{\int_Y} e^{\frac{i}{2\hbar}\langle y, T_1 y\rangle} e^{-\frac{i}{\hbar}V(y+z)} g(y+z)dydz, \end{aligned} \tag{4.56}$$

where T_1 and T_2 are defined by

$$T_1 y = (\pi_Y \circ T)(y), \qquad y \in Y,$$
$$T_2 z = (\pi_Z \circ T)(z), \qquad z \in Z,$$

and $C_T = (\det T)^{-1/2}(\det T1)^{1/2}(\det T_2)^{1/2}$. By assuming that $V, g \in \mathcal{F}(\mathcal{H})$, $V = \hat{\mu}$ and $g = \hat{\nu}$, and under some growth conditions on μ and ν, one has that the phase function

$$y \mapsto \frac{1}{2}\langle y, T_1 y\rangle - V(y+z), \qquad y \in Y,$$

of the oscillatory integral on Y

$$J(z, \hbar) = \widetilde{\int_Y} e^{\frac{i}{2\hbar}\langle y, T_1 y\rangle} e^{-\frac{i}{\hbar}V(y+z)} g(y+z)dy$$

has only one non degenerate stationary point $a(z) \in Y$. By applying then the theory developed for the non degenerate case one has

$$J(z, \hbar) = e^{\frac{i}{2\hbar}\langle a(z), T_1 a(z)\rangle} e^{-\frac{i}{\hbar}V(a(z)+z)} J^*(z, \hbar),$$

$$J^*(z, 0) = \left[\det\left(T_1 - \frac{\partial^2 V}{\partial^2 y}(a(z)+z)\right)\right]^{-1/2} g(a(z)+z).$$

As $I(\hbar) = \widetilde{\int_Z} e^{\frac{i}{\hbar}\Phi(z)} J^*(z, \hbar)dz$, where

$$\Phi(z) = \frac{1}{2}\langle z, T_2 z\rangle + \frac{1}{2}\langle a(z), T_1 a(z)\rangle - V(a(z)+z),$$

the main ingredient for the asymptotic behavior of $I(\hbar)$ comes from $J^*(z, 0)$. The phase function Φ has $z = 0$ as a unique degenerate critical point and, by applying the theory for asymptotic behavior for finite dimensional oscillatory integrals [165], one has to investigate the higher derivatives of Φ at 0. For example if $dim(Z) = 1$ and $\frac{\partial^3 V}{\partial^3 z}(0) \neq 0$ then

$$I(\hbar) \sim C\hbar^{-1/6}, \qquad \text{as} \quad \hbar \to 0.$$

More generally it is possible to handle other cases, taking into account the classification of different types of degeneracies (see, e.g., [45, 44]).

4.4 The semiclassical limit of quantum mechanics

The techniques and the results of the previous sections as well as the infinite dimensional oscillatory integral representation for the solution of the Schrödinger equation $\psi(t,x)$ (theorem 3.2) allow the rigorous study of the semiclassical limit of $\psi(t,x)$, i.e. the detailed behavior of the wave function and other quantum mechanical quantities when the Planck constant is regarded as a small parameter converging to 0.

This result is particularly important, as one of the most fascinating features of Feynman path integrals is their power to link, at least heuristically, quantum and classical mechanics. As the formal expression of the Feynman path integral representation for the solution of the Schrödinger equation

$$\psi(t,x) = \int_{\gamma(t)=x} e^{\frac{i}{\hbar}S_t(\gamma)}\psi(0,\gamma(0))d\gamma$$

suggests, according to the stationary phase method as $\hbar \downarrow 0$ the leading contribution to the asymptotic behavior of $\psi(t,x)$ should come from those paths γ which make stationary the action functional $S_t(\gamma)$. These, by Hamilton's least action principle, are exactly the classical orbits of the system.

The problem of the way quantum mechanics, and in particular the solution of the Schrödinger equation, approaches classical mechanics has been studied in different ways by several authors [221, 223, 119, 48, 57, 97, 102, 136, 135, 161, 252, 257, 281, 286] (we only mention some of them without any claim of completeness).

First rigorous results concerning the application of the stationary phase method for infinite dimensional oscillatory (Fresnel) integrals to the study of the semiclassical limit of the solution of the Schrödinger equation can be found in the pioneering paper [16], where it is assumed that the potential belongs to the class $\mathcal{F}(\mathbb{R}^d)$. By applying the general results concerning the infinite dimensional stationary phase method and the theory of Lagrangian manifolds, in the spirit of the geometrical analysis of oscillatory integrals and their asymptotics made in [110], the authors provide an alternative (Feynman path integral) derivation of Maslov's results on the semiclassical asymptotics of the solution of the Schrödinger equation [223]. Part of these results are generalized in [7, 5], by means of slightly different methods, to the case where the potential is the sum of a quadratic function and a smooth function in $\mathcal{F}(\mathbb{R}^d)$, providing not only the leading term, but also the higher order terms of the asymptotic expansion as well as a good control on the remainder. The authors consider a particular but physically relevant form for the initial wave function:

$$\psi_0(x) = e^{\frac{i}{\hbar}S_0(x)}\phi_0(x), \tag{4.57}$$

where S_0 is real and $S_0, \phi_0 \in \mathcal{F}(\mathbb{R}^d)$ are independent of $\hbar$. This initial data corresponds to an initial particle distribution with density $\rho_0(x) = |\phi_0|^2(x)$ and to a limiting value of the density of the probability current $J_{\hbar=0} = S_0'(x)\rho_0(x)/m$, giving an initial particle flux associated to the velocity field $S_0'(x)/m$ (S_0' stands for the gradient of S_0).

In [7, 5] the authors consider the Schrödinger equation

$$i\hbar\frac{\partial}{\partial t}\psi(t,x) = -\frac{\hbar^2}{2m}\Delta\psi(t,x)+\frac{1}{2}x\Omega^2 x\psi(t,x)+V_0(x)\psi(t,x), \quad t>0, x\in\mathbb{R}^d, \tag{4.58}$$

(where Ω^2 is a positive symmetric linear operator in $\mathbb{R}^d$ and $V_0 \in \mathcal{F}(\mathbb{R}^d)$) and provide first of all an infinite dimensional oscillatory integral representation of the solution of the initial value problem, which is similar to that presented in theorem 3.2.

Let us assume that $\det(\cos(t\Omega)) \neq 0$ and let $\beta_{t,x}$ be the unique solution of the boundary value problem

$$\ddot\beta(s) + \Omega^2\beta(s) = 0, \qquad 0 \le s \le t,$$

$$\dot\beta(0) = 0, \quad \beta(t) = x.$$

Let $\mathcal{H}_t$ denote the Cameron-Martin Hilbert space, that is the Sobolev space of absolutely continuous functions $\gamma : [0,t] \to \mathbb{R}^d$, such that $\gamma(t) = 0$, with square integrable weak derivative $\dot\gamma$:

$$\int_0^t |\dot\gamma(s)|^2 ds < \infty,$$

endowed with the inner product

$$\langle\gamma_1,\gamma_2\rangle = \int_0^t \dot\gamma_1(s)\cdot\dot\gamma_2(s)ds.$$

Let $L : \mathcal{H}_t \to \mathcal{H}_t$ be the self-adjoint linear operator defined by equation (3.5). Analogously to what has been done in theorem 3.2, by assuming that the initial datum is of the form (4.57), it is possible to prove [16, 7] that the solution of the Schrödinger equation is given by the infinite dimensional oscillatory integral:

$$\psi^\hbar(t,x) = e^{-\frac{i}{\hbar}x\tan(\Omega t)\Omega x}\widetilde{\int_{\mathcal{H}_t}} e^{\frac{i}{2\hbar}\langle\gamma,(I-L)\gamma\rangle}e^{-\frac{i}{\hbar}\int_0^t V_0(\gamma(s)+\beta_{t,x}(s))ds}$$
$$e^{\frac{i}{\hbar}S_0(\gamma(0)+\beta_{t,x}(0))}\phi_0(\gamma(0)+\beta_{t,x}(0))d\gamma, \tag{4.59}$$

where the superscript $\hbar$ stresses the dependence on the variable $\hbar \in \mathbb{R}^+$. Equation (4.59) is more convenient than representation (3.16) for the study of the semiclassical asymptotics of $\psi^\hbar(t,x)$ as $\hbar \to 0$, as it points out the role of the classical path $\beta_{t,x}$.

For the study of the asymptotic behavior of the integral (4.59) one has first of all to determine the stationary points of the phase functional $\Phi : \mathcal{H}_t \to \mathbb{R}$, which is given by

$$\Phi(\gamma) = \frac{1}{2}\|\gamma\| - \frac{1}{2}\int_0^t \gamma(s)\Omega^2\gamma(s)ds - x\tan(\Omega t)\Omega x - \int_0^t V_0(\gamma(s)+\beta_{t,x}(s))ds + S_0(\gamma(0)+\beta_{t,x}(0)). \quad (4.60)$$

Under the assumption that $V_0, S_0 \in C^2(\mathbb{R}^d)$, it is possible to compute the first and the second Fréchet derivative of the functional Φ:

$$\Phi'(\gamma)(\delta) = \langle\gamma,\delta\rangle - \int_0^t \gamma(s)\Omega^2\delta(s)ds - \int_0^t V_0'(\gamma(s)+\beta_{t,x}(s))\delta(s)ds + S_0'(\gamma(0)+\beta_{t,x}(0))\delta(0), \qquad \gamma,\delta \in \mathcal{H}_t \quad (4.61)$$

$$\Phi''(\gamma)(\delta_1,\delta_2) = \langle\delta_1,\delta_2\rangle - \int_0^t \delta_1(s)\Omega^2\delta_2(s)ds + S_0''(\gamma(0)+\beta_{t,x}(0))\delta_1(0)\delta_2(0) - \int_0^t V_0''(\gamma(s)+\beta_{t,x}(s))\delta_1(s)\delta_2(s)ds, \qquad \gamma,\delta_1,\delta_2 \in \mathcal{H}_t . \quad (4.62)$$

By means of Eq. (4.61) it is not difficult to verify the following result:

Lemma 4.4. [5] *let $V_0, S_0 \in C^2(\mathbb{R}^d)$. Then the function $\Phi : \mathcal{H}_t \to \mathbb{R}$ defined by Eq. (4.60) is of class C^2. Moreover, for $\gamma \in \mathcal{H}_t$, $\Phi'(\gamma) = 0$ iff γ is a solution of*

$$\ddot\gamma(s) + \Omega^2\gamma(s) + V_0'(\gamma(s)+\beta_{t,x}(s)) = 0, \qquad 0 \le s \le t, \quad (4.63)$$

$$\dot\gamma(0) = S_0'(\gamma(0)+\beta_{t,x}(0)), \quad \gamma(t) = 0.$$

Equivalently, $\chi(s) := \gamma(s) + \beta_{t,x}(s)$ is a solution of

$$\ddot\chi(s) + \Omega^2\chi(s) + V_0'(\chi(s)) = 0, \qquad 0 \le s \le t, \quad (4.64)$$

$$\dot\chi(0) = S_0'(\chi(0)), \quad \chi(t) = x.$$

Remark 4.7. One can easily recognize in the function χ the "classical path".

For the study of the degeneracy of the critical points γ_c, one has to determine the spectrum of Hessian of the phase function evaluated in γ_c. Let the operator $A : \mathcal{H}_t \to \mathcal{H}_t$ be defined by

$$I + A = \Phi''(\gamma),$$

i.e. for $\delta_1, \delta_2 \in \mathcal{H}_t$

$$\langle \delta_1, A\delta_2 \rangle = -\int_0^t \delta_1(s)\Omega^2\delta_2(s)ds - \int_0^t V_0''(\chi(s))\delta_1(s)\delta_2(s)ds + S_0''(\chi(0))\delta_1(0)\delta_2(0), \quad (4.65)$$

(χ being the solution of Eq. (4.63)).
The following lemma reduces the computation of the Fredholm determinant of the operator $\Phi''(\gamma)$, to the solution of a finite dimensional Cauchy problem.

Lemma 4.5. *The linear operator $A : \mathcal{H}_t \to \mathcal{H}_t$ defined by Eq. (4.65) is uniquely determined by the following conditions.*
For $\eta \in \mathcal{H}_t$, $A\eta$ is the unique solution to the following boundary value problem:

$$(\ddot{A\eta})(s) = p(s)\eta(s), \qquad 0 \le s \le t,$$

$$(\dot{A\eta})(0) = -Q\eta(0), \qquad (A\eta)(t) = 0,$$

where

$$p(s) = \Omega^2 + V_0''(\chi(s)),$$

$$Q = S_0''(\chi(0)).$$

Moreover

$$\det(I + A) = \det \Phi''(\gamma) = \det K(t),$$

where $K(s)$ is a (matrix valued) solution to the following second order equation

$$\ddot{K}(s) + p(s)K(s) = 0, \qquad s > 0, \qquad (4.66)$$

$$K(0) = I, \qquad \dot{K}(0) = Q,$$

and $\gamma \in \mathcal{H}_t$, $\chi = \gamma + \beta_{t,x}$.

Proof. For a detailed proof see [5], theorem 2.1, which is based on an idea of [109] and to the Hadamard Factorization Theorem. □

Equation (4.66) is a second order linear equation with non-constant coefficients, hence at a first glance its solution could seem an ardours task. On the other hand the following argument provides a simple technique for the construction of the solution.

Let $y(s, x_0)$, for $s \geq 0$, be the unique solution to

$$\ddot{y}(s) + \Omega^2 y(s) + V_0'(y(s)) = 0, \qquad s \geq 0, \tag{4.67}$$

$$y(0) = x_0, \quad \dot{y}(0) = S_0'(x_0).$$

Let

$$Y(s) = Y(s, x_0) := \frac{\partial y}{\partial x_0}(s, x_0).$$

It is simple to verify that $Y(s)$, for $s \geq 0$, is the unique solution to

$$\frac{d^2}{ds^2} Y(s) + (\Omega^2 + V_0''(y(s, x_0)))Y(s) = 0,$$

$$Y(0) = I, \quad \dot{Y}(0) = S_0''(x_0).$$

This implies the following result.

Lemma 4.6. *If* $\gamma \in \mathcal{H}_t$, $\chi = \gamma + \beta_{t,x}$. *Then*

$$\det \Phi''(\gamma) = \det \left(\frac{\partial y}{\partial x_0}(t, x_0) \right).$$

As lemma 4.6 states, given a critical point γ_c of the phase function Φ, the non degeneracy condition, i.e. $\det \Phi''(\gamma_c) \neq 0$ is equivalent to the non vanishing of the quantity $\det \left(\frac{\partial y}{\partial x_0}(t, x_0) \right)$. For the importance of this concept, let us introduce the definition of *focal* and *nonfocal point.*

Definition 4.6. A time-space point (t, x) is called *nonfocal* if for any $x_0 \in \mathbb{R}^d$ and for any solution $y(\cdot, x_0)$ to Eq. (4.67) such that $y(t, x_0) = x$ the following holds:

$$\det \left(\frac{\partial y}{\partial x_0}(t, x_0) \right) \neq 0.$$

If this condition is not satisfied for some $x_0 \in \mathbb{R}^d$, the time-space point (t, x) is called *focal.*

Moreover, by Morse theorem (see [227], theorem 15.1), the index of the operator $\Phi''(\gamma)$, with γ solution of Eq. (4.63), is equal to the Morse-Maslov index of the curve χ solution of Eq. (4.64) (see also [42]).

The following result is an application of theorem 4.4 to the study of the asymptotics of the oscillatory integral (4.59) and handles the case where the phase function has a unique stationary point.

Theorem 4.6. [5] *Let $\psi^\hbar(t,x)$ (given by the oscillatory integral (4.59)) be the unique solution of the Schrödinger equation (4.58) and initial condition ψ_0 of the form (4.57). Assume that*

$$V_0 = \hat{\mu}_0, \quad S_0 = \hat{\sigma}_0, \quad \phi_0 = \hat{\nu}_0 \in \mathcal{F}(\mathbb{R}^d),$$

and for some $K_i, \epsilon_i > 0, i = 1, 2, 3,$

$$\int_{\mathbb{R}^d} |y|^j d|\mu_i|(y) \leq K_i \frac{j!}{\epsilon_i^j}, \qquad j \in \mathbb{N},$$

with $\mu_1 = \mu_0$, $\mu_2 = \sigma_0$, $\mu_3 = \nu_0$.
Let $L : \mathcal{H}_t \to \mathcal{H}_t$ be the operator defined by Eq. (3.5). Assume that

$$t \neq \left(n + \frac{1}{2}\right)\frac{\pi}{\Omega_j}, \qquad n \in \mathbb{N}, j + 1, \ldots, d, \tag{4.68}$$

where Ω_j, for $j = 1, \ldots, d$ are the eigenvalues of the operator Ω. Then $(I - L)^{-1}$ exists and it is bounded. Let us put

$$K = K_1 + K_2, \qquad \epsilon = \frac{min\{\epsilon_1, \epsilon_2\}}{\sqrt{t}}.$$

let us assume that

$$2(3 + 2\sqrt{2})K\|(I - L)^{-1}\|\epsilon^{-2} < 1. \tag{4.69}$$

Then there exists a unique path $\bar{\gamma} \in \mathcal{H}_t$ making stationary the phase function Φ, i.e.

$$\Phi'(\bar{\gamma}) = 0,$$

where Φ is given by Eq. (4.60) and its Frechet derivative by Eq. (4.61). Let $x \in \mathbb{R}^d$ and put $\chi := \bar{\gamma} + \beta_{t,x}$, i.e.

$$\chi(s) = \bar{\gamma}(s) + \beta_{t,x}(s), \qquad 0 \leq s \leq t$$

then χ is a solution of the boundary value problem (4.64). Let $m(\chi)$ denote the Maslov (or Morse) index of the curve χ. then, as $\hbar \downarrow 0$, the following asymptotic formula for the integral (4.59) holds:

$$\begin{aligned}\psi^\hbar(t,x) &= e^{\frac{i}{\hbar}\Phi(\bar{\gamma})} e^{-\frac{i\pi}{2}m} \left|\det \frac{\partial y(t,x_0)}{\partial x_0}\Big|_{x_0=\chi(0)}\right|^{-1/2} \phi_0(\chi(0)) + O(\hbar) \\ &= e^{\frac{i}{\hbar}\left(\int_0^t |\dot{\chi}(s)|^2 ds/2 - \int_0^t \chi(s)\Omega^2\chi(s)ds/2 - \int_0^t V_0(\chi(s)ds + S_0(\chi(0))\right)} \\ &\quad e^{-\frac{i\pi}{2}m} \left|\det \frac{\partial y(t,x_0)}{\partial x_0}\Big|_{x_0=\chi(0)}\right|^{-1/2} \phi_0(\chi(0)) + O(\hbar)\end{aligned} \tag{4.70}$$

where $y(t,x_0)$ is the unique solution of the Cauchy problem (4.67).

Proof. By lemma 3.1, if condition (4.68) is satisfied, the operator L given by Eq. (3.5) is such that $(I-L)^{-1}$ exists and it is bounded. In this case we have the the matrix $\cos(\Omega t)$ is non singular and $\tan(\Omega t)$ is well defined.

Let us put

$$W(\gamma) = \int_0^t V_0(\gamma(s) + \beta_{t,x}(s))ds - S_0(\gamma(0) + \beta_{t,x}(0)), \qquad \gamma \in \mathcal{H}_t. \quad (4.71)$$

By reasoning as in the proof of lemma 3.3, it is easy to verify that $W = \hat{\mu}$, with $\mu \in \mathcal{M}(\mathcal{H}_t)$. Moreover the following holds:

$$\int_{\mathcal{H}_t} \|\gamma\|^j d|\mu|(\gamma) \leq t^{j/2} \int_{\mathbb{R}^d} |y|^j d|\sigma_0|(y) + \frac{t^{j/2+1}}{j/2+1} \int_{\mathbb{R}^d} |y|^j d|\mu_0|(y),$$

and

$$\int_{\mathcal{H}_t} \|\gamma\|^j d|\mu|(\gamma) \leq K \frac{j!}{\epsilon^j}. \quad (4.72)$$

Since the integral $\psi^\hbar(t,x)$ can be written as:

$$\psi^\hbar(t,x) = e^{-\frac{i}{\hbar} x \tan(\Omega t)\Omega x} \widetilde{\int_{\mathcal{H}_t}} e^{\frac{i}{2\hbar}\langle \gamma, (I-L)\gamma \rangle} e^{-\frac{i}{\hbar}W(\gamma)} g(\gamma) d\gamma, \quad (4.73)$$

with

$$g(\gamma) = \phi_0(\gamma(0) + \beta_{t,x}(0)), \quad (4.74)$$

and inequality (4.69) holds by assumption, we can apply theorem 4.4 and conclude that there exists a unique stationary point $\bar{\gamma} \in \mathcal{H}_t$ of the phase function and

$$\psi^\hbar(t,x) = e^{\frac{i}{\hbar}\Phi(\bar{\gamma})} (\det \Phi''(\bar{\gamma}))^{-1/2} g(a) + O(\hbar) \quad (4.75)$$

where

$$\Phi(\bar{\gamma}) = \int_0^t |\dot{\chi}(s)|^2 ds/2 - \int_0^t \chi(s)\Omega^2\chi(s)ds/2 - \int_0^t V_0(\chi(s))ds + S_0(\chi(0)).$$

The final result follows from lemma 4.6 □

Analogously to what is done in theorem 4.4, where the complete asymptotic expansion of the oscillatory integral $I(\hbar)$ in powers of $\hbar$ is provided as well as a good control on the remainder, it is possible to compute not only the first term in the expansion of $\psi^\hbar(t,x)$. Indeed it is possible to prove (see [5, 7] in the case where the phase function has a unique stationary point $\bar{\gamma}$, that

$$\psi^\hbar(t,x) = e^{\frac{i}{\hbar}\Phi(\bar{\gamma})} I^*(\hbar),$$

with $I^*(\hbar)$ being a C^∞ function of $\hbar \in \mathbb{R}$, and

$$I^*(\hbar) = \sum_{j=0}^{n-1} \alpha_j \hbar^j + R_n(\hbar),$$

where the coefficients α_j of the expansions depend on Ω as well as the derivatives of V_0, S_0, ϕ_0, and

$$|R_n(\hbar)| \leq C_n |\hbar|^n$$

for suitable coefficients C_n and $|\hbar|$ sufficiently small.

The assumption (4.69) in theorem 4.6 assures that the phase function Φ has a unique stationary point. Following [5], it is worthwhile to give also some weaker conditions assuring that the set of critical points is not empty and finite.

Lemma 4.7. *Let us assume that the operator $(I - L)$ is invertible and the functions V_0, S_0 are C^1 with bounded derivative. Then the number of non degenerate stationary points of the phase function Φ is finite.*

Lemma 4.8. *Let us assume that the functions V_0, S_0 are bounded C^1 functions with bounded derivatives. Then there exists at least a stationary point of the phase function Φ*

For a proof of both lemmas we refer to the original paper [5]. The next theorem handles the case where the phase function Φ has a finite number of non degenerate stationary points and can be seen as an application of theorem 4.5 (see also theorems 3.7, 4.1 and 5.4 in [7]).

Theorem 4.7. [5] *Let us assume that*

$$V_0 = \hat{\mu}_0, \quad S_0 = \hat{\sigma}_0, \quad \phi_0 = \hat{\nu}_0 \in \mathcal{F}(\mathbb{R}^d),$$

and for some $K_i, \epsilon_i > 0, i = 1, 2, 3$,

$$\int_{\mathbb{R}^d} |y|^j d|\mu_i|(y) \leq K_i \frac{j!}{\epsilon_i^j}, \qquad j \in \mathbb{N},$$

with $\mu_1 = \mu_0$, $\mu_2 = \sigma_0$, $\mu_3 = \nu_0$.
Let $L : \mathcal{H}_t \to \mathcal{H}_t$ be the operator defined by Eq. (3.5). Assume that

$$t \neq \left(n + \frac{1}{2}\right)\frac{\pi}{\Omega_j}, \qquad n \in \mathbb{N}, j + 1, \ldots, d, \tag{4.76}$$

where Ω_j, for $j = 1, \ldots, d$ are the eigenvalues of the operator Ω. Then $(I - L)^{-1}$ exists and it is bounded.
Let the point (t, x) be nonfocal.

Then there exists a finite number of solutions $\chi_1, \dots, \chi_n$ to the problem 4.64 and the solution $\psi^\hbar(t,x)$ of the Schrödinger equation (4.58) has the following asymptotic representation as $\hbar \downarrow 0$:

$$\psi^\hbar(t,x) = \sum_{j=1}^{n} e^{\frac{i}{\hbar}\left(\int_0^t |\dot{\chi}_j(s)|^2 ds/2 - \int_0^t \chi_j(s)\Omega^2\chi_j(s)ds/2 - \int_0^t V_0(\chi_j(s)ds + S_0(\chi_j(0))\right)}$$

$$e^{-\frac{i\pi}{2}m_j}\left|\det \frac{\partial y_j(t,x_0)}{\partial x_0}\right|^{-1/2}\phi_0(\chi_j(0)) + O(\hbar). \quad (4.77)$$

If some critical point of the phase function is degenerate, or, in other words, if the time-space point (t,x) is focal, then the problem can be solved by reducing the study of the degeneracy to a finite dimensional subspace of $\mathcal{H}_t$, in the way described at the end of section 4.3.

Particular examples are handled in [5, 6]. In the next section we shall describe the application of this technique to the study of the trace of the Schrödinger group. We present briefly here some results concerning the solution of the Schrödinger equation.

Let us assume that the space dimension d is less or equal then 3 and that the following inequalities are satisfied

$$\det \sin(\Omega t) \neq 0 \qquad \det \cos(\Omega t) \neq 0. \quad (4.78)$$

By lemma 3.1, the second condition assures that the operator $(I - L)$ is invertible. Let us consider the solution of the Schrödinger equation given in terms of the infinite dimensional oscillatory integral $\psi^\hbar(t,x)$ (Eq. (4.59)).

Let us introduce the subspaces of $\mathcal{H}_t$:

$$Y = \{\gamma \in \mathcal{H}_t : \gamma(0) = 0\},$$
$$Z = \{\gamma \in \mathcal{H}_t : \gamma(s) = \sin\Omega(s-t)z \text{ for some } z \in \mathbb{R}^d\}. \quad (4.79)$$

It is easy to verify that $(I - L)Y \subset Z^\perp$. Let us denote by Π_Y and Π_Z the orthogonal projections of $\mathcal{H}_t$ onto Y and Z and by $T_1 : Y \to Y$ and $T_2 : Z \to Z$ the linear operators defined by

$$T_1 := \Pi_Y \circ T_{|Y}, \qquad T_2 := \Pi_Z \circ T_{|Z},$$

where $T : \mathcal{H}_t \to \mathcal{H}_t$ is given by $T = I - L$.

By applying the Fubini theorem 2.7 to the oscillatory integral (4.59), or equivalently (4.73), we have

$$\widetilde{\int_{\mathcal{H}_t}} e^{\frac{i}{2\hbar}\langle\gamma,T\gamma\rangle} e^{-\frac{i}{\hbar}W(\gamma)} g(\gamma)d\gamma$$

$$= C_T \widetilde{\int_Z} e^{\frac{i}{2\hbar}\langle\zeta,T_2\zeta\rangle} \widetilde{\int_Y} e^{\frac{i}{2\hbar}\langle\eta,T_1\eta\rangle} e^{-\frac{i}{\hbar}W(\zeta+\eta)} g(\zeta+\eta)d\eta d\zeta \quad (4.80)$$

where $W : \mathcal{H}_t \to \mathbb{R}$ is given by Eq. (4.71) and $g : \mathcal{H}_t \to \mathbb{C}$ by Eq. (4.74), while the constant C_T is given by

$$C_T = (\det T)^{-1/2}(\det T_1)^{1/2}(\det T_2)^{1/2}$$

and it is equal to

$$C_T = 2\big(\sin^2 \Omega t(\Omega t)^{-1}(2\Omega t + \sin 2\Omega t)^{-1}\big)^{1/2}.$$

As $\eta(0) = 0$, equation (4.80) becomes

$$\widetilde{\int_{\mathcal{H}_t}} e^{\frac{i}{2\hbar}\langle\gamma, T\gamma\rangle} e^{-\frac{i}{\hbar}W(\gamma)} g(\gamma) d\gamma$$
$$= C_T \widetilde{\int_Z} e^{\frac{i}{2\hbar}\langle\zeta, T_2\zeta\rangle} e^{\frac{i}{\hbar}S(\zeta)} \Big(\widetilde{\int_Y} e^{\frac{i}{2\hbar}\langle\eta, T_1\eta\rangle} e^{-\frac{i}{\hbar}V(\zeta+\eta)} d\eta\Big) g(\zeta) d\zeta \quad (4.81)$$

where $S(\zeta) = S_0(\zeta(0) + \beta_{t,x}(0))$, and $V(\zeta + \eta) = \int_0^t V_0(\zeta(s) + \eta(s) + \beta_{t,x}(s))ds$. The key point is the fact that the integral over Z is a finite dimensional oscillatory integral and the classical results of the stationary phase method for finite dimensional oscillatory integrals apply [164].

By assuming that the functions $V_\zeta : Y \to \mathbb{R}$, with $\zeta \in Z$, given by

$$V_\zeta(\eta) := \int_0^t V_0(\zeta(s) + \eta(s) + \beta_{t,x}(s))ds, \qquad \eta \in Y,$$

satisfy assumptions analogous to Eq. (4.40) and Eq. (4.42), i.e.

$$\int_Y \|\eta\|^j d|\mu_\zeta|(\eta) \le L\frac{j!}{\epsilon^j},$$

$$\frac{L\|T_1^{-1}\|}{\epsilon^2} 2(3 + 2\sqrt{2}) < 1,$$

uniformly in $\zeta \in Z$, one has that the phase function $\Phi_\zeta : Y \to \mathbb{R}$, given by

$$\Phi_\zeta(\eta) = \frac{1}{2}\langle\eta, T_1\eta\rangle - V(\eta + \zeta), \qquad \eta \in Y,$$

has a unique non degenerate stationary point $a(\zeta) \in Y$. Moreover

$$\widetilde{\int_Y} e^{\frac{i}{2\hbar}\langle\eta, T_1\eta\rangle} e^{-\frac{i}{\hbar}V(\zeta+\eta)} d\eta = e^{\frac{i}{\hbar}\Phi_\zeta(a(\zeta))} J^*(\hbar, \zeta),$$

with J^* being a C^∞ function of $\hbar \in \mathbb{R}$ such that

$$J^*(0, \zeta) = (\det \Phi''_\zeta(a(\zeta)))^{-1/2}.$$

The solution of the Schrödinger equation is then given in terms of the following oscillatory integral on a d dimensional space Z:

$$\psi^\hbar(t,x) = C_T e^{-\frac{i}{\hbar}x\tan(\Omega t)\Omega x} \widetilde{\int_Z} e^{\frac{i}{2\hbar}\langle\zeta,T_2\zeta\rangle} e^{\frac{i}{\hbar}S(\zeta)} e^{\frac{i}{\hbar}\Phi_\zeta(a(\zeta))} J^*(\hbar,\zeta)g(\zeta)d\zeta \tag{4.82}$$

with a phase function $\Phi : Z \to \mathbb{R}$ equal to

$$\Phi(\zeta) = \frac{1}{2}\langle\zeta, T_2\zeta\rangle + S(\zeta) + \Phi_\zeta(a(\zeta)), \qquad \zeta \in Z \tag{4.83}$$

and the theory of asymptotic expansions of finite dimensional oscillatory integrals (with degenerate critical points) [110, 41] can be applied. In particular, if Φ has a degenerate critical point, i.e. if the time-space point (t,x) is focal, and if the space dimension d is less or equal then 3, then there exists a complete classification of the types of the possible degeneracies [144] and the asymptotics of the integral (4.82) is of the form.

$$\psi^\hbar(t,x) \sim \hbar^{-\beta} \qquad \hbar \downarrow 0,$$

where $\beta \in \mathbb{Q}$, $\beta > 0$, is called Coxeter number and depends on the type of degeneracy [43].

4.5 The trace formula

The study of the connections between quantum and classical quantities is very old and goes back to the very origin of the quantum theory: it is sufficient to think for instance to the Bohr quantization rules, relating the spectrum of the energy operator to the volume in the phase space enclosed by the classical periodic orbits of the system. The interest in this kind of relations has been renewed in recent years since, according to the theory of quantum chaos, the study of the distribution of the energy eigenvalues of a given quantum mechanical system should reflect that the underlying classical system is integrable (resp. chaotic) [152]. Particularly interesting is an (heuristic) trace formula connecting the semiclassical asymptotics for $\hbar \downarrow 0$ of trace of the Schrödinger group $\mathrm{Tr}(e^{-\frac{i}{\hbar}Ht})$ to the classical periodic orbits of the system. One can look at this as a quantum analogue of Selberg trace formula, relating the trace of the heat kernel on manifolds of constant negative curvature with the periodic geodesics [77, 167].

The first rigorous Feynman path integral derivation of the trace formula for the Schrödinger group and the study of its singularities as a function of the time variable can be found in [4], see also [2, 3]. Part of those results are generalized in [8, 6], where it is shown that in this particular problem the degenerate stationary points of the phase function of the Feynman integral play a fundamental role. Recently in [33, 34] some interesting results on the degenerate case have been applied to the study of a trace formula for the heat semigroup with a polynomial potential. In the present section we give some details of the results of [8, 6].

Let V_0 be a bounded real function belonging to $\mathcal{F}(\mathbb{R}^d)$ and $\Omega^2 > 0$ a positive symmetric $d \times d$ matrix and let us consider an anharmonic oscillator Hamiltonian of the following form

$$H = -\frac{\hbar^2}{2}\Delta + \frac{1}{2}x\Omega^2 x + V_0(x),$$

with domain

$$D(H) = \{\psi \in W^{2,2}(\mathbb{R}^d), \int_{\mathbb{R}^d} |x|^4 |\psi(x)|^2 dx < \infty\}.$$

It is well known [246] that H is a self-adjoint operator on $L^2(\mathbb{R}^d)$, it has a pure point spectrum and, by applying theorem 10.5 in [263] and Tauberian theorem, it follows that for some $\beta > 0$ its eigenvalues λ_n satisfy the following inequality:

$$\liminf_{n\to\infty} \frac{\lambda_n}{n^\beta} > 0.$$

This relation implies that the trace of the unitary group generated by iH, i.e. $\mathrm{Tr}(e^{-\frac{i}{\hbar}Ht})$ is, as a function of t, a well defined distribution over $\mathbb{R}$.

Let $\mathcal{H}_{p,t}$ be the Hilbert space of periodic functions $\gamma \in H^1(0,t;\mathbb{R}^d)$ such that $\gamma(0) = \gamma(t)$, with norm

$$\|\gamma\|^2 = \int_0^t \dot\gamma(s)^2 ds + \int_0^t \gamma(s)^2 ds.$$

Let $\Phi : \mathcal{H}_{p,t} \to \mathbb{R}$ denote the phase function

$$\Phi(\gamma) = \frac{1}{2}\int_0^t \dot\gamma(s)^2 s - \int_0^t V_1(\gamma(s))ds, \qquad \gamma \in \mathcal{H}_{p,t}, \tag{4.84}$$

where

$$V_1(x) = \frac{1}{2}x\Omega^2 x + V_0(x), \qquad x \in \mathbb{R}^d.$$

Let us consider the infinite dimensional oscillatory integral on $\mathcal{H}_{p,t}$

$$I(t,\hbar) = \widetilde{\int_{\mathcal{H}_{p,t}}} e^{\frac{i}{\hbar}\Phi(\gamma)} d\gamma. \tag{4.85}$$

The integral (4.85) can also be written in the following form

$$I(t,\hbar) = \widetilde{\int_{\mathcal{H}_{p,t}}} e^{\frac{i}{2\hbar}\langle\gamma,(I-L)\gamma\rangle} e^{-\frac{i}{\hbar}V(\gamma)} d\gamma, \tag{4.86}$$

where $L : \mathcal{H}_{p,t} \to \mathcal{H}_{p,t}$ is a self-adjoint trace class operator, given by:

$$\langle\gamma, L\gamma\rangle = \int_0^t \gamma(s)\Omega^2\gamma(s)ds + \int_0^t |\gamma(s)|^2 ds, \qquad \gamma \in \mathcal{H}_{p,t} \tag{4.87}$$

and $V : \mathcal{H}_{p,t} \to \mathbb{R}$ is equal to

$$V(\gamma) = \int_0^t V_0(\gamma(s))ds, \qquad \gamma \in \mathcal{H}_{p,t}.$$

As $V_0 \in \mathcal{F}(\mathbb{R}^d)$, it is not difficult to verify that $I(t,\hbar)$ is well defined, provided that $(I - L) : \mathcal{H}_{p,t} \to \mathcal{H}_{p,t}$ is an invertible operator on $\mathcal{H}_{p,t}$. Moreover the following holds.

Theorem 4.8. [6] *Let us assume that the time variable t satisfies the following condition*

$$\det \sin\left(\frac{t}{2}\Omega\right) \neq 0.$$

Then the infinite dimensional oscillatory integral (4.85) *is well defined and the function* $t \mapsto I(t,\hbar)$ *is of class* C^∞ *on the subset* D_Ω *of the real line defined by*

$$D_\Omega = \{t : \det \sin\left(\frac{t}{2}\Omega\right) \neq 0\}.$$

Moreover the trace of the Schrödinger group $\mathrm{Tr}(e^{-\frac{i}{\hbar}Ht})$ *is a well defined distribution over* $(0,\infty)$, *a* C^∞ *function on* D_Ω *and it is equal to*

$$\mathrm{Tr}(e^{-\frac{i}{\hbar}Ht}) = \big(2(\cosh t - 1)\big)^{-d/2} I(t,\hbar). \tag{4.88}$$

Proof. For a detailed proof we refer to [6]. The definition and the regularity properties of the oscillatory integral $I(t,\hbar)$ can be proved by means of the general theory (see chapter 2, in particular theorem 2.5). The proof of equation (4.88) is based on an analytic continuation technique and on the proof of an analogous equality for the trace of the corresponding heat semigroup. □

According to Eq. (4.88), the study of the semiclassical limit of the trace of the Schrödinger group $\mathrm{Tr}(e^{-\frac{i}{\hbar}Ht})$ is reduced to the study of the asymptotic behavior of the integral in the limit $\hbar \downarrow 0$.

If $V_1 : \mathbb{R}^d \to \mathbb{R}$ is of class C^2, then one proves that the functional Φ is of class C^2 and a path $\gamma \in \mathcal{H}_{p,t}$ is a stationary point for Φ if and only if γ is a solution of the Newton equation

$$\ddot{\gamma}(s) + V_1'(\gamma(s)) = 0, \qquad s \in [0,t] \tag{4.89}$$

satisfying the periodic conditions

$$\gamma(0) = \gamma(t), \qquad \dot{\gamma}(0) = \dot{\gamma}(t). \tag{4.90}$$

Concerning the second Frechet derivative of the phase function Φ, it is possible to prove that, given a vector $\gamma \in \mathcal{H}_{p,t}$, the operator $\Phi''(\gamma) : \mathcal{H}_{p,t} \to \mathcal{H}_{p,t}$ is defined by $\Phi''(\gamma) = I + L_\gamma$, where $u = L_\gamma \phi$ if and only if u satisfies Eq. (4.90) and

$$\ddot{u}(s) - u(s) = \big(V_1''(\gamma(s)) + 1\big)\phi(s), \qquad 0 < s < t. \tag{4.91}$$

The symmetry of L_γ can be easily verified, moreover the regularity of the solutions of Eq. (4.91) implies that the range of L_γ is contained in $H^3(0,t;\mathbb{R}^d)$ and L_γ is a trace-class operator.

A first example of stationary path of the phase Φ can easily be found if the potential V_1 admits critical points c_j, $j = 1, \dots, m$. Indeed in this case the functions γ_{c_j}, given by

$$\gamma_{c_j}(s) = c_j, \qquad s \in [0,t],$$

is a periodic solution of Eq. (4.89). Moreover the Fredholm determinant of $\Phi''(\gamma_{c_j})$ is given by (see [6], proposition 2.11)

$$\det \Phi''(\gamma_{c_j}) = \det\Big((\cosh t - 1)^{-1}\big(\cos(t\sqrt{V_1''(c_j)}) - 1\big)\Big).$$

From this equality we can easily see that the stationary point γ_{c_j} is non degenerate if and only if $\cos(t\sqrt{V_1''(c_j)}) - 1 \neq 0$.

Besides the "trivial" solution of the equations (4.89) and (4.90), there can be also other stationary points of Φ, that are degenerate. Indeed the following holds

Theorem 4.9. *Let $\gamma \in \mathcal{H}_{p,t}$ be a non-constant solution of Eq. (4.89) and Eq. (4.90). Then* $\mathrm{Ker}(\Phi''(\gamma)) \neq \{0\}$. *In particular:*

$$\Phi''(\gamma)(\dot{\gamma}) = 0.$$

Proof. [111, 6] Let us introduce the C_0-group of linearly unitary transformations in $\mathcal{H}_{p,t}$:

$$G_\tau \gamma(s) = \tilde{\gamma}(s + \tau), \qquad s \in [0,t],\ \tau \in \mathbb{R},\ \gamma \in \mathcal{H}_{p,t},$$

where $\tilde{\gamma}$ is the unique extension of γ to a t-periodic continuous function on $\mathbb{R}$. One can see that Φ is invariant under the action of G_τ, i.e. $\Phi \circ G_\tau = \Phi$. Let us define $\gamma_\tau := G_\tau \gamma$ and by the chain rule we have:

$$\Phi'(\gamma_\tau) \circ G_\tau = d_\gamma(\Phi \circ G_\tau) = d_\gamma(\Phi) = 0.$$

From this we can conclude that $\Phi'(\gamma_\tau) = 0$. Hence

$$\lim_{s \to 0} \frac{1}{s} \Phi'(\gamma_s) - \Phi'(\gamma) = 0,$$

but the last limit is equal to

$$\Phi''(\gamma)\Big(\frac{d}{ds}\gamma(s)_{|s=0}\Big) = \Phi''(\gamma)(\dot{\gamma}).$$

□

The proof of theorem 4.9 shows that if $\gamma \in \mathcal{H}_{p,t}$ is a non-constant solution of Eq. (4.89) and Eq. (4.90), then γ is not only a degenerate stationary point, but it is also non isolated as for each $\tau \in [0, t]$ the path γ_τ is also a degenerate stationary point of Φ. In particular, the set of degenerate stationary points

$$\{\gamma_\tau,\ \tau \in [0, t]\}$$

is a compact manifold diffeomorphic to the circle S^1.

The following result is an extension to this setting of the classical Morse theorem about non-degenerate critical points.

Theorem 4.10. [6] *Let $\gamma \in \mathcal{H}_{p,t}$ be a degenerate stationary point of Φ satisfying the following condition*

$$\dim \operatorname{Ker} \Phi''(\gamma) = 1. \qquad (4.92)$$

then the set

$$\{\gamma_\tau,\ \tau \in [0, t]\}$$

is isolated within the set of all stationary points of Φ, i.e., there is $\epsilon > 0$ such that if $\phi \in \mathcal{H}_{p,t}$ satisfies $\|\phi - \gamma_\tau\| < \epsilon$ for some $\tau \in \mathbb{R}$, and $\Phi'(\phi) = 0$, then $\phi = \gamma_s$ for some $s \in \mathbb{R}$.

Proof. Let

$$Z = \operatorname{Ker} \Phi''(\gamma) = \{\alpha\dot{\gamma},\ \alpha \in \mathbb{R}\} \subset \mathcal{H}_{p,t}$$

and let $Y := Z^\perp$ be the orthogonal complement of Z in $\mathcal{H}_{p,t}$. By equation (4.91), the operator $\Phi''(\gamma)$ is compact and self-adjoint. By the Fredholm

alternative theorem it follows that $\Phi''(\gamma)$ maps injectively Y onto Y.
Let $F : Y \to \mathbb{R}$ be defined by

$$F(\psi) = \Phi(\gamma + \psi), \qquad \psi \in Y.$$

Denoted by $j : Y \to \mathcal{H}_{p,t}$ the natural embedding map and by $j^* : \mathcal{H}_{p,t} \to Y$ its dual operator, we have that $F''(0) = j^* \circ \Phi'' \circ j$ is a linear isomorphism of Y. By the implicit function theorem, there is an $\epsilon_0 > 0$ such that for any $\eta \in Y$, with $\|\eta\| < \epsilon_0$, we have $F'(\eta) \neq 0$. By the tubular neighborhood theorem [228], the set

$$\tilde{W}_{\epsilon_0} := \cup_{s \in \mathbb{R}} \{\gamma_s + G_s \eta : \eta \in Y, \|\eta\| < \epsilon_0\}$$

is an open neighborhood of the manifold $\hat{\gamma} := \{\gamma_\tau,\ \tau \in [0,t]\}$, and there exists an $\epsilon \in (0, \epsilon_0)$ such that the open neighborhood $W_\epsilon(\hat{\gamma})$

$$W_\epsilon(\hat{\gamma}) := \{\xi \in \mathcal{H}_{p,t} : dist(\xi, \hat{\gamma}) < \epsilon\}$$

is included in $\tilde{W}_{\epsilon_0}$. It is then possible to see that $\Phi' \neq 0$ on $W_\epsilon(\hat{\gamma})$. Indeed if $\Phi(\psi) = 0$ for $\psi = \gamma_s + G_s \eta$ for some $s \in \mathbb{R}$ and some $\eta \in Y$ such that $\|\eta\| < \epsilon$, then, since $\Phi \circ G_{-s} = \Phi$, we have that $\Phi'(G_{-s}(\gamma_s + G_s \eta)) = 0$. On the other hand, $G_{-s}(\gamma_s + G_s \eta) = \gamma + \eta$, $G_{-s}(\eta) \in Y$. Therefore $F'(\eta) = 0$ and this gives $\eta = 0$. □

The results stated so far concerning the stationary points of the phase function $\Phi : \mathcal{H}_{p,t} \to \mathbb{R}$ are valid for a general Φ of the form (4.84). Let us now consider the particular case where the potential is of the following form

$$V_1(x) = \frac{1}{2} x \Omega^2 x + V_0(x), \qquad x \in \mathbb{R}^d, \tag{4.93}$$

with V_0 real bounded and $V_0 \in \mathcal{F}(\mathbb{R}^d)$.

In [6] V_1 is also assumed to satisfy the following conditions:

(1) V_1 has a finite number critical points $c_1, \ldots, c_m$, and each of them is non degenerate, i.e.

$$\det V_1''(c_j) \neq 0, \qquad j = 1, \ldots m.$$

(2) $t > 0$ is such that the function γ_{c_j}, given by $\gamma_{c_j}(s) = c_j$, $s \in [0,t]$, is a non degenerate stationary point for Φ;
(3) any non constant t−periodic solution γ of Eq. (4.89) and Eq. (4.90) is a "non degenerate periodic solution", in the sense of [111], i.e. $dim\ ker\ \Phi''(\gamma) = 1$.

Then by theorems 4.9 and 4.10, the set M of stationary points of the phase function Φ is a disjoint union of the following form:

$$M = \{x_{c_1}, \dots, x_{c_m}\} \cup \bigcup_{k=1}^{r} M_k,$$

where x_{c_i}, $i = 1, \dots m$, are non degenerate and M_k are manifolds (diffeomorphic to S^1) of degenerate stationary points, on which the phase function is constant. Under some growth conditions on V_0 it is also possible to compute the asymptotic behavior as $\hbar \to 0$ of the trace of the Schrödinger group, or, equivalently (see Eq. (4.88)), of the infinite dimensional oscillatory integral $I(t, \hbar)$.

Theorem 4.11. [6]*Let us assume that the potential* V_1 *is of the form* (4.93), *with* $V_0 \in \mathcal{F}(\mathbb{R}^d)$, $V_0 = \hat{\mu}_0$, *satisfying*

$$\int_{\mathbb{R}^d} |y|^j d|\mu_0|(y) \le K \frac{j!}{\epsilon^j}, \qquad j \in \mathbb{N},$$

for some $K, \epsilon > 0$. *Then, as* $\hbar \downarrow 0$, *the trace of the Schrödinger group has the following asymptotic behavior*

$$\mathrm{Tr}(e^{-\frac{i}{\hbar}Ht}) = \sum_{j=1}^{m} e^{\frac{i}{\hbar}tV_1(c_j)} I_j^*(\hbar) + (2\pi i\hbar)^{-1/2} \Big[\sum_{k=1}^{r} e^{\frac{i}{\hbar}\Phi(b_k)} |M_k| I_k^{**}(\hbar) + O(\hbar) \Big]$$

where c_j *are the points in condition 1,* $b_k \in M_k$ *are all non constant* t-*periodic solutions of* (4.89) *and* (4.90) *as in condition (3),* $|M_k|$ *is the Riemannian volume of* M_k, I_j^* *and* I_k^{**} *are* C^∞ *functions of* $\hbar \in \mathbb{R}$ *such that, in particular,*

$$I_j^*(0) = \Big(\det \Big[2 \Big[\cos \Big(t\sqrt{V''(c_j)} \Big) - 1 \Big] \Big] \Big)^{-1/2},$$

$$I_k^{**}(0) = \Big(\frac{d}{d\epsilon} \det(R_\epsilon^k(t) - I)|_{\epsilon=1} \Big)^{-1/2},$$

where $R_\epsilon^k(t)$ *denotes the fundamental solution of*

$$\begin{cases} \ddot{x}(s) = -\epsilon V''(b_k(s))x(s), & s > 0, \\ x(0) = x_0, \quad \dot{x}(0) = y_0 \end{cases}$$

written as a first order system of $2d$ *equations for real valued functions.*

Proof. For a detailed proof we refer to the paper [6], we give here only some hints. By equation (4.88), one has to study the asymptotic behavior of the oscillatory integral $I(t, \hbar)$. By the condition on the potential V_1, the set of stationary points of the phase function is completely determined.

Moreover it is possible to reduce the study of the degeneracy to a finite dimensional subspace of the Hilbert space $\mathcal{H}_{p,t}$ and to apply the technique described at the end of section 4.3 (see also [7]).

Indeed it is possible to find a finite dimensional subspace $Z \subset \mathcal{H}_{p,t}$, such that, given $Y := T(Z)^{\perp}$ with $T = (I - L)$ and L is given by Eq. (4.87), one has $Z + Y = \mathcal{H}_{p,t}$, and the phase function when restricted to y

$$\Phi_z : Y \to \mathbb{R}, \qquad \Phi_z(y) = \Phi(z + y),$$

has a unique stationary point, denoted by $a(z)$. Moreover the function $a : Z \to Y$, $z \mapsto a(z)$, is of class C^∞ and bounded with all its derivatives. By the Fubini theorem 2.7 for oscillatory integrals, $I(t, \hbar)$ is given by

$$I(t,\hbar) = C_T \widetilde{\int_Z} \Big(\widetilde{\int_Y} e^{\frac{i}{\hbar}\Phi_z(y)} dy \Big) dz,$$

with $C_T = (\det T)^{-1/2} (\det T_{|Y})^{1/2} (\det T_{|Z})^{1/2}$.

By theorem 4.4, the integral over Y is given by

$$\widetilde{\int_Y} e^{\frac{i}{\hbar}\Phi_z(y)} dy = e^{\frac{i}{\hbar}\Phi_z(a(z))} J^*(\hbar, z),$$

with J^* being a C^∞ function of both $\hbar \in (-1, 1)$ and $z \in Z$, such that

$$J^*(0, z) = \det \Big(\frac{\partial^2}{\partial y^2} \Phi_z(a(z)) \Big)^{-1/2}.$$

Therefore, as $\hbar \to 0$,

$$I(t, \hbar) = C_T \widetilde{\int_Z} \det \Big(\frac{\partial^2}{\partial y^2} \Phi_z(a(z)) \Big)^{-1/2} e^{\frac{i}{\hbar}\Phi_z(a(z))} dz + O(\hbar),$$

and the asymptotic behavior of $I(t, \hbar)$ is determined by the reduced phase function $\Psi : Z \to \mathbb{R}$, given by

$$\Psi(z) = \Phi(z + a(z)), \qquad z \in Z.$$

One can prove (see [7], lemma 5.3) that $\Psi'(z) = 0$ iff $\Phi'(z + a(z)) = 0$ and

$$\ker(\Phi''(z + a(z)) = (I_Z, a'(z))(\ker \Psi''(z)).$$

Denoting by Q the projection from $\mathcal{H}_{p,t}$ onto Z along y, i.e. $Q(z + y) = z$, the set N of stationary points of the phase Ψ is the projection $Q(M)$ of the stationary points of Φ:

$$N = \{d_1, \ldots, d_m\} \cup \bigcup_{k=1}^{r} Q(M_k),$$

with $d_1, \ldots, d_m$ are non degenerate and $Q(M_k)$ for $j = k, \ldots, r$ are compact and connected one dimensional manifolds such that $\dim \ker \Psi''(a) = 1$ for each $a \in \cup_{k=1}^r Q(M_k)$. The final results follows by applying the stationary phase method on finite dimensional oscillatory integrals (see [6]). □

Chapter 5

Open Quantum Systems

5.1 Feynman path integrals and open quantum systems

Our treatment of the Feynman path integral formulation of (non relativistic) quantum mechanics has started with the description of the dynamics of a pointwise particle, described by the Schrödinger equation (1.4). The generalization of the formalism developed so far to more complicated and more realistic quantum systems can be realized, on one hand, by increasing the number of degree of freedom and by considering more complicated Hamiltonian operators H. On the other hand in many situations occurring in quantum mechanics, one is not interested in the detailed description of the whole system (that, when the number of degrees of freedom increases, becomes a very difficult task), but only of a part of it. Let us think for instance to the interaction of a radiating atom with the electromagnetic field when one is interested only in the state of the atom and not in the emitted radiation. Another fundamental problem of this kind is the description of the process of quantum measurement. Indeed when a quantum system is submitted to the measurement of one of its observables, it interacts with a (macroscopic) measuring apparatus which is not of primary interest but, on the other hand, whose influence on the system, by the Heisenberg uncertainty principle, cannot be neglected.

The development of a formalism allowing the quantum description of the dynamic of a system interacting with an external "environment" is the task of the quantum theory of open systems [92, 116]. A possible approach is the description of the compound "system plus environment" as a whole, by means of a Schrödinger equation involving an interaction Hamiltonian. As a second step, the environment's degrees of freedom have to be traced out.

An alternative solution to the problem is provided by the path integral formalism. This chapter concerns a few instances of the descriptive power of Feynman path integrals when applied to the theory of open quantum systems. Let us briefly introduce some heuristic considerations which motivate a further detailed treatment.

The simplest example of a system subjected to an external influence is a classical one dimensional linearly forced harmonic oscillator, whose equation of motion is

$$m\ddot{x}(t) + kx(t) + f(t) = 0, \qquad t \in \mathbb{R}, \tag{5.1}$$

where $m, k \in \mathbb{R}^+$ and f is a "fluctuating force". The quantization of the classical system (5.1) in terms of Feynman path integrals provides for the Green function of the corresponding Schrödinger equation an heuristic formula of the form

$$G(0, y, t, x) = \int_{\substack{\gamma(t)=x \\ \gamma(0)=y}} e^{\frac{i}{\hbar}\left(\frac{m}{2}\int_0^t \dot{\gamma}(s)^2 ds - \frac{k}{2}\int_0^t \gamma(s)^2 ds - \int_0^t \gamma(s) f(s) ds\right)} D\gamma. \tag{5.2}$$

As a further step one can consider the external force f to be random, and be interested to the averaged value over f of the quantum dynamics given by Eq. (5.2). In order to obtain a physically meaningfull expression, the average has to be done on quantum observables, as for instance the transition amplitudes, rather than on mathematical objects, as the wave functions. Therefore let us consider two state vectors $\psi, \phi \in L^2(\mathbb{R}^d)$ and compute the probability of a transition from ψ to ϕ, i.e. $|\langle \phi, \psi(t) \rangle|^2$, where $\psi(t, x) = \int G_f(0, y, t, x)\psi(y)dy$,

$$|\langle \phi, \psi(t) \rangle_f|^2 = \int\int\int\int \phi(x')\bar{\phi}(x) G_f(0, y, t, x)\overline{G_f(0, y', t, x')}\psi(y) \bar{\psi}(y')dydxdy'dx', \tag{5.3}$$

where we have introduced the subscript in order to stress the f-dependence of the Green function. By inserting in (5.3) the Feynman path integral representation (5.2), we obtain an heuristic formula which represents the transition amplitude in terms of a double path integral:

$$|\langle \phi, \psi(t) \rangle_f|^2 = \int\int\int\int \phi(x')\bar{\phi}(x)\psi(y)\bar{\psi}(y') \int\int e^{\frac{i}{\hbar}\left(S(\gamma) - S(\gamma')\right)} e^{-\frac{i}{\hbar}\int_0^t \left(\gamma(s) - \gamma'(s)\right) f(s) ds} d\gamma d\gamma' dydxdy'dx', \tag{5.4}$$

where $\gamma(0) = y, \gamma(t) = x, \gamma'(0) = y', \gamma'(t) = x'$ and

$$S(\gamma) = \frac{m}{2}\int_0^t \dot{\gamma}(s)^2 ds - \frac{k}{2}\int_0^t \gamma(s)^2 ds.$$

By averaging the above quantity over the random force f we obtain:

$$\mathbb{E}_f|\langle\phi,\psi(t)\rangle_f|^2 = \int\int\int\int \phi(x')\bar{\phi}(x)\psi(y)\bar{\psi}(y')$$
$$\int\int e^{\frac{i}{\hbar}\left(S(\gamma)-S(\gamma')\right)}F(\gamma,\gamma')d\gamma d\gamma' dy dx dy' dx', \quad (5.5)$$

where the influence of the random fluctuating force f is modelled by the functional

$$F(\gamma,\gamma') = \mathbb{E}_f[e^{-\frac{i}{\hbar}\int_0^t\left(\gamma(s)-\gamma'(s)\right)f(s)ds}].$$

This, still heuristic, considerations show some key features of the Feynman path integral description of the quantum open systems, such as the introduction of a *double path integral* of the form (5.5) and the modellization of an interacting environment in terms of an *influence functional* $F(\gamma,\gamma')$ which couples the paths γ,γ'. This particular formalism was introduced by Feynman and Vernon in [121], where a formula similar to Eq. (5.5) was heuristically derived by modelling the environment in terms of a many body quantum system and by tracing out its degrees of freedom. This technique has been applied to the modellization of several complex quantum phenomena, including the quantum Brownian motion [63], i.e. the description of the dynamics of a quantum particle interacting with an macroscopic environment.

Let us consider for instance the time evolution of a quantum system made of two linearly interacting subsystems A and B. Let us assume that the state space of the system A is $L^2(\mathbb{R}^d)$ while the state space of the system B is $L^2(\mathbb{R}^N)$. Let the total Hamiltonian of the compound systems be of the form

$$H_{AB} = H_A + H_B + H_{INT} \quad (5.6)$$

$$H_A = -\frac{\Delta_{\mathbb{R}^d}}{2M} + \frac{1}{2}x\Omega_A^2 x + v_A(x), \quad x\in\mathbb{R}^d,$$

$$H_B = -\frac{\Delta_{\mathbb{R}^N}}{2m} + \frac{1}{2}R\Omega_B^2 R + v_B(R), \qquad R\in\mathbb{R}^N,$$

$$H_{INT} = xCR, \qquad x\in\mathbb{R}^d, R\in\mathbb{R}^N,$$

with $C:\mathbb{R}^N\to\mathbb{R}^d$ is a linear operator, Ω_A, resp. Ω_B, a symmetric positive $d\times d$, resp. $N\times N$, matrix, v_a, resp. v_B, real bounded functions. Let us assume that the quadratic part of the total potential, i.e. the function $x,R\mapsto \frac{1}{2}x\Omega_A^2 x + \frac{1}{2}R\Omega_B^2 R + xCR$ is positive definite (so that the total

Hamiltonian is bounded from below). Let us assume moreover that the density matrix of the compound system factorizes $\rho_{AB} = \rho_A \rho_B$ and has a smooth kernel $\rho_{AB}(x, y, R, Q) = \rho_A(x, y)\rho_B(R, Q)$. By writing the Feynman path integral representation for the evolution of the density matrix, we have

$$\rho_t(x, y, R, Q) = \left(e^{-\frac{i}{\hbar}H_{AB}t}\rho_0 e^{\frac{i}{\hbar}H_{AB}t}\right)(x, y, R, Q)$$
$$= \iiiint \rho_A(\gamma(0), \gamma'(0))\rho_B(\Gamma(0), \Gamma'(0))e^{\frac{i}{\hbar}\left(S_A(\gamma)-S_A(\gamma')\right)}e^{\frac{i}{\hbar}\left(S_B(\Gamma)-S_B(\Gamma')\right)}$$
$$e^{\frac{i}{\hbar}\left(S_{AB}(\gamma,\Gamma)-S_{AB}(\gamma',\Gamma')\right)}d\gamma d\gamma' d\Gamma d\Gamma' \quad (5.7)$$

where the integral is taken over the path $\gamma, \gamma' : [0, t] \to \mathbb{R}^d$ and $\Gamma, \Gamma' : [0, t] \to \mathbb{R}^N$ such that $\gamma(t) = x$, $\gamma'(t) = y$, $\Gamma(t) = R$, $\Gamma'(t) = Q$. S_A, S_B and S_{AB} are given by:

$$S_A(\gamma) = \int_0^t \left(\frac{M}{2}\dot{\gamma}(s)^2 - \frac{1}{2}\gamma(s)\Omega_A^2\gamma(s) - v_A((\gamma(s)))\right)ds,$$
$$S_B(\Gamma) = \int_0^t \left(\frac{m}{2}\dot{\Gamma}(s)^2 - \frac{1}{2}\Gamma(s)\Omega_B^2\Gamma(s) - v_B((\Gamma(s)))\right)ds,$$
$$S_{AB}(\gamma, \Gamma) = -\int_0^t \gamma(s)C\Gamma(s)ds\,.$$

By tracing over the coordinates of the system B, one obtains the following (heuristic) Feynman path integral representation for the reduced density matrix $\rho_A^r(t)$ of the system A:

$$\rho_A^r(t)(x, y) = \int \rho_t(x, y, R, R)dR$$
$$= \iint \rho_A(\gamma(0), \gamma'(0))e^{\frac{i}{\hbar}\left(S_A(\gamma)-S_A(\gamma')\right)}F(\gamma, \gamma')d\gamma d\gamma' \quad (5.8)$$

where

$$F(\gamma, \gamma') = \int \int_{\Gamma(t)=R} \int_{\Gamma'(t)=R} \rho_B(\Gamma(0), \Gamma'(0))e^{\frac{i}{\hbar}\left(S_B(\Gamma)-S_B(\Gamma')\right)}$$
$$e^{\frac{i}{\hbar}\left(S_{AB}(\gamma,\Gamma)-S_{AB}(\gamma',\Gamma')\right)}d\Gamma d\Gamma' dR. \quad (5.9)$$

In section 5.2 we shall give a rigorous mathematical meaning to the double Feynman path integral (5.8) in terms of a well defined infinite dimensional oscillatory integral and analyze the Caldeira-Leggett model [63] for the description of the quantum Brownian motion.

Another fundamental problem of quantum theory is the description of the process of quantum measurement, i.e. the interaction of a physical system with a (macroscopic) measuring apparatus allowing an observer to obtain the result of the measurement of a physical quantity. In the traditional formulation of quantum mechanics, the continuous time evolution described by the Schrödinger equation (1.4) is valid if the quantum system is "undisturbed". On the other hand we should not forget that all the informations we can have on the state of a quantum particle are the result of some measurement process. When the particle interacts with the measuring apparatus, its time evolution is no longer continuous: the state of the system after the measurement is the result of a random and discontinuous change, the so-called "collapse of the wave function", which cannot be described by the ordinary Schrödinger equation. Quoting Dirac [107], after the introduction of the Planck constant $\hbar$ the concept of "large" and "small" are no longer relative: it is "microscopic"[1] one object such that the influence on the measuring apparatus on it cannot be neglected.

Let us recall the main features of the traditional quantum description of the measurement of an observables $\mathcal{A}$, represented by a self-adjoint operator $A : D(A) \subset \mathcal{H} \to \mathcal{H}$ on a complex separable Hilbert space $\mathcal{H}$, whose unitary vectors represent the states of the system. Let us consider for simplicity the case where the operator A is bounded and its spectrum is discrete. Let $\{a_i\}_{i\in\mathbb{N}} \subset \mathbb{R}$ and $\{\psi_i\}_{i\in\mathbb{N}} \subset \mathcal{H}$ be the corresponding eigenvalues and eigenvectors. According to the traditional mathematical formulation by Von Neumann the consequences of the measurement are:

(1) the *decoherence* of the state of the quantum system.

Because of the interaction with the measuring apparatus, the initial pure state ψ of the system becomes a mixed state, described by the density operator $\rho^{prior}(t) = \sum_i w_i P_{\psi_i}$, where P_{ψ_i} denotes the projector operator onto the eigenspace which is spanned by the vector ψ_i and $w_i = |\langle\psi_i, \psi\rangle|^2$. By considering another observable $\mathcal{B}$ (represented by a bounded self-adjoint operator $B : \mathcal{H} \to \mathcal{H}$), its expectation value at time t, after the measurement of the observable $\mathcal{A}$ (but without the information of the result of the measurement of $\mathcal{A}$), is given by

$$\mathbb{E}(B)_t^{prior} = \mathrm{Tr}[\rho^{prior}(t)B].$$

The existence of the trace is assumed. The transformation mapping ψ to the so-called "prior state" $\rho^{prior}(t)$ is named "prior dynamics" or

[1]It would be more correct the word "quantum" as there exist also macroscopic quantum systems, but they were unknown at Dirac's time.

non selective dynamics.

(2) The so-called "collapse of the wave function".
After the reading of the result of the measurement (i.e. the real number a_i) the state of the system is the corresponding eigenstate of the measured observable:

$$\rho(t)^{post}_{a_i} = P_{\psi_i}.$$

The expectation value of another observable $\mathcal{B}$ of the system at time t (taking into account the information about the value of the measurement of $\mathcal{A}$) is given by:

$$\mathbb{E}^{post}(B|A = a_i)_t = \mathrm{Tr}[\rho^{post}_{a_i}(t)B] = \langle\psi_i, B\psi_i\rangle.$$

The transformation mapping the initial state ψ to one of the so-called "posterior states" $\rho^{post}_{a_i}(t)$ is called "posterior dynamics" or selective dynamics and depends on the result a_i of the measurement of $\mathcal{A}$.

As it is suggested by the collapse of the wave function, the non selective dynamics maps pure states into mixed states, while the selective one maps pure states into pure states. The relation between the posterior state and the prior state is given by:

$$\rho^{prior}(t) = \sum_i P(A = a_i)\rho^{post}_{a_i}(t),$$

where $P(A = a_i)$ denotes the probability that the outcome of the measurement of $\mathcal{A}$ is the eigenvalue a_i and it is given by

$$P(A = a_i) = |\langle\psi_i, \psi\rangle|^2.$$

We remark that

$$\mathbb{E}(B)^{prior}_t = \sum_i \mathbb{E}^{post}(B|A = a_i)P(A = a_i)\,. \tag{5.10}$$

There are several efforts to include the process of measurement into the traditional quantum theory and to deduce from its laws, instead of postulating, both the process of decoherence (see point 1) and the collapse of the wave function (point 2). In particular the aim of the *quantum theory of measurement* is a description of the process of measurement taking into account the properties of the measuring apparatus, which is handled as a quantum system, and its interaction with the system submitted to the measurement [92, 62]. Even if also this approach is not completely satisfactory (also in this case one has to postulate the collapse of the state of the compound system "measuring apparatus plus observed system"), it is able to give a better description of the process of measurement.

An interesting result of the quantum theory of measurement is the so-called "Zeno effect", which seems to forbid a satisfactory description of continuous measurements. Indeed if a sequence of "ideal"[2] measurements of an observable A with discrete spectrum is performed and the time interval between two measurements is sufficiently small, then the observed system does not evolve. In other words a particle whose position is continuously monitored cannot move. This result is in apparent contrast with the experience: indeed in a bubble chamber repeated measurements of the position of microscopical particles are performed without "freezing" their state. For a detailed description of the quantum Zeno paradox see for instance [229, 80, 240].

An heuristic Feynman path integral description of the process of quantum measurement has been proposed by several authors, in particular by M.B. Mensky [225, 226], who considers the selective dynamics of a particle whose position is continuously observed. According to Mensky the state of the particle at time t if the observed trajectory is the path $\omega(s)_{s\in[0,t]}$ is given by the "restricted path integrals"

$$\psi(t,x,\omega) = \text{“}\int_{\{\gamma(t)=x\}} e^{\frac{i}{\hbar}S_t(\gamma)}e^{-\lambda\int_0^t(\gamma(s)-\omega(s))^2ds}\phi(\gamma(0))D\gamma,\text{”} \quad (5.11)$$

where $\lambda \in \mathbb{R}^+$ is a real positive parameter which is proportional to the accuracy of the measurement. Heuristically formula (5.11) suggests that, as an effect of the correction term $e^{-\lambda\int_0^t(\gamma(s)-\omega(s))^2ds}$ due to the measurement, the paths γ giving the main contribution to the integral (5.11) are those closer to the observed trajectory ω.

As one looks to the observed trajectory ω in Eq. (5.11) as a random variable and to the state $\psi(t,x,\omega)$, $t \geq 0$ as a stochastic process, it is natural to think at Eq. (5.11) as the Feynman path integral implementation of a quantum stochastic dynamics. Indeed in the physical and in the mathematical literature a class of stochastic Schrödinger equations giving a phenomenological description of quantum measurements has been proposed by several authors, see for instance [59, 50, 51, 105, 226, 138]. Let us consider for instance Belavkin equation, a stochastic Schrödinger equation describing the selective dynamics of a $d-$dimensional particle submitted to the measurement of one of its (possible $M-$dimensional vector) observables, described

[2] A measurement is called *ideal* if the correlation between the state of the measuring apparatus and the state of the system after the measurement is maximal.

by the self-adjoint operator R on $L^2(\mathbb{R}^d)$:

$$\begin{cases} d\psi(t,x) = -\frac{i}{\hbar}H\psi(t,x)dt - \frac{\lambda}{2}R^2\psi(t,x)dt + \sqrt{\lambda}R\psi(t,x)dB(t) \\ \psi(0,x) = \psi_0(x) \qquad (t,x) \in [0,T] \times \mathbb{R}^d. \end{cases} \tag{5.12}$$

H is the quantum mechanical Hamiltonian, B is an $M-$dimensional Brownian motion on a probability space $(\Omega, \mathcal{F}, \mathbb{P})$ (see section A.2), $dB(t)$ is the Ito differential and $\lambda > 0$ is a coupling constant, which is proportional to the accuracy of the measurement. In the particular case of the description of the continuous measurement of position one has $R = x$ (the multiplication operator), so that equation (5.12) assumes the following form:

$$\begin{cases} d\psi(t,x) = -\frac{i}{\hbar}H\psi(t,x)dt - \frac{\lambda}{2}x^2\psi(t,x)dt + \sqrt{\lambda}x\psi(t,x)dB(t) \\ \psi(0,x) = \psi_0(x) \qquad (t,x) \in [0,T] \times \mathbb{R}^d. \end{cases} \tag{5.13}$$

Belavkin derives equation (5.12) by modeling the measuring apparatus (but it is better to say "the informational environment") by means of a one-dimensional bosonic field and by assuming a particular form for the interaction Hamiltonian between the field and the system on which the measurement is performed. The resulting dynamics is such that there exists a family of mutually commuting Heisenberg operators of the compound system, denoted by $X(t)_{t\in[0,T]}$, such that:

$$[X(t), X(s)] = 0, \qquad s,t \in [0,T],$$

(on a dense domain in $L^2(\mathbb{R}^d)$). In this description the concept of trajectory of X is meaningful, even from a quantum mechanical point of view. Furthermore the "non-demolition principle" is fulfilled: the measurement of any future Heisenberg operator $Z(t)$ of the system is compatible with the measurement of the trajectory of X up to time t, that is:

$$[Z(t), X(s)] = 0, \qquad s < t,$$

(on a dense domain in $L^2(\mathbb{R}^d)$). The measured observable R is connected to the operator X by the following relation

$$X(t) = R(t) + \lambda(B_t + B_t^+), \tag{5.14}$$

(where $(B_t + B_t^+)$ is a quantum Brownian motion [166]). Equation (5.14) shows how the measurement of $X(t)$ gives some (indirect and not precise) information on the value of R, overcoming the problems of quantum Zeno paradox. Indeed we are dealing with "unsharp" in spite of "ideal" measurements.

For an alternative derivation of Eq. (5.13), we point out to the reader the appendix A of [207], where a physically intuitive model of unsharp continuous quantum measurement is proposed, as well as a Feynman path integral representation of the state ψ analogous to Eq. (5.11).

In section 5.3 we shall realize, in terms of infinite dimensional oscillatory integrals, a Feynman path integral representation of the solution of Belavkin equation (5.13):

$$\psi(t,x) = \widetilde{\int} e^{\frac{i}{2\hbar}\int_0^t |\dot\gamma(s)|^2 ds - \lambda \int_0^t |\gamma(s)+x|^2 ds} e^{-\frac{i}{\hbar}\int_0^t V(\gamma(s)+x)ds} e^{\int_0^t \sqrt{\lambda}(\gamma(s)+x)\cdot dB(s)} \psi_o(\gamma(0)+x) d\gamma \quad (5.15)$$

and therefore we shall give a rigorous mathematical meaning to Mensky's heuristic formula (5.11).

5.2 The Feynman-Vernon influence functional

Let us introduce a particular type of infinite dimensional oscillatory integral on a real separable Hilbert space $\mathcal{H}$ which will be used in the mathematical definition of double Feynman path integrals of the form (5.8).

Definition 5.1. A Borel measurable function $f : \mathcal{H}\times\mathcal{H} \to \mathbb{C}$ is called *Fresnel integrable* if for any sequence $\{P_n\}$ of projectors onto n-dimensional subspaces of $\mathcal{H}$, such that $P_n \le P_{n+1}$ and $P_n \to 1$ strongly as $n \to \infty$ (1 being the identity operator in $\mathcal{H}$), the finite dimensional oscillatory integrals (suitably normalized)

$$(2\pi\hbar)^{-n} \int^{\circ}_{P_n\mathcal{H}} \int^{\circ}_{P_n\mathcal{H}} e^{\frac{i}{2\hbar}\langle P_n x, P_n x\rangle} e^{-\frac{i}{2\hbar}\langle P_n y, P_n y\rangle} f(P_n x, P_n y) d(P_n x) d(P_n y)$$

are well defined (in the sense of definition 2.1) and the limit

$$\lim_{n\to\infty} (2\pi\hbar)^{-n} \int^{\circ}_{P_n\mathcal{H}} \int^{\circ}_{P_n\mathcal{H}} e^{\frac{i}{2\hbar}\langle P_n x, P_n x\rangle} e^{-\frac{i}{2\hbar}\langle P_n y, P_n y\rangle} f(P_n x, P_n y) d(P_n x) d(P_n y)$$

exists and is independent of the sequence $\{P_n\}$. In this case the limit is denoted by

$$\widetilde{\int\int} e^{\frac{i}{2\hbar}\langle x,x\rangle} e^{-\frac{i}{2\hbar}\langle y,y\rangle} f(x,y) dx dy.$$

For the integrals of definition 5.1 it is possible to prove a result analogous to theorem 2.5.

Theorem 5.1. *Let $L : \mathcal{H} \to \mathcal{H}$ be a trace-class operator, such that $I - L$ is invertible, and let $f : \mathcal{H} \times \mathcal{H} \to \mathbb{C}$ be the Fourier transform of a complex bounded variation measure μ_f on $\mathcal{H} \times \mathcal{H}$. Then the integral*

$$\widetilde{\int\int} e^{\frac{i}{2\hbar}\langle x,x\rangle} e^{-\frac{i}{2\hbar}\langle y,y\rangle} e^{-\frac{i}{2\hbar}\langle x-y, L(x+y)\rangle} f(x,y)dxdy$$

is well defined and is equal to

$$\frac{1}{\det(I-L)} \int_{\mathcal{H}}\int_{\mathcal{H}} e^{-\frac{i\hbar}{2}\langle \alpha+\beta,(I-L)^{-1}(\alpha-\beta)\rangle} d\mu_f(\alpha,\beta)\,,$$

where $\det(I-L)$ *is the Fredholm determinant of* $I - L$.

Proof. Let us consider a sequence $\{P_n\}$ of projectors onto n-dimensional subspaces of $\mathcal{H}$, such that $P_n \le P_{n+1}$ and $P_n \to 1$ strongly as $n \to \infty$. The finite dimensional approximations of the oscillatory integral

$$\widetilde{\int\int} e^{\frac{i}{2\hbar}\langle x,x\rangle} e^{-\frac{i}{2\hbar}\langle y,y\rangle} e^{-\frac{i}{2\hbar}\langle x-y, L(x+y)\rangle} f(x,y)dxdy$$

are given by

$$\frac{1}{(2\pi\hbar)^n} \int_{P_n\mathcal{H}}^{\circ}\int_{P_n\mathcal{H}}^{\circ} e^{\frac{i}{2\hbar}\langle x_n-y_n,(I_n-L_n)(x_n+y_n)\rangle} f(x_n,y_n)dx_n dy_n\,,$$

where $x_n = P_n x$, $x \in \mathcal{H}$, and $I_n - L_n = I|_{P_n\mathcal{H}} - P_n L P_n$. The finite dimensional approximations are defined by the following sequence of regularized integrals:

$$\lim_{\epsilon\to 0}\frac{1}{(2\pi\hbar)^n} \int_{P_n\mathcal{H}}\int_{P_n\mathcal{H}} e^{\frac{i}{2\hbar}\langle x_n-y_n,(I_n-L_n)(x_n+y_n)\rangle} \phi(\epsilon x_n,\epsilon y_n) f(x_n,y_n)dx_n dy_n\,,$$

with $\phi \in \mathcal{S}(\mathbb{R}^n \times \mathbb{R}^n)$, $\phi(0) = 1$.

Since $I - L$ is invertible, for any sequence $\{P_n\}$ of projectors there exists an $\bar{n}$ such that for any $n \ge \bar{n}$ the operator $P_n(I-L)P_n$ is invertible and thus $\det(I_n - L_n) \neq 0$. Hence, for $n \ge \bar{n}$, by introducing the new variables $z_n = x_n - y_n$ and $w_n = x_n + y_n$, and by Fubini theorem, the integral can be written as

$$\lim_{\epsilon\to 0}\frac{1}{(4\pi\hbar)^n} \int_{P_n\mathcal{H}}\int_{P_n\mathcal{H}}\Bigg(\int_{P_n\mathcal{H}}\int_{P_n\mathcal{H}} e^{i\langle \alpha,\frac{z_n+w_n}{2}\rangle + i\langle \beta,\frac{w_n-z_n}{2}\rangle} e^{\frac{i}{2\hbar}\langle z_n,(I_n-L_n)w_n\rangle}$$

$$\times\, \phi_T(\epsilon z_n,\epsilon w_n)\, dz_n dw_n\Bigg) d\mu_n(\alpha,\beta)\,,$$

where $\mu_n \in \mathcal{F}(P_n\mathcal{H} \times P_n\mathcal{H})$ is defined as $\mu_n = \mu \circ P_n$ and $\phi_T \in \mathcal{S}(P_n\mathcal{H} \times P_n\mathcal{H})$ as

$$\phi_T(z_n,w_n) = \phi\left(\frac{z_n+w_n}{2}, \frac{w_n-z_n}{2}\right).$$

If we write the function ϕ_T in terms of its Fourier transform and apply Fubini again, the integrals over the variables z_n and w_n become

$$\int_{P_n\mathcal{H}}\int_{P_n\mathcal{H}} e^{i\langle\alpha,\frac{z_n+w_n}{2}\rangle+i\langle\beta,\frac{w_n-z_n}{2}\rangle} e^{\frac{i}{2\hbar}\langle z_n,(I_n-L_n)w_n\rangle}\phi_T(\epsilon z_n,\epsilon w_n)\,dz_n dw_n$$
$$=\left(\frac{\hbar}{\pi}\right)^n \det(I_n-L_n)^{-1}\int_{P_n\mathcal{H}}\int_{P_n\mathcal{H}} e^{\frac{-i\hbar}{2}\langle\alpha+\beta-2\epsilon\gamma_n,(I_n-L_n)^{-1}(\alpha-\beta-2\epsilon\delta_n)\rangle}$$
$$\widehat{\phi}_T(\gamma_n,\delta_n)d\gamma_n d\delta_n\,.$$

Lebesgue's dominated convergence theorem allows us to exchange limit and integrals. Thus, by taking into account that

$$\int_{P_n\mathcal{H}}\int_{P_n\mathcal{H}} \widehat{\phi}_T(\gamma_n,\delta_n)d\gamma_n d\delta_n=(2\pi)^{2n}\phi_T(0,0),$$

we obtain

$$\det(I_n-L_n)^{-1}\int_{P_n\mathcal{H}}\int_{P_n\mathcal{H}} e^{-\frac{i\hbar}{2}\langle\alpha+\beta,(I_n-L_n)^{-1}(\alpha-\beta)\rangle}d\mu_n(\alpha,\beta)\,.$$

The statement follows by taking the limit $n\to\infty$, since $\det(I_n-L_n)$ converges to $\det(I-L)$ (see section 2.4). □

The next result is a straightforward consequence of Theorem 5.1.

Corollary 5.1. *Under the assumptions of Theorem 5.1, the functional*

$$f\in\mathcal{F}(\mathcal{H}\times\mathcal{H})\mapsto \widetilde{\int\int} e^{\frac{i}{2\hbar}\langle x,x\rangle}e^{-\frac{i}{2\hbar}\langle y,y\rangle}e^{-i\langle x-y,L(x+y)\rangle}f(x,y)dxdy$$

is continuous in the $\mathcal{F}(\mathcal{H}\times\mathcal{H})$-norm.

For the applications that will follow, it is convenient to introduce the following Fubini-type theorem on the change of order of integration between oscillatory integrals and Lebesgue integrals.

Let $\{\mu_\alpha|\,\alpha\in\mathbb{R}^d\}$ be a family in $\mathcal{M}(\mathcal{H})$. We denote by $\int_{\mathbb{R}^d}\mu_\alpha d\alpha$ the measure defined by

$$f\mapsto\int_{\mathbb{R}^d}\int_{\mathcal{H}} f(x)d\mu_\alpha(x)d\alpha,\qquad f\in C_0(\mathcal{H}),$$

whenever it exists.

Theorem 5.2. *Let $L:\mathcal{H}\to\mathcal{H}$ be as in the assumptions of Theorem 5.1 and let $\mu:\mathbb{R}^d\to\mathcal{M}(\mathcal{H}\times\mathcal{H})$, $\alpha\mapsto\mu_\alpha$, be a continuous map such that*

$$\int_{\mathbb{R}^d}|\mu_\alpha|d\alpha<\infty.$$

Further, for any $\alpha \in \mathbb{R}^d$, let $f_\alpha \in \mathcal{F}(\mathcal{H} \times \mathcal{H})$ be given by

$$f_\alpha(x, y) = \hat{\mu}_\alpha(x, y), \qquad (x, y) \in \mathcal{H} \times \mathcal{H}.$$

Then $\int_{\mathbb{R}^d} f_\alpha d\alpha \in \mathcal{F}(\mathcal{H} \times \mathcal{H})$ and

$$\int_{\mathbb{R}^d} \widetilde{\int_{\mathcal{H}}} \widetilde{\int_{\mathcal{H}}} e^{\frac{i}{2\hbar}\langle x,x\rangle} e^{-\frac{i}{2\hbar}\langle y,y\rangle} e^{-\frac{i}{2\hbar}\langle x-y, L(x+y)\rangle} f_\alpha(x, y) dx dy d\alpha$$
$$= \widetilde{\int_{\mathcal{H}}} \widetilde{\int_{\mathcal{H}}} e^{\frac{i}{2\hbar}\langle x,x\rangle} e^{-\frac{i}{2\hbar}\langle y,y\rangle} e^{-\frac{i}{2\hbar}\langle x-y, L(x+y)\rangle} \int_{\mathbb{R}^d} f_\alpha(x, y) d\alpha dx dy\,. \quad (5.16)$$

Proof. Since f_α is assumed to be the Fourier transform of μ_α, by Fubini's theorem

$$\int_{\mathbb{R}^d} f_\alpha d\alpha = \int_{\mathbb{R}^d} \int_{\mathcal{H}\times\mathcal{H}} e^{i\langle k,x\rangle + i\langle h,y\rangle} d\mu_\alpha(k, h) d\alpha$$
$$= \int_{\mathcal{H}\times\mathcal{H}} e^{i\langle k,x\rangle + i\langle h,y\rangle} \int_{\mathbb{R}^d} d\mu_\alpha(k, h) d\alpha\,,$$

so that $\int_{\mathbb{R}^d} f_\alpha d\alpha \in \mathcal{F}(\mathcal{H} \times \mathcal{H})$.
By applying Theorem 5.1 to the left hand side of (5.16), we obtain

$$\int_{\mathbb{R}^d} \widetilde{\int_{\mathcal{H}}} \widetilde{\int_{\mathcal{H}}} e^{\frac{i}{2\hbar}\langle x,x\rangle} e^{-\frac{i}{2\hbar}\langle y,y\rangle} e^{-\frac{i}{2\hbar}\langle x-y, L(x+y)\rangle} f_\alpha(x, y) dx dy d\alpha$$
$$= \det(I - L)^{-1} \int_{\mathbb{R}^d} \int_{\mathcal{H}} \int_{\mathcal{H}} e^{-\frac{i\hbar}{2}\langle k+h, (I-L)^{-1}(k-h)\rangle} d\mu_\alpha(k, h) d\alpha\,.$$

By the usual Fubini theorem the latter is equal to

$$\det(I - L)^{-1} \int_{\mathcal{H}} \int_{\mathcal{H}} e^{-\frac{i\hbar}{2}\langle k+h, (I-L)^{-1}(k-h)\rangle} \int_{\mathbb{R}^d} d\mu_\alpha(k, h) d\alpha\,.$$

But this expression, by theorem 5.1, is equal to the r.h.s. of (5.16). □

We can now apply these results to the mathematical realization of the Feynman-Vernon influence functional in the Caldeira-Leggett model [120, 63, 11].

Let us consider the Cameron-Martin space $\mathcal{H}_t$, i.e. the Hilbert space of absolutely continuous paths $\gamma : [0, t] \to \mathbb{R}$, such that $\gamma(t) = 0$ and $\int_0^t |\dot{\gamma}(s)|^2 ds < \infty$, endowed with the inner product

$$\langle \gamma_1, \gamma_2 \rangle = \int_0^t \dot{\gamma}_1(s) \dot{\gamma}_2(s) ds\,.$$

Let $L : \mathcal{H}_t \to \mathcal{H}_t$ be the trace-class symmetric operator on $\mathcal{H}_t$ given by

$$(L\gamma)(s) = \int_s^t ds' \int_0^{s'} ds'' \gamma(s''), \qquad \gamma \in \mathcal{H}_t\,.$$

Further let $\mathcal{H}_t^d = \oplus_{i=1}^d \mathcal{H}_t$ and let $L^d : \mathcal{H}_t^d \to \mathcal{H}_t^d$ denote the operator defined by

$$L^d = L^{(1)} \otimes L^{(2)} \otimes \cdots \otimes L^{(d)},$$

where $L^{(j)} = I \otimes \cdots \otimes I \otimes L \otimes I \cdots \otimes I$ with L acting on the jth space. Given a positive symmetric $d \times d$ matrix Ω, let $L_\Omega : \mathcal{H}_t^d \to \mathcal{H}_t^d$ be the trace-class symmetric operator on $\mathcal{H}_t^d$ defined by

$$(L_\Omega \gamma)(s) = \int_s^t ds' \int_0^{s'} (\Omega^2 \gamma)(s'')ds'', \qquad \gamma \in \mathcal{H}_t^d. \tag{5.17}$$

Clearly, $L_\Omega \gamma = L^d \Omega^2 \gamma$, for all $\gamma \in \mathcal{H}_t^d$. One can easily verify that

$$\langle \gamma_1, L_\Omega \gamma_2 \rangle = \int_0^t \gamma_1(s) \Omega^2 \gamma_2(s) ds.$$

By lemma 3.1 and lemma 3.2, if $t \neq [(n + 1/2)\pi]/\Omega_j$, for any $n \in \mathbb{N}$ and any eigenvalue Ω_j of Ω, one has that the operator $I - L_\Omega$ is invertible with

$$(I - L_\Omega)^{-1} \gamma(s) = \gamma(s) - \Omega \int_s^t \sin(\Omega(s' - s))\gamma(s')ds'$$
$$+ \sin(\Omega(t - s)) \int_0^t (\cos \Omega t)^{-1} \Omega \cos(\Omega s')\gamma(s')ds' \tag{5.18}$$

and

$$\det(I - L_\Omega) = \det(\cos(\Omega t))$$

(see lemmas 3.1 and 3.2).

Let $v \in \mathcal{F}(\mathbb{R}^d)$ be a real bounded function and let H be the quantum Hamiltonian, given on smooth vectors $\psi \in \mathcal{S}(\mathbb{R}^d)$ by

$$H\psi(x) = -\frac{\Delta}{2}\psi(x) + \frac{1}{2} x \Omega^2 x \psi(x) + v(x)\psi(x), \qquad x \in \mathbb{R}^d.$$

Let $U_t = e^{-\frac{i}{\hbar}Ht}$ be the unitary group generated by H. As we have seen in chapter 3, given an initial datum $\psi_0 \in \mathcal{F}(\mathbb{R}^d)$ and assuming that $t \neq [(n + 1/2)\pi]/\Omega_j$, $\forall n \in \mathbb{N}$, the solution of the Schrödinger equation $\psi(t) = U_t \psi_0$ is given by an infinite dimensional oscillatory integral on the Cameron Martin space $\mathcal{H}_t$:

$$\psi(t, x) = e^{-\frac{i}{2\hbar} x \Omega^2 x t} \widetilde{\int_{\mathcal{H}_t^d}} e^{\frac{i}{2\hbar}\langle \gamma, (I - L_\Omega)\gamma \rangle} e^{-\frac{i}{\hbar}\int_0^t x \Omega^2 \gamma(s) ds} e^{-\frac{i}{\hbar}\int_0^t v(\gamma(s) + x) ds}$$
$$\psi_0(\gamma(0) + x) d\gamma.$$

(see theorem 3.2). This result can be generalized to the Feynman path integral representation of the time evolution for a mixed state, represented by a density matrix.

Theorem 5.3. *Let ρ_0 be a density matrix operator on $L^2(\mathbb{R}^d)$, such that ρ_0 admits a regular kernel $\rho_0(x,y)$, $x,y \in \mathbb{R}^d$. Let us assume moreover that ρ_0 admits a decomposition into pure states of the form $\rho_0(x,y) = \sum_i \lambda_i e_i(x)\bar{e}_i(y)$, with $\lambda_i > 0$, $\sum_i \lambda_i = 1$, $\langle e_i, e_j\rangle_{L^2(\mathbb{R}^d)} = \delta_{ij}$, and $e_i(x) = \hat{\mu}_i(x)$, satisfying*

$$\sum_i \lambda_i |\mu_i|^2 < \infty. \tag{5.19}$$

Let $t \neq [(n+1/2)\pi]/\Omega_j$, $\forall n \in \mathbb{N}$. Then the density matrix operator ρ_t at time t, $\rho_t = U_t\rho_0 U_t^\dagger$, admits a smooth kernel $\rho_t(x,y)$, which is given by the infinite dimensional oscillatory integral

$$\begin{aligned} e^{-\frac{i}{2\hbar}(x\Omega^2 x - y\Omega^2 y)t} \widetilde{\int_{\mathcal{H}_t^d}} \widetilde{\int_{\mathcal{H}_t^d}} & e^{\frac{i}{2\hbar}\langle\gamma,(I-L_\Omega)\gamma\rangle} e^{-\frac{i}{2\hbar}\langle\gamma',(I-L_\Omega)\gamma'\rangle} \\ & e^{-\frac{i}{\hbar}\int_0^t (x\Omega^2\gamma(s) - y\Omega^2\gamma'(s))ds} e^{-\frac{i}{\hbar}\int_0^t (v(\gamma(s)+x) - v(\gamma'(s)+y))ds} \\ & \rho_0(\gamma(0)+x, \gamma'(0)+y) d\gamma d\gamma' . \end{aligned} \tag{5.20}$$

Proof. By decomposing ρ into pure states, by Corollary 5.1, and by condition (5.19), the integral (5.20) is equal to

$$\sum_i \lambda_i \Big(e^{-\frac{i}{2\hbar}x\Omega^2 xt} \widetilde{\int_{\mathcal{H}_t^d}} e^{\frac{i}{2\hbar}\langle\gamma,(I-L_\Omega)\gamma\rangle} e^{-\frac{i}{\hbar}\int_0^t x\Omega^2\gamma(s)ds} e^{-\frac{i}{\hbar}\int_0^t v(\gamma(s)+x)ds} e_i(\gamma(0)+x) d\gamma \Big)$$

$$\Big(e^{\frac{i}{2\hbar}y\Omega^2 yt} \widetilde{\int_{\mathcal{H}_t^d}} e^{-\frac{i}{2\hbar}\langle\gamma',(I-L_\Omega)\gamma'\rangle} e^{\frac{i}{\hbar}\int_0^t y\Omega^2\gamma'(s)ds} e^{\frac{i}{\hbar}\int_0^t v(\gamma'(s)+y)ds} e_i^*(\gamma(0)+y) d\gamma \Big)$$

$$= \sum_i \lambda_i \Big(e^{-\frac{i}{2\hbar}x\Omega^2 xt} \widetilde{\int_{\mathcal{H}_t^d}} e^{\frac{i}{2\hbar}\langle\gamma,(I-L_\Omega)\gamma\rangle} e^{-\frac{i}{\hbar}\int_0^t x\Omega^2\gamma(s)ds} e^{-\frac{i}{\hbar}\int_0^t v(\gamma(s)+x)ds} e_i(\gamma(0)+x) d\gamma \Big)$$

$$\overline{\Big(e^{-\frac{i}{2\hbar}y\Omega^2 yt} \widetilde{\int_{\mathcal{H}_t^d}} e^{\frac{i}{2\hbar}\langle\gamma',(I-L_\Omega)\gamma'\rangle} e^{-\frac{i}{\hbar}\int_0^t y\Omega^2\gamma'(s)ds} e^{-\frac{i}{\hbar}\int_0^t v(\gamma'(s)+y)ds} e_i(\gamma(0)+y) d\gamma \Big)} .$$

This is equal to

$$\sum_i \lambda_i U_t e_i(x) \overline{(U_t e_i)}(y) = \rho_t(x,y) ,$$

(where for $z \in \mathbb{C}$, $\bar{z}$ denotes the conjugate of the complex number z). □

Heuristically, expression (5.20) can be written as a double Feynman path integral:

$$\text{“}\int\int e^{\frac{i}{\hbar}(S_t(\gamma+x)-S_t(\gamma'+y))}\rho_0(\gamma(0)+x,\gamma'(0)+y)d\gamma d\gamma'\text{”}.$$

Let us consider now the time evolution of a quantum system made of two linearly interacting subsystems A and B. Let us assume that A is d–dimensional, B is N–dimensional and the quantum mechanical Hamiltonian of the compound system is given by Eq. (5.6). Let us assume that the density matrix of the compound system factorizes as $\rho_{AB}=\rho_A\rho_B$ and has a regular kernel $\rho_{AB}(x,y,R,Q)=\rho_A(x,y)\rho_B(R,Q)$. We are going to see how it is possible to construct an infinite dimensional oscillatory integral realization for the Feynman path integral (5.8) representing the reduced density operator at time t, namely

$$\int\left(e^{-\frac{i}{\hbar}H_{AB}t}\rho_{AB}e^{\frac{i}{\hbar}H_{AB}t}\right)(x,y,R,R)dR\,.$$

Heuristically:

$$\text{“}\int\int_{\substack{\gamma(t)=x\\ \Gamma(t)=R}}\int_{\substack{\gamma'(t)=y\\ \Gamma'(t)=R}} e^{\frac{i}{\hbar}(S_A(\gamma)+S_B(\Gamma)+S_{INT}(\gamma,\Gamma)-S_A(\gamma')-S_B(\Gamma')-S_{INT}(\gamma',\Gamma'))}$$
$$\times\rho_A(\gamma(0),\gamma'(0))\rho_B(\Gamma(0),\Gamma'(0))D\gamma D\gamma' D\Gamma D\Gamma' dR\text{”},\qquad(5.21)$$

where γ, resp. Γ, represents a generic path in the configuration space of the system, resp. of the reservoir, and

$$\begin{aligned}&S_A(\gamma)+S_B(\Gamma)+S_{INT}(\gamma,\Gamma)\\&=\int_0^t\left(\frac{M}{2}\dot\gamma^2(s)-\frac{M}{2}\gamma(s)\Omega_A^2\gamma(s)-v_A(\gamma(s))\right)ds\\&\quad+\int_0^t\left(\frac{m}{2}\dot\Gamma^2(s)-\frac{m}{2}\Gamma(s)\Omega_B^2\Gamma(s)-v_B(\Gamma(s))\right)ds-\int_0^t\gamma(s)C\Gamma(s)ds\,.\end{aligned}$$

If we rescale γ via $\gamma\to\gamma/\sqrt{M}$ and Γ via $\Gamma\to\Gamma/\sqrt{m}$, formula (5.21) becomes

$$\text{“}\int\int_{\substack{\gamma(t)=\sqrt{M}x\\ \Gamma(t)=\sqrt{m}R}}\int_{\substack{\gamma'(t)=\sqrt{M}y\\ \Gamma'(t)=\sqrt{m}R}} e^{\frac{i}{2\hbar}\int_0^t\left(\dot\gamma^2(s)-\gamma(s)\Omega_A^2\gamma(s)-2v_A\left(\frac{\gamma(s)}{\sqrt{M}}\right)\right)ds}$$
$$e^{\frac{i}{2\hbar}\int_0^t\left(\dot\Gamma^2(s)-\Gamma(s)\Omega_B^2\Gamma(s)-2v_B\left(\frac{\Gamma(s)}{\sqrt{m}}\right)\right)ds}$$
$$e^{-\frac{i}{\hbar}\int_0^t\gamma(s)\frac{C}{\sqrt{mM}}\Gamma(s)ds}e^{-\frac{i}{2\hbar}\int_0^t\left(|\dot\gamma'|^2(s)-\gamma'(s)\Omega_A^2\gamma'(s)-2v_A\left(\frac{\gamma'(s)}{\sqrt{M}}\right)\right)ds}$$
$$e^{-\frac{i}{2\hbar}\int_0^t\left((\dot\Gamma')^2(s)-\Gamma'(s)\Omega_B^2\Gamma'(s)-2v_B\left(\frac{\Gamma'(s)}{\sqrt{m}}\right)\right)ds}e^{\frac{i}{\hbar}\int_0^t\gamma'(s)\frac{C}{\sqrt{mM}}\Gamma'(s)ds}$$
$$\rho_A\left(\frac{\gamma(0)}{\sqrt{M}},\frac{\gamma'(0)}{\sqrt{M}}\right)\rho_B\left(\frac{\Gamma(0)}{\sqrt{m}},\frac{\Gamma'(0)}{\sqrt{m}}\right)D\gamma D\gamma' D\Gamma D\Gamma' dR\text{”}.$$

Let

$$\Omega^2_{AB} = \begin{pmatrix} \Omega^2_A & C' \\ C'^T & \Omega^2_B \end{pmatrix}, \tag{5.22}$$

where $C' = C/\sqrt{Mm}$, and define for simplicity $L_A\gamma = L_{\Omega_A}\gamma$, $L_B\Gamma = L_{\Omega_B}\Gamma$, and $L_{AB}(\gamma,\Gamma) = L_{\Omega_{AB}}(\gamma,\Gamma)$.

Formula (5.21) can be made completely rigorous under suitable assumptions. In the following we shall assume without loss of generality that $m = M = 1$. The result in the general case can be obtained by replacing C, v_A, v_B, ρ_0^A, and ρ_0^B by $C' = C/\sqrt{mM}$, $v'_A(\cdot) = v_A(\cdot/\sqrt{M})$, $v'_B(\cdot) = v_B(\cdot/\sqrt{m})$, $\rho'_A(\cdot) = \rho_A(\cdot/\sqrt{M})$ and $\rho'_B(\cdot) = \rho_B(\cdot/\sqrt{m})$ respectively.

Theorem 5.4. *Let ρ_A and ρ_B be two density matrix operators on $L^2(\mathbb{R}^d)$ and $L^2(\mathbb{R}^N)$, respectively, with regular kernels $\rho_A(x,x')$ and $\rho_B(R,R')$ and such that they decompose into sums of pure states:*

$$\rho_A = \sum_i w_i^A P_{\psi_i^A}, \qquad \rho_B = \sum_j w_j^B P_{\psi_j^B}, \tag{5.23}$$

with $\psi_i^A \in \mathcal{F}(\mathbb{R}^d)$, $\psi_j^B \in \mathcal{F}(\mathbb{R}^N)$, and

$$\sum_{i,j} w_i^A w_j^B |\mu_i^A|^2 |\mu_j^B|^2 < \infty. \tag{5.24}$$

Further let $\rho_B \in \mathcal{S}(\mathbb{R}^N \times \mathbb{R}^N)$. Let t satisfy the following assumptions

$$t \neq [(n+1/2)\pi]/\Omega_j^A, \qquad n \in \mathbb{N}, \quad j = 1\dots d, \tag{5.25}$$

$$t \neq [(n+1/2)\pi]/\Omega_j^B, \qquad n \in \mathbb{N}, \quad j = 1\dots N, \tag{5.26}$$

$$t \neq [(n+1/2)\pi]/\lambda_j, \qquad n \in \mathbb{N}, \quad j = 1\dots d+N, \tag{5.27}$$

with $\Omega_j^A, \Omega_j^B, \lambda_j$ being respectively the eigenvalues of $\Omega_A, \Omega_B, \Omega_{AB}$. Let us assume moreover that the determinant of the $d \times d$ left upper block of the matrix $\cos(\Omega_{AB}t)$ is non-vanishing.

Then the kernel $\rho_R(t,x,y)$ of the reduced density operator of the system A evaluated at time t is given by

$$\begin{aligned}
&\rho_R(t,x,y) \\
&= e^{-\frac{it}{2\hbar}x\Omega_A^2 x} e^{\frac{it}{2\hbar}y\Omega_A^2 y} \widetilde{\int}_{\mathcal{H}_t^d} \widetilde{\int}_{\mathcal{H}_t^d} e^{\frac{i}{2\hbar}\langle\gamma,(I_d-L_A)\gamma\rangle} e^{-\frac{i}{2\hbar}\langle\gamma',(I_d-L_A)\gamma'\rangle} \\
&\qquad e^{-\frac{i}{\hbar}\int_0^t (x\Omega_A^2\gamma(s)ds - y\Omega_A^2\gamma'(s))ds} e^{-\frac{i}{\hbar}\int_0^t (v_A(\gamma(s)+x) - v_A(\gamma(s)+y))ds} \\
&\qquad F(\gamma,\gamma',x,y)\rho_A(\gamma(0)+x,\gamma'(0)+y)d\gamma d\gamma',
\end{aligned} \tag{5.28}$$

where $F(\gamma,\gamma',x,y)$ is the influence functional

$$F(\gamma,\gamma',x,y)$$

$$= \int_{\mathbb{R}^N} e^{-\frac{it}{\hbar}xCR} e^{+\frac{it}{\hbar}yCR} e^{-\frac{i}{\hbar}\int_0^t(\gamma(s)-\gamma'(s))CRds}$$

$$\widetilde{\int}_{\mathcal{H}_t^N}\widetilde{\int}_{\mathcal{H}_t^N} e^{\frac{i}{2\hbar}\langle\Gamma,(I_N-L_B)\Gamma\rangle} e^{-\frac{i}{2\hbar}\langle\Gamma',(I_N-L_B)\Gamma'\rangle} e^{-\frac{i}{\hbar}\langle\Gamma,L^N C^T\gamma\rangle} e^{\frac{i}{\hbar}\langle\Gamma',L^N C^T\gamma'\rangle}$$

$$e^{-\frac{i}{\hbar}\int_0^t R\Omega_B^2(\Gamma(s)-\Gamma'(s))ds} e^{-\frac{i}{\hbar}\int_0^t(xC\Gamma(s)-yC\Gamma'(s))ds}$$

$$e^{-\frac{i}{\hbar}\int_0^t\left(v_B(\Gamma(s)+R)-v_B(\Gamma'(s)+R)\right)ds}$$

$$\rho_B(\Gamma(0)+R,\Gamma'(0)+R)d\Gamma d\Gamma' dR\,.$$

Proof. The proof of the present theorem involves a large amount of computation. We give here only the main steps and refer to [11] for more details.

First of all one has to prove that the functional $(\gamma,\gamma')\mapsto F(\gamma,\gamma',x,y)$ is well defined for any $\gamma,\gamma'\in\mathcal{H}_t^d$, $x,y\in\mathbb{R}^d$ and is Fresnel integrable in the sense of Definition 5.1. By decomposing the mixed state ρ_B into pure states according to (5.23), the influence functional can be written as

$$\int_{\mathbb{R}^N}\sum_j w_j^B\psi_j^B(x,\gamma;R)\overline{\psi_j^B(y,\gamma';R)}dR\,,$$

where $\psi_j^B(x,\gamma)$ is the solution of the Schrödinger equation on $L^2(\mathbb{R}^N)$ with Hamiltonian

$$H=-\frac{1}{2}\Delta_R+\frac{1}{2}R\Omega_B^2R+v_B(R)+(x+\gamma(t))CR=H_B+(x+\gamma(t))CR$$

and initial state ψ_j^B. In particular,

$$\int_0^t\gamma(s)C\Gamma(s)ds=\langle L^N C^T\gamma,\Gamma\rangle\,.$$

Because of the unitarity of the evolution operator, $\|\psi_j^B(x,\gamma)\|_{L^2(\mathbb{R}^N)}=1$ for any $x\in\mathbb{R}^d$, $\gamma\in\mathcal{H}_t^d$ and, by Schwartz inequality,

$$\sum_j w_j^B\int_{\mathbb{R}^N}\psi_j^B(x,\gamma;R)\overline{\psi_j^B(y,\gamma';R)}dR$$

$$\leqslant\sum_j w_j^B\|\psi_j^B(x,\gamma)\|_{L^2(\mathbb{R}^N)}\|\psi_j^B(y,\gamma')\|_{L^2(\mathbb{R}^N)}=\sum_j\omega_j^B=1\,.$$

Thus we can conclude that $F(\gamma,\gamma',x,y)$ is well defined for any $x,y\in\mathbb{R}^d$ and $\gamma,\gamma'\in\mathcal{H}_t^d$.

By exploiting the assumptions on $v_B \in \mathcal{F}(\mathbb{R}^n)$ and on ρ_B and by a large amount of explicit computation, it is possible to see that $F(\gamma, \gamma', x, y)$ has the following form:

$$F(\gamma, \gamma', x, y) = e^{-\frac{i}{2\hbar}\langle(\gamma-\gamma'), A(\gamma+\gamma')\rangle} f(\gamma, \gamma')$$

with $f \in \mathcal{F}(\mathcal{H}_t^d \oplus \mathcal{H}_t^d)$ and the operator $A : \mathcal{H}_t^d \to \mathcal{H}_t^d$ is given by

$$\langle \gamma, A\gamma' \rangle = -\int_0^t C^T \gamma(s) \Omega_B^{-1} \int_0^s \sin(\Omega_B(s-r)) C^T \gamma'(r) dr ds.$$

Hence one has to verify that the operator $(I - L_A - A) : \mathcal{H}_t^d \to \mathcal{H}_t^d$ is invertible. In fact a vector $\gamma \in \mathcal{H}_t^d$ belongs to the kernel of the operator $I - L_A - A$, if it satisfies the following equation for all $s \in [0, t]$:

$$\begin{aligned} &\gamma(s) - \int_s^t ds' \int_0^{s'} \Omega_A^2 \gamma(s'') ds'' \\ &+ \int_s^t ds' \int_0^{s'} ds'' \int_0^{s''} C\Omega_B^{-1} \sin(\Omega_B(s''-r)) C^T \gamma(r) dr = 0\,, \end{aligned} \tag{5.29}$$

with $\gamma(t) = 0$. By differentiating twice, (5.29) becomes

$$\begin{cases} \ddot{\gamma}(s) + \Omega_A^2 \gamma(s) - \int_0^s C\Omega_B^{-1} \sin(\Omega_B(s-r)) C^T \gamma(r) dr = 0\,, \\ \gamma(t) = 0 = \dot{\gamma}(0)\,. \end{cases}$$

By further differentiating (5.29), one can see that its solution, if it exists, is a C^∞-function and its odd derivatives, evaluated for $s = 0$, vanish, while the even derivatives satisfy the following relation:

$$\gamma^{2(M+2)}(0) + \Omega_A^2 \gamma^{2(M+1)}(0) - \sum_{k=0}^{M} (-1)^k C\Omega_B^{2k} C^T \gamma^{2(M-k)}(0) = 0\,.$$

By induction it is possible to prove that $\gamma^{2M}(0) = (-1)^M [\Omega_{AB}^{2M}]_{d\times d} \gamma(0)$, where $[\Omega_{AB}^{2M}]_{d\times d}$ denotes the $d \times d$ left upper block of the M-th power of the matrix Ω_{AB}^2. One can conclude that the solution of equation (5.29) is of the form $\gamma(s) = [\cos(\Omega_{AB} s)]_{d\times d} \gamma(0)$. By imposing the condition $\gamma(t) = 0$, one concludes that if $\det([\cos(\Omega_{AB} t)]_{d\times d}) \neq 0$ then equation (5.29) cannot admit nontrivial solutions and the operator $I - L_A - A$ is invertible. Hence, by theorem 5.1, we can finally conclude that the influence functional is a Fresnel integrable function.

The second step is the proof that the reduced density operator $\rho_R(t, x, y)$ of the system A is given by the infinite dimensional oscillatory integral (5.28). this results follows from a regularization procedure and by theorems 5.3, 5.2 and corollary 5.1 (see [11] for further details). □

This result can now be applied to the Caldeira-Leggett model [63], describing the influence of a heat bath on a quantum Brownian particle. The heat bath is described by a finite number of oscillators and thus $v_B = 0$. Further the model presumes that the environment is initially in equilibrium at temperature T, i.e. that its initial density $\rho_B(R,Q)$ is a product of Gaussian functions: $\rho_B(R,Q) = \prod_{j=1}^N \rho_B^{(j)}(R_j, Q_j, 0)$, where

$$\rho_B^{(j)}(R_j, Q_j, 0) = \sqrt{\frac{m\Omega_j^B}{\pi\hbar\coth(\hbar\Omega_j^B/2kT)}} \; e^{-\left(\frac{m\Omega_j^B}{2\hbar\sinh(\hbar\Omega_j^B/kT)}\left((R_j^2+Q_j^2)\cosh\frac{\hbar\Omega_j^B}{kT}-2R_jQ_j\right)\right)}$$

with Ω_j^B, $j = 1 \ldots N$, the eigenvalues of the matrix Ω_B.

By a direct computation and by exploiting the results of theorem 5.4 the influence functional becomes

$$\begin{aligned} F(\gamma,\gamma',x,y) \qquad\qquad (5.30) \\ = e^{\frac{i}{2\hbar}\int_0^t C^T(\gamma(s)+x-\gamma'(s)-y)\Omega_B^{-1}\int_0^s \sin(\Omega_B(s-r))C^T(\gamma(r)+x+\gamma'(r)+y)drds} \\ e^{-\frac{1}{2\hbar}\int_0^t C^T(\gamma(s)+x-\gamma'(s)-y)\Omega_B^{-1}\coth\left(\frac{\hbar\Omega_B}{2kT}\right)\int_0^s \cos(\Omega_B(s-r))C^T(\gamma(r)+x-\gamma'(r)-y)drds}, \end{aligned}$$

which yields the result heuristically derived in [121, 63].

In Theorem 5.4 the allowed values of the time variable t are restricted by conditions (5.25), (5.26), and (5.27), as well as by $\det[\cos(\Omega_{AB}t)]_{d\times d} \neq 0$. Since the influence functional (5.30) is well defined also for the excluded values of t, we can extend the formula "by continuity" to all times.

5.3 The stochastic Schrödinger equation

In order to realize the heuristic Feynman path integrals (5.11) and (5.15), it is necessary to generalize definition 2.4 and theorem (2.5) to complex-valued phase functions.

Let $\mathcal{H}$ be a real separable Hilbert space and let us denote by $\mathcal{H}^\mathbb{C}$ its complexification. An element $x \in \mathcal{H}^\mathbb{C}$ is a couple of vectors $x = (x_1, x_2)$, with $x_1, x_2 \in \mathcal{H}$, or with a different notation $x = x_1 + ix_2$. The multiplication of the vector $x \in \mathcal{H}^\mathbb{C}$ for the pure imaginary scalar $i = \sqrt{-1}$ is given by $ix = (-x_2, x_1)$. A vector $y \in \mathcal{H}$ can be seen as the element $(y, 0) \in \mathcal{H}^\mathbb{C}$. With an abuse of notation, let us denote with A the extension to $\mathcal{H}^\mathbb{C}$ of a linear operator $A : D(A) \subseteq \mathcal{H} \to \mathcal{H}$:

$$A : D^\mathbb{C}(A) \subseteq H^\mathbb{C} \to H^\mathbb{C}, \qquad D^\mathbb{C}(A) = D(A) + iD(A),$$

$$Ax = A(x_1, x_2) = (Ax_1, Ax_2).$$

Let $dim(\mathcal{H}) = 1$, i.e. $\mathcal{H} = \mathbb{R}$, $\mathcal{H}^{\mathbb{C}} = \mathbb{C}$. Then, for any $f \in \mathcal{F}(\mathbb{R})$, $f = \hat{\mu}_f$, and any complex constant $\alpha \in \mathbb{C}$, $\alpha \neq 0$, $\mathrm{Im}(\alpha) \geq 0$, one can easily prove the following equality

$$\widetilde{\int_{\mathbb{R}}} e^{\frac{i\alpha}{2\hbar}x^2} f(x)dx = \alpha^{-1/2} \int_{\mathbb{R}} e^{\frac{-i\hbar}{2\alpha}x^2} d\mu_f(x)\,. \tag{5.31}$$

The proof is completely similar to the proof of theorem 2.2.
More generally, given $\alpha \in \mathbb{C}$, $\alpha \neq 0$, $\mathrm{Im}(\alpha) > 0$ and $\beta \in \mathbb{R}$

$$\widetilde{\int_{\mathbb{R}}} e^{\frac{i\alpha}{2\hbar}x^2} e^{\beta x} f(x)dx = \alpha^{-1/2} \int_{\mathbb{R}} e^{\frac{-i\hbar}{2\alpha}(x-i\beta)^2} d\mu_f(x)\,. \tag{5.32}$$

Such a result can be generalized to the infinite dimensional case [22, 23, 13]:

Theorem 5.5. *Let $\mathcal{H}$ be a real separable Hilbert space, let $y \in \mathcal{H}$ be a vector in $\mathcal{H}$ and let L_1 and L_2 be two self-adjoint, trace class commuting operators on $\mathcal{H}$ such that $I + L_1$ is invertible and L_2 is non negative. Let moreover $f : \mathcal{H} \to \mathbb{C}$ be the Fourier transform of a complex bounded variation measure μ_f on $\mathcal{H}$:*

$$f(x) = \hat{\mu}_f(x), \qquad f(x) = \int_{\mathcal{H}} e^{i\langle x,k\rangle} d\mu_f(k).$$

Then the function $g : \mathcal{H} \to \mathbb{C}$ given by

$$g(x) = e^{\frac{i}{2\hbar}\langle x,Lx\rangle} e^{\langle y,x\rangle} f(x)$$

(L being the operator on the complexification $\mathcal{H}^{\mathbb{C}}$ of the real Hilbert space $\mathcal{H}$ given by $L = L_1 + iL_2$) is Fresnel integrable (in the sense of definition 2.4) and its Fresnel integral

$$\widetilde{\int_{\mathcal{H}}} e^{\frac{i}{2\hbar}\langle x,(I+L)x\rangle} e^{\langle y,x\rangle} f(x)dx$$

can be explicitly computed by means of the following Parseval type equality:

$$\widetilde{\int_{\mathcal{H}}} e^{\frac{i}{2\hbar}\langle x,(I+L)x\rangle} e^{\langle y,x\rangle} f(x)dx$$
$$= \det(I+L)^{-1/2} \int_{\mathcal{H}} e^{\frac{-i\hbar}{2}\langle k-iy,(I+L)^{-1}(k-iy)\rangle} d\mu_f(k)\,. \tag{5.33}$$

Proof. First of all one can notice that both sides of Eq. (5.33) are well defined. Indeed one can easily prove that $(I+L): H_{\mathbb{C}} \to H_{\mathbb{C}}$ is invertible, if $(I+L_1)$ is invertible and that $\det(I+L)$ exists as L is trace class. On the other hand the function on $f_1 : \mathcal{H} \to \mathbb{C}$

$$f_1(x) = e^{-\frac{1}{2\hbar}\langle x, L_2 x\rangle} e^{\langle y,x\rangle} f(x)$$

where $y \in \mathcal{H}$ and $f \in \mathcal{F}(\mathcal{H})$, $f(x) = \hat{\mu}_f(x)$, $\mu_f \in \mathcal{M}(\mathcal{H})$ is the Fourier transform of a complex bounded variation measure μ on $\mathcal{H}$. In fact μ is the convolution of μ_f and the measure ν, with

$$\nu(dx) = e^{\frac{\hbar}{2}\langle y, L_2^{-1} y\rangle - i\hbar\langle y, L_2^{-1} x\rangle} \mu_{L_2}(dx),$$

where μ_{L_2} is the Gaussian measure on $\mathcal{H}$ with covariance operator $L_2/\hbar$. By theorem 2.5 the Fresnel integral of the function g is well defined and can be explicitly computed:

$$\begin{aligned}
\widetilde{\int_{\mathcal{H}}} e^{\frac{i}{2\hbar}\langle x,(I+L)x\rangle} e^{\langle y,x\rangle} f(x)dx &= \widetilde{\int_{\mathcal{H}}} e^{\frac{i}{2\hbar}\langle x,(I+L_1)x\rangle} e^{-\frac{1}{2\hbar}\langle x,L_2 x\rangle} e^{\langle y,x\rangle} f(x)dx \\
&= \det(I+L_1)^{-1/2} \int_{\mathcal{H}} e^{\frac{-i\hbar}{2}\langle x,(I+L_1)^{-1}x\rangle} \mu_f * \nu(dx) \\
&= \det(I+L_1)^{-1/2} \int_{\mathcal{H}}\int_{\mathcal{H}} e^{\frac{-i\hbar}{2}\langle x+z,(I+L_1)^{-1}(x+z)\rangle} \mu_f(dz)\nu(dx) \\
&= \det(I+L_1)^{-1/2} \int_{\mathcal{H}}\int_{\mathcal{H}} e^{\frac{-i\hbar}{2}\langle x+z,(I+L_1)^{-1}(x+z)\rangle} \\
&\qquad e^{\frac{\hbar}{2}\langle y,L_2^{-1}y\rangle - i\hbar\langle y,L_2^{-1}x\rangle} \mu_{L_2}(dx)\mu_f(dz)\,. \quad (5.34)
\end{aligned}$$

Equation (5.33) can be proved by taking the finite dimensional approximation of the last line of equation (5.34) and of the right hand side of (5.33) and showing they coincide (see [13] for more details). □

Let us consider now Belavkin equation (5.13) describing the posterior dynamics of a quantum particle, whose position is continuously observed. Equation (5.13) can also be written in the Stratonovich equivalent form:

$$\begin{cases} d\psi = -\frac{i}{\hbar}H\psi dt - \lambda|x|^2\psi dt + \sqrt{\lambda}x\psi \circ dB(t) \\ \psi(0,x) = \psi_0(x) \qquad\qquad t \geq 0,\ x \in \mathbb{R}^d \end{cases} \quad (5.35)$$

The existence and uniqueness of a strong solution of equations (5.13) and (5.35)[3] is proved in [138]. We shall prove that it can be represented by an infinite dimensional oscillatory integral on a suitable Hilbert space.

[3] A strong solution for the stochastic equation (5.35) is a predictable process with values in $\mathcal{H} = L^2(\mathbb{R}^d)$, such that:
$\psi(t) \in D(-i/\hbar H - \lambda|x|^2)$ **P**-a.s.

Let us consider the Cameron Martin space $\mathcal{H}_t$ and let $\mathcal{H}_t^{\mathbb{C}}$ be its complexification. Let $L : \mathcal{H}_t^{\mathbb{C}} \to \mathcal{H}_t^{\mathbb{C}}$ be the linear operator on $\mathcal{H}_t^{\mathbb{C}}$ defined by

$$\langle \gamma_1, L\gamma_2 \rangle = -a^2 \int_0^t \gamma_1(s) \cdot \gamma_2(s) ds,$$

where $a^2 = -2i\lambda\hbar$. The j–th component of $L\gamma$, $L\gamma = (L\gamma_1, \ldots, L\gamma_d)$, is given by

$$(L\gamma)_j(s) = 2i\lambda\hbar \int_s^t ds' \int_0^{s'} \gamma_j(s'')ds'' \qquad j = 1, \ldots, d. \tag{5.37}$$

As we have seen in section 3.1, the operator $iL : \mathcal{H}_t \to \mathcal{H}_t$ is self-adjoint with respect to the $\mathcal{H}_t$-inner product, it is trace-class and the Fredholm determinant of $(I + L)$ is given by:

$$\det(I + L) = \cos(at).$$

Moreover $(I + L)$ is invertible and its inverse is given by:

$$[(I + L)^{-1}\gamma]_j(s) = \gamma_j(s) - a \int_s^t \sin[a(s' - s)]\gamma_j(s')ds'$$

$$+ \sin[a(t - s)] \int_0^t [\cos at]^{-1} a \cos(as')\gamma_j(s')ds', \qquad j = 1, \ldots, d.$$

For any $\omega \in C_t$, let us introduce the vector $l \in \mathcal{H}_t$ defined by

$$\langle l, \gamma \rangle = -\sqrt{\lambda} \int_0^t \omega(s) \cdot \dot{\gamma}(s)ds = \sqrt{\lambda} \int_0^t \gamma(s) \cdot dB(s), \tag{5.38}$$

$$l(s) = \sqrt{\lambda} \int_s^t \omega(\tau)d\tau.$$

With this notation, we can apply the theorem 5.5 and prove that, under suitable assumptions on the potential V and the initial wave function ψ_0,

---

$\mathbf{P}\Big(\int_0^T (\|\psi(t)\|^2 + \|(-i/\hbar H - \lambda|x|^2)\psi\|^2)\, dt < \infty \Big) = 1$

$\mathbf{P}\Big(\int_0^T \||x|\psi(t)\, dt\|^2 < \infty \Big) = 1$ and

$\mathbf{P}$ a.s. for all $t \in [0, T]$:

$$\begin{cases} d\psi = -\frac{i}{\hbar} H\psi dt - \lambda|x|^2\psi dt + \sqrt{\lambda} x \cdot \psi \circ dB(t) & t \geq 0,\ x \in \mathbb{R}^d \\ \psi(0, x) = \psi_0(x)\,. \end{cases} \tag{5.36}$$

the heuristic expression (5.15) can be realized as the infinite dimensional oscillatory integral with complex phase on the Cameron-Martin space $\mathcal{H}_t$:

$$C(t,x,\omega)\widetilde{\int_{\mathcal{H}_t}} e^{\frac{i}{2\hbar}\langle\gamma,(I+L)\gamma\rangle}e^{\langle l,\gamma\rangle}e^{-2\lambda x\cdot\int_0^t\gamma(s)ds}e^{-\frac{i}{\hbar}\int_0^t V(\gamma(s)+x)ds}\psi_0(\gamma(0)+x)d\gamma \tag{5.39}$$

where $C(t,x,\omega)=e^{-\lambda|x|^2+\sqrt{\lambda}x\cdot\omega(t)}$ is a constant depending on t, $x\in\mathbb{R}^d$, $\omega\in C_t$. Indeed the integrand $\exp(\frac{i}{2\hbar}\Phi)$ in (5.15), where

$$\Phi(\gamma)\equiv\int_0^t|\dot\gamma(s)|^2ds+2i\hbar\lambda\int_0^t|\gamma(s)+x|^2ds-2i\hbar\int_0^t\sqrt{\lambda}(\gamma(s)+x)\cdot dB(s),$$

can be rigorously defined as the functional on the Cameron Martin space $\mathcal{H}_t$ given by

$$\Phi(\gamma)=\langle\gamma,(I+L)\gamma\rangle-2i\hbar\langle l,\gamma\rangle-2\hbar\int_0^t a^2x\cdot\gamma(s)ds-a^2|x|^2t-2i\hbar\sqrt{\lambda}x\cdot\omega(t),$$

where L is the operator (5.37) and l is the vector (5.38).
By means of theorem 5.5 one can compute the integral (5.39) in terms of an absolutely convergent integral on $\mathcal{H}_t$. Moreover it is possible to prove it represents the solution of Belavkin equation (5.35) (see [13]).

Theorem 5.6. *Let V and ψ_0 be Fourier transforms of complex bounded variation measures on $\mathbb{R}^d$. Then there exist a (strong) solution to the Stratonovich stochastic differential equation (5.35) and it is given by the infinite dimensional oscillatory integral with complex phase* (5.39).

Remark 5.1. The result can be extended to general initial vectors $\psi_0\in L^2(\mathbb{R}^d)$, using the fact that $\mathcal{F}(\mathbb{R}^d)$ is dense in $L^2(\mathbb{R}^d)$.

Proof. The proof is divided into 3 steps: in the first two we consider the case $V\equiv 0$. First of all we deal with an approximated problem and we find a representation for its solution via an infinite dimensional oscillatory integral, then we show that the sequence of approximated solutions converges in a suitable sense to the solution of problem (5.35). In the final step we introduce the potential V and show that the right hand side of (5.39) is in fact the solution of the equation (5.35).
1. The solution of the approximated problem.
We approximate the trajectory $t\to\omega(t)$ of the Wiener process by a sequence of smooth curves. More precisely we consider the sequence of functions[4]

$$n\int_{t-\frac{1}{n}}^t\omega(s)d\,s\equiv\omega_n(t),\qquad n\in\mathbb{N}.$$

[4]Here we denote, as usual, the trajectory of the Wiener process $B(t)$ as $\omega(t)$.

We have $\omega_n \to \omega$ uniformly on $[0,T]$, more precisely

$$\sup_{s\in[0,T]} |w_n(s) - w(s)| \to 0 \quad \text{as } n \to \infty \qquad \mathbb{P} \text{ a.s.}$$

Let us consider the sequence of approximated problems:

$$\begin{cases} d\psi_n = -\frac{i}{\hbar}H\psi_n dt - \lambda|x|^2\psi_n dt + \sqrt{\lambda}x \cdot \psi_n dB_n(t) \\ \psi_n(0,x) = \psi_0(x) \end{cases} \tag{5.40}$$

where $dB_n(t)$ is an ordinary differential, i.e. $dB_n(t) = \dot{\omega}_n(t)dt$, and we can also write:

$$\begin{cases} \dot{\psi}_n = -\frac{i}{\hbar}H\psi_n - \lambda|x|^2\psi_n + \sqrt{\lambda}x \cdot \psi_n\dot{\omega}_n(t) \\ \psi_n(0,x) = \psi_0(x) \end{cases} \tag{5.41}$$

which can be recognized as a family of Schrödinger equations, with a complex potential, labeled by the random parameter $\omega \in \Omega$.
Now we compute a representation of the solution of (5.41) by means of an infinite dimensional oscillatory integral with complex phase, under suitable assumptions on the (real) potential V and on the initial datum $\psi_n(0,x,\omega) = \psi_0(x)$.
We can write equation (5.41) in the following form:

$$\begin{cases} \dot{\psi}_n = -\frac{i}{\hbar}(\frac{-\hbar^2\Delta}{2m} - i\lambda\hbar|x|^2)\psi_n - \frac{i}{\hbar}V\psi_n + \sqrt{\lambda}x \cdot \psi_n\dot{\omega}_n(t) \\ \psi_n(0,x) = \psi_0(x) \end{cases} \tag{5.42}$$

so that we can recognize in it the Schrödinger equation for an anharmonic oscillator with a complex potential, i.e.

$$\begin{cases} \dot{\psi}_n = -\frac{i}{\hbar}(\frac{-\hbar^2\Delta}{2m} + \frac{a^2}{2}|x|^2)\psi_n - \frac{i}{\hbar}U\psi_n \\ \psi_n(0,x) = \psi_0(x) \end{cases} \tag{5.43}$$

where $a^2 = -2i\lambda\hbar$ and $U = U(t,x,\omega) = V(x) + i\hbar\sqrt{\lambda}x \cdot \dot{\omega}_n(t)$.
We introduce the sequence of vectors $l_n \in \mathcal{H}_t$ defined by

$$\langle l_n, \gamma\rangle = \sqrt{\lambda}\int_0^t \gamma(s) \cdot \dot{\omega}_n(s)ds = -\sqrt{\lambda}\int_0^t \omega_n(s) \cdot \dot{\gamma}(s)ds,$$

which is given by

$$l_n(s) = \sqrt{\lambda}\int_s^t \omega_n(\tau)d\tau. \tag{5.44}$$

First of all let us consider equation (5.35) with H replaced by the free Hamiltonian $H = -\hbar^2\Delta/2$. The following result holds:

Lemma 5.1. *Let $\psi_0 \in \mathcal{S}(\mathbb{R}^d)$. Then the solution of the Cauchy problem:*

$$\begin{cases} \dot{\psi}_n(t,x) = \frac{i\hbar}{2}\Delta\psi_n(t,x) - \lambda|x|^2\psi_n(t,x) + \sqrt{\lambda}x \cdot \dot{\omega}_n(t)\psi_n(t,x) \\ \psi_n(0,x) = \psi_0(x), \qquad x \in \mathbb{R}^d \end{cases} \tag{5.45}$$

is given by:

$$\psi_n(t,x) = \widetilde{\int_{\mathcal{H}_t}} e^{\frac{i}{2\hbar}\int_0^t |\dot{\gamma}(s)|^2 ds - \lambda \int_0^t |\gamma(s)+x|^2 ds} e^{\sqrt{\lambda}\int_0^t (\gamma(s)+x)\cdot\dot{\omega}_n(s) ds} \psi_0(\gamma(0)+x) d\gamma \tag{5.46}$$

(where the right hand side is interpreted as the infinite dimensional oscillatory integral of $\psi_0(\gamma(0)+x)e^{\langle l_n,\gamma\rangle}$ with complex quadratic phase function $\langle\gamma,(I+L)\gamma\rangle/\hbar$, with $\mathcal{H}_t$ the Cameron-Martin space, l_n the vector defined by (5.44) and L the operator defined by (5.37).)

Proof. Formula (5.46) can be realized as

$$\widetilde{\int_{\mathcal{H}_t}} e^{\frac{i}{2\hbar}\int_0^t |\dot{\gamma}(s)|^2 ds - \lambda \int_0^t |\gamma(s)+x|^2 ds} e^{\sqrt{\lambda}\int_0^t (\gamma(s)+x)\cdot\dot{\omega}_n(s) ds} \psi_0(\gamma(0)+x) d\gamma$$

$$= e^{\frac{-ia^2|x|^2 t}{2\hbar} + \sqrt{\lambda}x\cdot\omega_n(t)} \widetilde{\int_{\mathcal{H}_t}} e^{\frac{i}{2\hbar}\langle\gamma,(I+L)\gamma\rangle} e^{\langle l_n,\gamma\rangle} \int_{\mathbb{R}^d} e^{i\alpha\cdot x} e^{i\langle b(\alpha,x),\gamma\rangle} \tilde{\psi}_0(\alpha) d\alpha d\gamma$$

where $b(\alpha,x) \in \mathcal{H}_t$, precisely:

$$b(\alpha,x)(s) = \alpha(t-s) - \frac{xa^2}{2\hbar}(t^2-s^2),$$

One can directly verify that the function

$$f(\gamma) \equiv \int_{\mathbb{R}^d} e^{i\alpha\cdot x} e^{i\langle b(\alpha,x),\gamma\rangle} \tilde{\psi}_0(\alpha) d\alpha, \qquad \gamma \in \mathcal{H}_t$$

is the Fourier transform of a measure $\mu \in \mathcal{M}(\mathcal{H}_t)$, that is:

$$\mu(d\gamma) = \int_{\mathbb{R}^d} e^{i\alpha\cdot x} \tilde{\psi}_0(\alpha) \delta_{b(\alpha,x)}(d\gamma) d\alpha$$

so we can apply theorem 5.5 and the integral (5.46) is equal to:

$$e^{\frac{-ia^2|x|^2 t}{2\hbar} + \sqrt{\lambda}\cdot\omega_n(t)} \int_{\mathbb{R}^d} e^{i\alpha\cdot x} \det(I+L)^{-1/2} \\ e^{\frac{-i\hbar}{2}\langle b(\alpha,x)-il_n,(I+L)^{-1}(b(\alpha,x)-il_n)\rangle} \tilde{\psi}_0(\alpha) d\alpha.$$

By simple calculations we get the final result:

$$\psi_n(t,x) = \int_{\mathbb{R}^d} G_n(t,x,y)\psi_0(y) dy,$$

where $G_n(t,x,y)$ is given by:

$$G_n(t,x,y) \equiv \frac{1}{\sqrt{2\pi i\hbar}}\sqrt{\frac{a}{\sin(at)}}e^{\sqrt{\lambda}x\cdot\omega_n(t)-\frac{\sqrt{\lambda}ax}{\sin(at)}\cdot\int_0^t\omega_n(s)\cos(as)ds}$$

$$e^{\frac{i\hbar\lambda}{2}\int_0^t|\omega_n(s)|^2ds}e^{\frac{i\hbar\lambda}{2}(-a\int_0^t\omega_n(s)\cdot\int_s^t\omega_n(s')\sin[a(s'-s)]ds'ds)}$$

$$\cdot e^{\frac{i\hbar\lambda}{2}(-a\int_0^t\sin(as)\omega_n(s)ds\cdot\int_0^t\cos(as)\omega_n(s)ds-a\cot(at)|\int_0^t\cos(as)\omega_n(s)ds|^2)}$$

$$e^{\frac{i}{2\hbar}(\cot(at)(|x|^2+|y|^2)-\frac{2x\cdot y}{\sin(at)})}\cdot e^{a\sqrt{\lambda}y\cdot(\cot(at)\int_0^t\cos(as)\omega_n(s)ds+\int_0^t\sin(as)\omega_n(s)ds)} \tag{5.47}$$

which is, as one can easily directly verify, the fundamental solution to the approximate Cauchy problem (5.40). □

2. The convergence of the sequence of approximated solutions.
We will prove the following result:

Lemma 5.2. *The following equation*

$$\begin{cases} d\psi = -\frac{i}{\hbar}H\psi dt - \lambda|x|^2\psi dt + \sqrt{\lambda}x\cdot\psi\circ dB(t) & t>0 \\ \psi(0,x)=\psi_0(x), \quad \psi_0\in S(\mathbb{R}^d), \end{cases} \tag{5.48}$$

with $H=-\hbar^2\Delta/2$, has a unique strong solution given by the Feynman path integral

$$\psi(t,x) = \widetilde{\int} e^{\frac{i}{2\hbar}\int_0^t|\dot\gamma(s)|^2ds-\lambda\int_0^t|\gamma(s)+x|^2ds}e^{\sqrt{\lambda}\int_0^t(\gamma(s)+x)\cdot dB(s)}$$

$$\psi_0(\gamma(0)+x)d\gamma$$

rigorously realized as the infinite dimensional oscillatory integral with complex phase on the Hilbert space $\mathcal{H}_t$

$$e^{-\lambda|x|^2+\sqrt{\lambda}x\cdot\omega(t)}\widetilde{\int_{\mathcal{H}_t}} e^{\frac{i}{2\hbar}\langle\gamma,(I+L)\gamma\rangle}e^{\langle l,\gamma\rangle}e^{-2\lambda x\cdot\int_0^t\gamma(s)ds}\psi_0(\gamma(0)+x)d\gamma.$$

Moreover it can be represented by the process

$$\psi(t,x) = \int_{\mathbb{R}^d} G(t,x,y)\psi_0(y)dy$$

where

$$G(t,x,y) = \frac{1}{\sqrt{2\pi i\hbar}}\sqrt{\frac{a}{\sin(at)}}e^{\sqrt{\lambda}x\cdot\omega(t)-\frac{\sqrt{\lambda}ax}{\sin(at)}\cdot\int_0^t\cos(as)\omega(s)ds}$$

$$e^{\frac{i\hbar\lambda}{2}(-a\int_0^t\omega(s)\cdot\int_s^t\omega(s')\sin[a(s'-s)]ds'ds)}$$

$$\cdot e^{\frac{i\hbar\lambda}{2}(-a\int_0^t\sin(as)\omega(s)ds\cdot\int_0^t\cos(as)\omega(s)ds-a\cot(at)|\int_0^t\cos(as)\omega(s)ds|^2)}$$

$$e^{\frac{i}{2\hbar}\left(\cot(at)(|x|^2+|y|^2)-\frac{2x\cdot y}{\sin(at)}\right)}e^{a\sqrt{\lambda}y\cdot\frac{1}{\sin(at)}(\int_0^t\cos[a(s-t)]\omega(s)ds)}.$$

Proof. Let us consider the sequence of approximated solutions

$$\psi_n(t,x) = \int_{\mathbb{R}^d} G_n(t,x,y)\psi_0(y)dy.$$

Using the dominated convergence theorem we have that:

$$\mathbf{P}\Big(\lim_{n\to\infty}\int_{\mathbb{R}^d}|\psi_n(t,x)-\tilde{\psi}(t,x)|^2dx \to 0\Big) = 1 \tag{5.49}$$

with $\tilde{\psi}(t,x) = \int_{\mathbb{R}} G(t,x,y)\psi_0(y)dy$, as:

$$\lim_{n\to\infty}|G_n(t,x,y)-G(t,x,y)| \to 0$$

for all $t \in [0,T]$ and $x, y \in \mathbb{R}^d$. Moreover, one can see by a direct computation that $a = \sqrt{-2i\hbar\lambda}$ can be chosen is such a way that:

$$|\int_{\mathbb{R}^d} G_n(t,x,y)\psi_0(y)dy|^2 \le C(t)e^{P(t,x)}\|\psi_0(y)\|^2, \tag{5.50}$$

where $P(t,x)$ is a second order polynomial with negative leading coefficient and $C(t)$ and $P(t,x)$ are continuous functions of the variable $t \in [0,T]$. Applying the Itô formula to the limit process $\tilde{\psi}(t)$ we see that it verifies equation (5.48) for every (t,x,y). Since the kernel $G(t,x,y)$ is adapted to the filtration of the Brownian motion by construction, it follows that the solution is predictable. By direct computation and using estimates analogous to (5.50) one can verify that $\tilde{\psi}$ is a strong solution. On the other hand every $\psi_n(t,x)$ is equal to

$$\widetilde{\int_{\mathcal{H}_t}} e^{\frac{i}{2\hbar}\int_0^t|\dot{\gamma}(s)|^2ds-\lambda\int_0^t|\gamma(s)+x|^2ds}e^{\sqrt{\lambda}\int_0^t(\gamma(s)+x)\cdot\dot{\omega}_n(s)ds}\psi_0(\gamma(0)+x)d\gamma$$

$$= e^{\frac{-ia^2|x|^2t}{2\hbar}+\sqrt{\lambda}x\cdot\omega_n(t)}\widetilde{\int_{\mathcal{H}_t}} e^{\frac{i}{2\hbar}\langle\gamma,(I+L)\gamma\rangle}e^{\langle l_n,\gamma\rangle}e^{-i\int_0^t a^2x\cdot\gamma(s)ds}\psi_0(\gamma(0)+x)d\gamma$$

$$= e^{\frac{-ia^2|x|^2t}{2\hbar}+\sqrt{\lambda}x\cdot\omega_n(t)}\det(I+L)^{-1/2}\int_{\mathcal{H}_t} e^{\frac{-i\hbar}{2}\langle\gamma-il_n,(I+L)^{-1}(\gamma-il_n)\rangle}\mu(d\gamma)$$

where $\mu(d\gamma)$ is the measure on $\mathcal{H}_t$ whose Fourier transform is the function $\gamma \to e^{-i\int_0^t a^2x\cdot\gamma(s)ds}\psi_0(\gamma(0)+x)$.

We have $\|l_n - l\|^2_{\mathcal{H}_t} \to 0$ as $n \to \infty$, where $l(s) = \sqrt{\lambda}\int_s^t \omega(r)dr$. Therefore, by the Lebesgue's dominated convergence theorem, we have that, for every $x \in \mathbb{R}^d$:

$$\lim_{n\to\infty} e^{\frac{-ia^2|x|^2t}{2\hbar}+\sqrt{\lambda}x\cdot\omega_n(t)}\det(I+L)^{-1/2}\int_{\mathcal{H}_t} e^{\frac{-i\hbar}{2}\langle\gamma-il_n,(I+L)^{-1}(\gamma-il_n)\rangle}\mu(d\gamma)$$

$$= e^{\frac{-ia^2|x|^2t}{2\hbar}+\sqrt{\lambda}x\omega(t)}\det(I+L)^{-1/2}\int_{\mathcal{H}_t} e^{\frac{-i\hbar}{2}\langle\gamma-il,(I+L)^{-1}(\gamma-il)\rangle}\mu(d\gamma). \tag{5.51}$$

Therefore, taking into account the uniqueness of the pointwise limit, we have shown that:

$$\psi(t,x) = \int_{\mathbb{R}} G(t,x,y)\psi_0(y)dy$$
$$= \widetilde{\int_{\mathcal{H}_t}} e^{\frac{i}{2\hbar}\int_0^t |\dot\gamma(s)|^2 ds - \lambda\int_0^t |\gamma(s)+x|^2 ds} e^{\int_0^t (\gamma(s)+x)\cdot dB(s)} \psi_0(\gamma(0)+x)d\gamma. \tag{5.52}$$

□

Remark 5.2. The result can be extended by continuity to all $\psi_0 \in L^2(\mathbb{R}^d)$, using the density of $S(\mathbb{R}^d)$ in $L^2(\mathbb{R}^d)$.

3. The proof of Feynman-Kac-Ito formula by means of Dyson expansion. Let us consider now the general case where $H = -\hbar^2\Delta/2 + V$ and complete the proof of theorem 5.6. We follow here the technique by Elworthy and Truman [114].

We set for $t > 0$, $x \in \mathbb{R}^d$:

$$\Theta(t,0)\psi_0(x) = \widetilde{\int_{\mathcal{H}_t}} e^{\frac{i}{2\hbar}\int_0^t |\dot\gamma(s)|^2 ds - \lambda\int_0^t |\gamma(s)+x|^2 ds} e^{-\frac{i}{\hbar}\int_0^t V(\gamma(s)+x)ds}$$
$$\cdot e^{\sqrt{\lambda}\int_0^t (\gamma(s)+x)\cdot dB(s)} \psi_0(\gamma(0)+x)d\gamma \tag{5.53}$$

and

$$\Theta_0(t,0)\psi_0(x) = \widetilde{\int_{\mathcal{H}_t}} e^{\frac{i}{2\hbar}\int_0^t |\dot\gamma(s)|^2 ds - \lambda\int_0^t |\gamma(s)+x|^2 ds} e^{\sqrt{\lambda}\int_0^t (\gamma(s)+x)\cdot dB(s)}$$
$$\psi_0(\gamma(0)+x)d\gamma. \tag{5.54}$$

Then we have:

$$\Theta(t,0)\psi_0(x) = e^{\frac{-ia^2|x|^2 t}{2\hbar} + \sqrt{\lambda}x\cdot\omega(t)} \widetilde{\int_{\mathcal{H}_t}} e^{\frac{i}{2\hbar}\langle\gamma,(I+L)\gamma\rangle} e^{\langle l,\gamma\rangle} e^{-i\int_0^t a^2 x\cdot\gamma(s)ds}$$
$$\cdot e^{-\frac{i}{\hbar}\int_0^t V(x+\gamma(s))ds} \psi_0(\gamma(0)+x)d\gamma. \tag{5.55}$$

Let $\mu_0(\psi)$ be the measure on $\mathcal{H}_t$ such that its Fourier transform evaluated in $\gamma \in \mathcal{H}_t$ is $\psi_0(\gamma(0)+x)$.

For $0 \le u \le t$ let $\mu_u(V,x)$, $\nu_u^t(V,x)$ and $\eta_u^t(x)$ be the measures on $\mathcal{H}_t$, whose Fourier transforms when evaluated at $\gamma \in \mathcal{H}_t$ are respectively $V(x+\gamma(u))$, $\exp\left(-\frac{i}{\hbar}\int_u^t V(x+\gamma(s))ds\right)$, and $\exp\left(\frac{i}{\hbar}\int_u^t a^2 x\gamma(s)ds\right)$. We shall often write $\mu_u \equiv \mu_u(V,x)$, $\nu_u^t \equiv \nu_u^t(V,x)$ and $\eta_u^t \equiv \eta_u^t(x)$. If $\{\mu_u : a \le u \le b\}$ is a family in $\mathcal{M}(\mathcal{H}_t)$, we shall let $\int_a^b \mu_u du$ denote the measure on $\mathcal{H}_t$ given by:

$$f \to \int_a^b \int_{\mathcal{H}_t} f(\gamma)\mu_u(d\gamma)du,$$

whenever it exists.

Then, since for any continuous path γ

$$\exp\Big(-\frac{i}{\hbar}\int_0^t V(\gamma(s)+x)ds\Big) = \\ 1-\frac{i}{\hbar}\int_0^t V(\gamma(u)+x)\exp\Big(-\frac{i}{\hbar}\int_u^t V(\gamma(s)+x)ds\Big)du, \tag{5.56}$$

we have

$$\nu_0^t = \delta_0 - \frac{i}{\hbar}\int_0^t (\mu_u * \nu_u^t)du \tag{5.57}$$

where δ_0 is the Dirac measure at $0 \in \mathcal{H}_t$.
By the Parseval-type equality:

$$\Theta(t,0)\psi_0(x) = e^{\frac{-ia^2|x|^2t}{2\hbar}+\sqrt{\lambda}x\cdot\omega(t)}\det(I+L)^{-1/2} \\ \cdot\int_{\mathcal{H}_t} e^{\frac{-i\hbar}{2}\langle\alpha-il,(I+L)^{-1}(\alpha-il)\rangle}(\eta_0^t*\nu_0^t*\mu_0(\psi))(d\alpha) \tag{5.58}$$

Applying to this equality (5.57) we obtain:

$$\Theta(t,0)\psi_0(x) = \\ e^{\frac{-ia^2|x|^2t}{2\hbar}+\sqrt{\lambda}x\cdot\omega(t)}\det(I+L)^{-1/2}\int_{\mathcal{H}_t} e^{\frac{-i\hbar}{2}\langle\alpha-il,(I+L)^{-1}(\alpha-il)\rangle}(\eta_0^t*\mu_0(\psi))(d\alpha) \\ -\frac{i}{\hbar}\int_0^t e^{\frac{-ia^2|x|^2t}{2\hbar}+\sqrt{\lambda}x\cdot\omega(t)}\det(I+L)^{-1/2} \\ \cdot\int_{\mathcal{H}_t} e^{\frac{-i\hbar}{2}\langle\alpha-il,(I+L)^{-1}(\alpha-il)\rangle}(\eta_0^t*\mu_u(V,x)*\nu_u^t*\mu_0(\psi))(d\alpha)du \\ = \Theta_0(t,0)\psi_0(x) - \frac{i}{\hbar}\int_0^t \widetilde{\int_{\mathcal{H}_t}} e^{\frac{i}{2\hbar}\int_0^t|\dot\gamma(s)|^2ds-\lambda\int_0^t|\gamma(s)+x|^2ds}e^{-\frac{i}{\hbar}\int_u^t V(\gamma(s)+x)ds} \\ e^{\sqrt{\lambda}\int_0^t(\gamma(s)+x)\cdot dB(s)}V(\gamma(u)+x)\psi_0(\gamma(0)+x)d\gamma du.$$

By the Fubini theorem for oscillatory integrals (see 2.7, we get that

$$\widetilde{\int_{\mathcal{H}_t}} e^{\frac{i}{2\hbar}\int_0^t|\dot\gamma(s)|^2ds-\lambda\int_0^t|\gamma(s)+x|^2ds}e^{-\frac{i}{\hbar}\int_u^t V(\gamma(s)+x)ds} \\ e^{\sqrt{\lambda}\int_0^t(\gamma(s)+x)\cdot dB(s)}V(\gamma(u)+x)\psi_0(\gamma(0)+x)d\gamma \\ = \widetilde{\int_{\mathcal{H}_{u,t}}} e^{\frac{i}{2\hbar}\int_u^t|\dot\gamma_2(s)|^2ds-\lambda\int_u^t|\gamma_2(s)+x|^2ds}e^{-\frac{i}{\hbar}\int_u^{tu} V(\gamma_2(s)+x)ds}\cdot \\ e^{\sqrt{\lambda}\int_u^t(\gamma_2(s)+x)\cdot dB(s)}V(\gamma_2(u)+x)\widetilde{\int_{\mathcal{H}_{0,u}}} e^{\frac{i}{2\hbar}\int_0^u|\dot\gamma_1(s)|^2ds-\lambda\int_0^u|\gamma_1(s)+\gamma_2(u)+x|^2ds}\cdot \\ e^{\sqrt{\lambda}\int_0^u(\gamma_1(s)+\gamma_2(u)+x)\cdot dB(s)}\psi_0(\gamma_1(0)+\gamma_2(u)+x)d\gamma_1 d\gamma_2.$$

Here $\gamma_1 \in \mathcal{H}_{0,u}$ and $\gamma_2 \in \mathcal{H}_{u,t}$ are the integration variables, where we have denoted by $\mathcal{H}_{r,s}$ the Cameron-Martin space of paths $\gamma : [r,s] \to \mathbb{R}^d$.
Finally we have:

$$\Theta(t,0)\psi_0(x) = \Theta_0(t,0)\psi_0(x) - i\int_0^t \Theta(t,u)(V\Theta_0(u,0)\psi_0)(x)du \qquad (5.59)$$

Now the iterative solution of the latter integral equation is the Dyson series for $\Theta(t,0)$, which coincides with the corresponding power series expansion of the solution of the stochastic Schrödinger equation, which converges strongly in $L^2(\mathbb{R}^d)$. The equality holds pointwise. On the other hand, following [138], it is possible to prove that the problem (5.48) has a strong solution that verifies (5.59) in the L^2 sense, therefore $\Theta(t,0)\psi_0$ coincides with the solution $\psi(t)$. This concludes the proof of theorem 5.6. □

It is also possible to consider Belavkin's equation describing the continuous measurement of the momentum p of a $d-$dimensional quantum particle:

$$\begin{cases} d\psi(t,x) = -\frac{i}{\hbar}H\psi(t,x)dt + \frac{\lambda\hbar^2}{2}\Delta\psi(t,x)dt - i\sqrt{\lambda}\hbar\nabla\psi(t,x)dB(t) \\ \psi(0,x) = \psi_0(x) \qquad (t,x) \in [0,T]\times\mathbb{R}^d\,. \end{cases} \qquad (5.60)$$

The stochastic term plays the role of a complex random potential depending on the momentum of the particle. In this case one has to use the phase space Feynman path integrals described in section 3.3. More precisely by means of a infinite dimensional oscillatory integral with complex phase on the space of paths in phase space one can give a rigorous mathematical meaning to the following heuristic expression:

$$\psi(t,x) = \widetilde{\int} e^{\frac{i}{\hbar}(\int_0^t(\dot{q}(s)p(s)-\frac{1}{2m}p(s)^2)ds - \lambda\int_0^t p(s)^2 ds}e^{-\frac{i}{\hbar}\int_0^t V(q(s)+x)ds} \cdot e^{\sqrt{\lambda}\int_0^t p(s)\cdot dB(s)}\psi_0(\gamma(0)+x)dqdp. \qquad (5.61)$$

See [14] for a detailed study of this problem.

Chapter 6

Alternative Approaches to Feynman Path Integration

Since the first introduction of Feynman path integrals [122], several approaches for their mathematical definition and several applications have been proposed, both in the mathematical and in the physical literature. The present chapter is a brief survey of this topic, without any claim of completeness.

6.1 Analytic continuation of Wiener integrals

One of the first attempts to the rigorous mathematical realization of Feynman path integrals involves analytic continuation of Gaussian Wiener integrals. The first rigorous results can be found in Cameron's paper in 1960 [64] and where further developed in a series of papers [65, 66, 68, 67, 69–71, 181, 182, 184, 183, 185, 186, 175]. As Cameron proved, it is not possible to define a Wiener measure W_λ on the space of continuous paths $C([0,t])$ with a complex covariance $\lambda \in \mathbb{C}$, as, unless $\lambda \in \mathbb{R}^+$, it would have infinite total variation. Therefore the expression

$$\int_{C([0,t])} f(\omega) dW_\lambda(\omega),$$

is meaningless. On the other hand, for $\lambda \in \mathbb{R}^+$ and for a Borel function f on $C_t \equiv C([0,t])$ satisfying suitable conditions, the following formula holds [64]:

$$\int_{C_t} f(\omega) dW_\lambda(\omega) = \int_{C_t} f(\sqrt{\lambda}\omega) dW(\omega). \tag{6.1}$$

If λ is complex, the left hand side or Eq. (6.1) is not well defined, but the right hand side can still be meaningfull, provided that the function f has suitable analyticity and measurability properties. In particular, for

$\lambda = i$, the right hand side of Eq. (6.1) is the natural candidate for an "analytic Wiener integral" or "analytic Feynman integral". The class of functions which are "integrable" in this sense does not substantially differ from the Fresnel class $\mathcal{F}(\mathcal{H}_t)$ which can be handled by means of the infinite dimensional oscillatory integral approach (see [175]).

Concerning the application of analytically continued Wiener Gaussian integrals to the Feynman path integral representation for the solution of the Schrödinger equation, the fundamental idea is the extension to the complex case of the Wiener integral representation for the solution of the heat equation

$$\begin{cases} -\frac{\partial}{\partial t}u = -\frac{1}{2}\Delta u + V(x)u \\ u(0,x) = u_0(x), \qquad x \in \mathbb{R}^d, \end{cases} \tag{6.2}$$

i.e. the Feynman-Kac formula:

$$u(t,x) = \int_{C_t} e^{-\int_0^t V(\omega(s)+x)ds} u_0(\omega(t)+x)dW(\omega). \tag{6.3}$$

Indeed, by introducing in Eq. (6.2) a real parameter λ, related to the time t

$$-\lambda\hbar\frac{\partial}{\partial t}u = -\frac{1}{2m}\hbar^2\Delta u + V(x)u$$
$$u(t,x) = \int_{C_t} e^{-\frac{1}{\lambda\hbar}\int_0^t V(\sqrt{\hbar/(m\lambda)}\omega(s)+x)ds} u_0(\sqrt{\hbar/(m\lambda)}\omega(t)+x)dW(\omega),$$

or the Planck constant $\hbar$ [108]

$$\lambda\frac{\partial}{\partial t}u = \frac{1}{2m}\lambda^2\Delta u + V(x)u$$
$$u(t,x) = \int_{C_t} e^{\frac{1}{\lambda}\int_0^t V(\sqrt{\lambda/m}\,\omega(s)+x)ds} u_0(\sqrt{\lambda/m}\,\omega(t)+x)dW(\omega),$$

or the mass m [235, 191, 10]

$$\frac{\partial}{\partial t}u = \frac{1}{2\lambda}\Delta u - iV(x)u,$$
$$u(t,x) = \int_{C_t} e^{-i\int_0^t V(\sqrt{1/\lambda}\omega(s)+x)ds} u_0(\sqrt{1/\lambda}\omega(t)+x)dW(\omega),$$

and by substituting respectively $\lambda = -i$, $\lambda = i\hbar$, or $\lambda = -im$ (when the resulting Wiener integral is well defined), one gets, at least heuristically, Schrödinger equation (with $\hbar = 1$ in the latter case) and its solution. These procedures can be made completely rigorous under suitable conditions on the potential V and the initial datum u_0. For further results and details, see also for instance [81, 82, 87, 176, 177, 180, 190, 193, 218, 230, 234, 264, 271, 272] and [274–278, 281, 282, 294].

The class of classical potentials V which can be handled by means of these methods does not substantially differ from those of the type "quadratic plus Fourier transform of measure". It is worthwhile to point out however that Nelson [235] handles potentials which are singular at the origin, Doss [108] can deal with some polynomially growing potentials, while Albeverio, Brzeźniak and Haba [10] study the case of potentials that are Laplace transform of measures and can have an exponential growth at infinity (for analogous results in the framework of the white noise calculus see [210]).

Furthermore, this particular approach allows the study of the semiclassical limit $\hbar \downarrow 0$ of the solution of the Schrödinger equation [47, 48, 57] by means of (a modified version of) the Laplace method for the study of the asymptotic of the Wiener integrals [256, 56, 58, 112, 113, 192, 241, 242, 252, 93–96, 40, 24].

It is worthwhile to recall the alternative way to define Feynman integrals by means of convergent Wiener integrals proposed by Daubechies and Klauder [90, 91, 88, 89, 204]. The authors study the matrix elements of the unitary evolution operator $U(t) = e^{-\frac{i}{\hbar}Ht}$ between two coherent states $\phi_{q',p'}, \phi_{q'',p''}$, defined by

$$\phi_{q,p} = e^{i(pQ-qP)}\phi_0,$$

where Q, P are respectively the quantum position and momentum operators and ϕ_0 is the ground state of an harmonic oscillator. Formally the matrix elements $\langle \phi_{q'',p''}, U(t)\phi_{q',p'}\rangle$, in the case $\hbar = 1$ and $d = 1$, should be given by the phase space Feynman path integral:

$$\langle \phi_{q'',p''}, U(t)\phi_{q',p'}\rangle = \mathcal{N} \int e^{\frac{i}{2}\int_0^t (p(s)\dot{q}(s)-q(s)\dot{p}(s))ds - i\int_0^t H(p(s),q(s))ds} dpdq, \tag{6.4}$$

where $(q(s), p(s))_{s\in[0,t]}$ represents a generic path in the phase space, while $H : \mathbb{R}^2 \to \mathbb{R}$ is defined as the matrix element of the quantum Hamiltonian operator, i.e.

$$H(q,p) = \langle \phi_{q,p}, H\phi_{q,p}\rangle, \qquad (q,p) \in \mathbb{R}^2,$$

and $\mathcal{N}$ represents a normalization constant. The heuristic expression (6.4) is defined by inserting into the integrand an extra factor

$$e^{-\frac{1}{2\nu}\int_0^t (\dot{p}(s)^2+\dot{q}(s)^2)ds},$$

representing formally the density of a Wiener measure (on the space of paths $(q(s), p(s))_{s\in[0,t]}$ in the phase space) with diffusion constant $\nu > 0$. The exponent $\int_0^t (p(s)\dot{q}(s) - q(s)\dot{p}(s))ds$ is replaced by $\int_0^t (p(s)dq(s) - q(s)dp(s))$

and interpreted as a (Ito or Stratonovich) stochastic integral. In this way, for $\nu > 0$, the resulting expression

$$\int e^{\frac{i}{2}\int_0^t (p(s)dq(s)-q(s)dp(s))-i\int_0^t H(p(s),q(s))ds} dW_\nu(p,q) \qquad (6.5)$$

is a well defined Wiener integral. The main result is the proof that, by replacing in Eq. (6.5) the function $H:\mathbb{R}^2 \to \mathbb{R}$ with $h:\mathbb{R}^2 \to \mathbb{R}$, given by

$$h(p,q) = \exp\big(-\frac{1}{2}(\partial_p^2 + \partial_q^2)\big)H(p,q), \qquad (p,q) \in \mathbb{R}^2, \qquad (6.6)$$

the matrix elements $\langle \phi_{q'',p''}, U(t)\phi_{q',p'}\rangle$ are given by the following limit

$$\langle \phi_{q'',p''}, U(t)\phi_{q',p'}\rangle = \lim_{\nu\to\infty} 2\pi e^{\nu t/2} \int e^{\frac{i}{2}\int_0^t (p(s)dq(s)-q(s)dp(s))-i\int_0^t h(p(s),q(s))ds} dW_\nu(p,q),$$

where W_ν is the product of two independent Wiener measures (one in p and one in q) with diffusion constant ν pinned at p', q' for $s=0$ and at p'', q'' for $s=t$. This formula is valid for all self-adjoint Hamiltonian operators H on $L^2(\mathbb{R})$ for which the linear span of the harmonic oscillator eigenstates is a core and such that

$$H = \int h(p,q) P_{p,q} \frac{dpdq}{2\pi},$$

where $h:\mathbb{R}^2 \to \mathbb{R}$ is given by Eq. (6.6) and $P_{p,q}: L^2(\mathbb{R}) \to L^2(\mathbb{R})$ for $(p,q) \in \mathbb{R}^2$ is the projection operator

$$P_{p,q}(\psi) = \phi_{q',p'}\langle \phi_{q',p'}, \psi\rangle, \qquad \psi \in L^2(\mathbb{R}).$$

One has also to impose that for all $\alpha > 0$ the bound

$$\int_{\mathbb{R}^2} |h(p,q)|^2 e^{-\alpha(p^2+q^2)} dpdq < \infty$$

is satisfied. The class of Hamiltonian operator satisfying these condition includes all Hamiltonians that are polynomial in P and Q. The same technique has been also applied to systems with spin.

The key point of the whole procedure is the "duplication" of the path integral by working on the phase space rather than in the configuration space. Indeed if one tries to apply the same regularization technique on a configuration space path integral and to define an heuristic integral of the form $\int e^{\frac{i}{2}\int_0^t \dot{q}(s)^2 ds} dq$ by inserting the term $e^{-\frac{1}{2\nu}\int_0^t \dot{q}(s)^2 ds}$, $\nu > 0$, one has to face with Cameron's result [64] on the non existence of a Wiener measure with complex covariance (in this case $i + 1/\nu$), or equivalently with the non integrability of the function $e^{\frac{i}{2}\int_0^t \dot{q}(s)^2 ds}$.

6.2 The sequential approach

An alternative approach to the rigorous mathematical definition of Feynman path integrals which is very close to Feynman's original derivation of formula (1.6) is the "sequential approach", which has already been briefly discussed in the introduction (see Eqs. (1.7)-(1.9)).

The starting point is the Lie-Kato-Trotter product formula [279, 78, 79, 235], that, under suitable assumptions on the potential V, allows one to write the unitary evolution operator $e^{-\frac{i}{\hbar}tH}$, whose generator is the Hamiltonian $H = H_0 + V$, $H_0 = -\frac{\hbar^2}{2m}\Delta$, in terms of the following strong operator limit:

$$e^{-\frac{i}{\hbar}tH} = s - \lim_{n\to\infty}\left(e^{-\frac{i}{\hbar}\frac{t}{n}V}e^{-\frac{i}{\hbar}\frac{t}{n}H_0}\right)^n. \tag{6.7}$$

Let us consider a vector ϕ in in Schwartz space $S(\mathbb{R}^d)$. For $V = 0$, the vector $e^{-\frac{i}{\hbar}tH_0}\phi$ can be expressed in terms of the Green function of $e^{-\frac{i}{\hbar}tH_0}$:

$$e^{-\frac{i}{\hbar}tH_0}\phi(x) = \left(2\pi i\frac{\hbar}{m}t\right)^{-\frac{d}{2}}\int e^{im\frac{(x-y)^2}{2\hbar t}}\phi(y)dy. \tag{6.8}$$

On the other hand, if we drop the term H_0, the Hamiltonian operator H is a multiplication operator and $e^{-\frac{i}{\hbar}tH}\phi$ is simply given by:

$$e^{-\frac{i}{\hbar}tH}\phi(x) = e^{-\frac{i}{\hbar}tV(x)}\phi(x). \tag{6.9}$$

By substituting Eq. (6.8) and Eq. (6.9) into (6.7), one gets the following expression:

$$e^{-\frac{i}{\hbar}tH}\phi(x)$$
$$= \lim_{n\to\infty}\left(2\pi i\frac{\hbar}{m}\frac{t}{n}\right)^{-\frac{dn}{2}}\int\limits_{\mathbb{R}^{nd}} e^{-\frac{i}{\hbar}\sum_{j=1}^{n}\left[\frac{m}{2}\frac{(x_j-x_{j-1})^2}{\left(\frac{t}{n}\right)^2}-V(x_j)\right]\frac{t}{n}}\phi(x_0)$$
$$dx_0\dots dx_{n-1} \tag{6.10}$$

where $x_n = x$. The right hand side of Eq. (6.10) can be interpreted as the finite dimensional approximation of a path integral. Indeed, if γ is a continuous trajectory from $[0,t]$ to $\mathbb{R}^d$, with $\gamma(t) = x$, let us set $x_j := \gamma(jt/n)$, for $j = 0,\dots n$. The exponent in the integrand can be interpreted in terms of the Riemann sum of the classical action functional evaluated along the path γ:

$$S_t(\gamma) = \int_0^t\left(\frac{m}{2}\dot\gamma^2(s) - V(\gamma(s))\right)ds$$
$$= \lim_{n\to\infty}\sum_{j=1}^{n}\left[\frac{m}{2}\frac{(x_j-x_{j-1})^2}{\left(\frac{t}{n}\right)^2} - V(x_j)\right]\frac{t}{n}.$$

The sequential approach for the mathematical definition of the Feynman path integrals has in fact two versions. In the first one, the attention is focused on the definition of the evolution operator of the quantum system in terms of a strong operator limit, by means of a formula analogous to (6.7), which is a special case of the semigroup product formula

$$s-\lim_{n\to\infty}(F(t/n))^n=\exp(tF'(0)), \tag{6.11}$$

where $t\mapsto F(t)$ is a strongly continuous function from $\mathbb{R}$ (or $\mathbb{R}^+$) into the space of bounded linear operators on an Hilbert space $\mathcal{H}$, while $F'(0)$ has to be interpreted as some operator extension of the strong limit $s-\lim_{t\to 0}t^{-1}(F(t)-I)$. In particular if A, B are self-adjoint operators in $\mathcal{H}$ and $F(t)=e^{itA}e^{itB}$, one gets formally the Trotter product formula (see lemma 3.5):

$$s-\lim_{n\to\infty}(e^{itA/n}e^{itB/n})^n=e^{it(A+B)} \tag{6.12}$$

(where the sum $A+B$ has to be suitably interpreted). Nelson in 1964 [235] proved Eq. (6.12) in connection with the rigorous mathematical definition of Feynman path integrals, under the assumption that the potential V belongs to the class considered by Kato [196]. Some time later Friedman [127] studied formula (6.11) in connection with the description of continuous quantum observations. Feynman himself in [122] considered particular "ideal" quantum measurements of position, made to determine whether or not the trajectory of a particle lies in a certain space-time region. By substituting in formula (6.11) for $F(t)$ the operator $EP(t)E$, where $P(t)$ is a contraction semigroup in the Hilbert space $\mathcal{H}$ and E is an orthogonal projection, and by letting $P(t)=e^{-itH_0}$ and $E: L^2(\mathbb{R}^d)\to L^2(\mathbb{R}^d)$ be the orthogonal projection given by multiplication by the characteristic function of a suitable region $\mathcal{R}$ of $\mathbb{R}^d$, the limit

$$\lim_{n\to\infty}\|(Ee^{-itH_0/n}E)^n\phi\|^2$$

(if it exists) should give the probability that a continual observation during the time interval $[0,t]$ yields the result that the particle, whose initial state is the vector $\phi\in L^2(\mathbb{R}^d)$, lies constantly in the region $\mathcal{R}$.

The second version of the sequential approach has been extensively studied by Fujiwara and Kumano-Go [129–137, 209] (see also [169, 170]) and has the right hand side of Eq. (6.10) as a starting point. A formal path integral on a space of paths $\gamma:[0,t]\to\mathbb{R}^d$ of the form

$$\int e^{\frac{i}{\hbar}S(\gamma)}F(\gamma)d\gamma \tag{6.13}$$

is realized in terms of a "time slicing approximation". More precisely for any $n \in \mathbb{N}$ one considers a partition of the interval $[0, t]$ into n subintervals $t_0 = 0 < t_1 < \ldots < t_j < \ldots < t_n = t$ and for each $j = 0, \ldots, n$ a point $x_j \in \mathbb{R}^d$. The path γ in expression (1.6) is then approximated by a broken line path, passing, for each $j = 0, \ldots, n$, from the point x_j at time t_j. There are two possible approaches to this "time slicing approximation": one can connect the point x_j at time t_j with the point x_{j+1} at time t_{j+1} by means of a straight line path [283, 209, 134], i.e.

$$\gamma(\tau) = x_j + \frac{x_{j+1} - x_j}{t_{j+1} - t_j}(\tau - t_j), \qquad \tau \in [t_j, t_{j+1}],$$

or by means of a classical path [129], i.e. the (unique, for suitable V and if $|t_{j+1} - t_j|$ is sufficiently small) solution of the classical equation of motion

$$\begin{cases} m\ddot{\gamma}(\tau) = -\nabla V(\tau, \gamma(\tau)) \\ \gamma(t_j) = x_j, \\ \gamma(t_{j+1}) = x_{j+1} \,. \end{cases}$$

In particular the term $\frac{(x_j - x_{j-1})}{(\frac{t}{n})}$ is the (constant) velocity of the path connecting the points x_{j-1} and x_j in the time interval $\frac{t}{n}$.

An heuristic expression like Eq. (6.13) is then realized as the limit of the time slicing approximation for suitable functionals F on the path space. Indeed by denoting with γ_n the broken line path (straight resp. piecewise classical) associated to the partition $t_0 = 0 < t_1 < \ldots < t_j < \ldots < t_n = t$, and by Δ_j the amplitude of each time subinterval, i.e. $\Delta_j = |t_{j+1} - t_j|$, one defines

$$\int e^{\frac{i}{\hbar}S_t(\gamma)} F(\gamma) d\gamma := \lim_{|\Delta| \to 0} \prod_{j=1}^{n} \left(\frac{1}{2\pi i \hbar \Delta_j}\right)^{d/2} \int_{\mathbb{R}^{nd}} e^{\frac{i}{\hbar}S_t(\gamma_n)} F(\gamma_n) \prod_{j=1}^{n} dx_j, \tag{6.14}$$

(where $|\Delta| := \sup_j \Delta_j$) whenever the limit exists. The integrals on the right hand side do not converge absolutely and are meant as (finite dimensional) oscillatory integrals.

D. Fujiwara in the case of approximation with piecewise classical paths and D. Fujiwara and N. Kumano-go in the case of broken line paths prove the existence of the limit (6.14) for a suitable class of functionals F. They assume that the potential $V(t, x)$ is a real valued function of $(t, x) \in \mathbb{R} \times \mathbb{R}^d$ and for any multi-index α, $\partial_x^\alpha V(t, x)$ is continuous in $\mathbb{R} \times \mathbb{R}^d$. Moreover they assume that for any integer $k \geq 2$ there exists a positive constant A_k such that

$$|\partial_x^\alpha V(t, x)| \leq A_k, \qquad |\alpha| = k,$$

(this excludes polynomial behaviour at infinity, except for at most quadratic degree).

The so defined functional has some important properties. Integration by parts and Taylor expansion formula with respect to functional differentiation hold. The functional is invariant under orthogonal transformations and transforms naturally under translations. The fundamental theorem of calculus holds and it is possible to interchange the order of integration with Riemann-Stieltjes integrals as well as to interchange the order with a limit [134].

The semiclassical approximation as $\hbar \to 0$ of the integral has been detailed studied, providing not only the leading term, but also the other ones. We refer in particular to the review paper [136] for a complete treatment of the topic which provides also an intuitive explanation of the main ideas.

We point to the reader that a phase space Feynman path integral version of this approach has recently been implemented [137] by approximating a generic path $(q,p) : [0,t] \to \mathbb{R}^{2d}$ in the phase space of the system by means of piecewise bicharacteristic paths, i.e the solutions of the Hamilton equation of motion with particular boundary condition (i.e. the initial velocity and the final position are fixed).

Finally we recall that the time slicing approximation, in particular with piecewise polygonal paths, is extensively used in the physical literature not only as a tool for the definition of the Feynman integral, but also as a practical method of computation for particular solvable models [145–147, 153, 205, 255, 257, 139, 290].

6.3 White noise calculus

Another idea which has been largely implemented is the definition of the Feynman integral as an "infinite dimensional distribution". In other words, the heuristic expression $\int e^{\frac{i}{\hbar}S(\gamma)} f(\gamma)d\gamma$, which cannot be defined in terms of a Lebesgue integral, is realized as the distributional pairing between $e^{\frac{i}{\hbar}S}$ and a suitable function f.

The first proposal of the definition of the infinite dimensional oscillatory integrals in terms of a duality relation can be found in two papers by Ito [172, 173] and further developed in Albeverio and Høegh-Krohn' work [17, 16], where an infinite dimensional Fresnel integral on a real separable

Hilbert space $\mathcal{H}$

$$\widetilde{\int_{\mathcal{H}}} e^{\frac{i}{2\hbar}\langle\gamma,\gamma\rangle} f(\gamma)d\gamma$$

is defined for $f \in \mathcal{F}(\mathcal{H})$ in term of the Parseval type equality:

$$\widetilde{\int_{\mathcal{H}}} e^{\frac{i}{2\hbar}\langle\gamma,\gamma\rangle} f(\gamma)d\gamma = \int_{\mathcal{H}} e^{-\frac{i\hbar}{2}\langle\gamma,\gamma\rangle} d\mu_f(\gamma), \qquad f = \hat{\mu}_f. \tag{6.15}$$

The main difference with the infinite dimensional oscillatory integral approach described in chapter 2 is the fact that here Parseval equality (6.15) plays the role of the *definition* of the Fresnel integral, while the same equality in chapter 2 is a theorem.

The definition of the Feynman integrand $e^{\frac{i}{\hbar}S}$ as a special kind of infinite dimensional distribution was also realized in C. DeWitt-Morette's work [100, 101, 103, 102, 74, 73, 72, 75, 104]. This idea has been recently implemented in a mathematical rigorous setting by means of white noise calculus [162, 163, 118, 117, 150, 210, 214, 261, 270]. In white noise - Hida calculus the integral $\int e^{\frac{i}{2\hbar}\langle\gamma,\gamma\rangle} f(\gamma)d\gamma$ is realized as a distributional pairing on a well defined measure space. The idea can be simply explained in a finite dimensional setting. Indeed for $\mathcal{H} = \mathbb{R}^d$ and $\hbar = 1$, one has

$$\begin{aligned}
&(2\pi i)^{-d/2} \int_{\mathbb{R}^d} e^{\frac{i}{2}(x,x)} f(x)dx \\
&= (2\pi i)^{-d/2} \int_{\mathbb{R}^d} e^{\frac{i}{2}(x,x)+\frac{1}{2}(x,x)} f(x)e^{-\frac{1}{2}(x,x)}dx \\
&= \frac{1}{i^{d/2}} \int_{\mathbb{R}^d} e^{\frac{i}{2}(x,x)+\frac{1}{2}(x,x)} f(x)d\mu_G(x),
\end{aligned}$$

where the latter line can be interpreted as the distributional pairing of $i^{-d/2}e^{\frac{i}{2}(x,x)+\frac{1}{2}(x,x)}$ and f not with respect to Lebesgue measure dx on $\mathbb{R}^d$, but rather with respect to the centered Gaussian measure μ_G on $\mathbb{R}^d$ with covariance the identity.

This idea can be generalized to the infinite dimensional case by exploiting the fact that, even if Lebesgue measure does not exist, Gaussian measures are still well defined in this setting.

The first step is the construction of the underlying measure space, the infinite dimensional analogous of $(\mathbb{R}^d, \mu_G)$.

Let E be a real separable Hilbert space, with inner product $\langle\ ,\ \rangle$, and let $\mathcal{E}_1 \supset \mathcal{E}_2 \supset \ldots$ be vector subspaces of E, each $\mathcal{E}_p$ being a Hilbert space with inner product $\langle\ ,\ \rangle_p$ such that:

(1) $\mathcal{E} := \cap_p \mathcal{E}_p$ is dense in E and in each $\mathcal{E}_p$,

(2) $|u|_p \leq |u|_q$ for every $q \geq p$ and $u \in \mathcal{E}_q$,
(3) for every p, the Hilbert-Schmidt norm $\|i_{qp}\|_{HS}$ of the inclusion i_{qp} : $\mathcal{E}_q \to \mathcal{E}_p$ is finite for some $q \geq p$ and $\lim_{q\to\infty} \|i_{qp}\|_{HS} = 0$.

By identifying $E := \mathcal{E}_0$ with its dual $\mathcal{E}_0^*$, it is possible to construct the chain of spaces

$$\mathcal{E} := \cap_p \mathcal{E}_p \subset \cdots \subset \mathcal{E}_2 \subset \mathcal{E}_1 \subset \mathcal{E}_0 = E \simeq \mathcal{E}_0^* \subset \mathcal{E}_{-1} \subset \mathcal{E}_{-2} \cdots \subset \mathcal{E}^* := \cup_p \mathcal{E}_{-p},$$

where $\mathcal{E}_{-p} = \mathcal{E}_p^*$, $p \in \mathbb{Z}$. In a typical example one has an Hilbert-Schmidt operator K, with Hilbert-Schmidt norm $\|K\|_{HS} < 1$, $\mathcal{E}_p$ is taken as the range $Im(K^p)$ and

$$\langle u, v\rangle_p = \langle K^{-p}u, K^{-p}v\rangle. \tag{6.16}$$

According to Milnos' theorem there is a unique probability measure μ on the Borel $\sigma-$algebra (of the weak topology) on $\mathcal{E}^*$ such that for every $x \in \mathcal{E}$ the function on $\mathcal{E}^*$ given by

$$\phi \mapsto (\phi, x) := \phi(x), \qquad \phi \in \mathcal{E}^*$$

is a mean zero Gaussian variable of variance $|x|_0^2$. By unitary extension, for every $x \in E$ there is a μ-almost-everywhere defined Gaussian random variable $(\,\cdot\,, x)$ on $\mathcal{E}^*$ with mean 0 and variance $|x|_0^2$. This extends by complex linearity to complex Gaussian random variables corresponding to elements z of the complexification $E_{\mathbb{C}}$ of E.

The space $(\mathcal{E}^*, \mu)$ is then taken as the underlying measure space for the realization of the (at this level still heuristic) expression

$$\int_{\mathcal{E}^*} e^{\frac{i}{2}(x,x)+\frac{1}{2}(x,x)} f(x) d\mu(x), \tag{6.17}$$

i.e. $(\mathcal{E}^*, \mu)$ plays the role of the infinite dimensional analogue of $(\mathbb{R}^d, \mu_G)$.

Let us consider the symmetric Fock space $\mathcal{F}_s(E_{\mathbb{C}})$ on $E_{\mathbb{C}}$, i.e. the Hilbert space obtained by completing the symmetric tensor algebra over $E_{\mathbb{C}}$ with respect to the inner product given by

$$\langle\langle \sum_n u_n, \sum_m v_m \rangle\rangle_0 = \sum_n n! \langle u_n, v_n\rangle_0,$$

where u_n and v_n are n-tensors and $\langle\cdot,\cdot\rangle_0$ denotes the inner product on n-tensors induced by the inner product on $E_{\mathbb{C}}$. Analogously the inner products $\langle\cdot,\cdot\rangle_p$ produce inner products $\langle\langle\cdot,\cdot\rangle\rangle_p$ on $\mathcal{F}_s(E_{\mathbb{C}})$.

$\mathcal{F}_s(E_{\mathbb{C}})$ is unitary equivalent to $L^2(\mathcal{E}^*, \mu)$ by the Hermite-Ito-Segal isomorphism $\mathcal{I} : L^2(\mathcal{E}^*, \mu) \to \mathcal{F}_s(E_{\mathbb{C}})$, which is defined by

$$\mathcal{I}(e^{(\cdot,z)-(z,z)_0/2}) = \mathrm{Exp}(z)$$

for every $z \in E_{\mathbb{C}}$, where

$$(\cdot, z) : \mathcal{E}^* \to \mathbb{C} : \phi \mapsto \phi(z)$$

and

$$\text{Exp}(z) = 1 + z + \frac{z^{\otimes 2}}{2!} + \frac{z^{\otimes 3}}{3!} + \cdots \in \mathcal{F}_s(E_{\mathbb{C}}).$$

It allows to transfer the inner products $\langle\langle \cdot, \cdot \rangle\rangle_p$ from $\mathcal{F}_s(E_{\mathbb{C}})$ to $L^2(\mathcal{E}^*, \mu)$, denoted again with $\langle\langle \cdot, \cdot \rangle\rangle_p$.

Let us call *white noise distribution* over $\mathcal{E}^*$ any element of the completion $[\mathcal{E}_{-p}]$ of $L^2(\mathcal{E}^*, \mu)$ with respect to the dual norm $\langle\langle \cdot, \cdot \rangle\rangle_{-p}$, for some integers $p \geq 0$.

It is possible to construct the following chain of Hilbert spaces

$$\begin{aligned}[\mathcal{E}] &:= \cap_p [\mathcal{E}_p] \subset \cdots \subset [\mathcal{E}_2] \subset [\mathcal{E}_1] \subset [\mathcal{E}_0] \\ &= L^2(\mathcal{E}^*, \mu) \simeq [\mathcal{E}_0^*] \subset [\mathcal{E}_{-1}] \subset [\mathcal{E}_{-2}] \cdots \subset [\mathcal{E}^*] := \cup_p [\mathcal{E}_{-p}].\end{aligned}$$

The elements of $[\mathcal{E}]$ are taken to be test functions over $[\mathcal{E}^*]$, which is the corresponding space of distributions.

The aim of the white noise approach is the proof that the expression $e^{\frac{i}{2}(x,x)+\frac{1}{2}(x,x)}$ can be rigorously defined in terms of a white noise distribution. In particular in this framework it is possible to realize the Feynman path integral representation for the fundamental solution $G(t, x; 0, y)$ of the Schrödinger equation over $\mathbb{R}^d$

$$G(t, x; 0, y) = \int_{\substack{\gamma(t)=x \\ \gamma(0)=y}} e^{\frac{i}{\hbar}\int_0^t (\dot{\gamma}^2(s)/2) - V(\gamma(s)))ds} d\gamma, \qquad t > 0,\ x, y \in \mathbb{R}^d,$$

in terms of a white noise distributional pairing on a suitable "path space" $\mathcal{E}^*$ [163, 118, 210, 214]. We present here some results given in [210].

Let us consider the Hilbert space $E = L_d := L^2(\mathbb{R}) \otimes \mathbb{R}^d$, the nuclear space $\mathcal{E} = S_d := S(\mathbb{R}) \otimes \mathbb{R}^d$, i.e. the space of $d-$dimensional Schwartz test functions, and the corresponding dual space $\mathcal{E}^* = S'_d := S'(\mathbb{R}) \otimes \mathbb{R}^d$. Let μ be the Gaussian measure on the Borel σ-algebra of S'_d identified by its characteristic function

$$\int_{S'_d} e^{i\int X(s)f(s)ds} d\mu(X) = e^{-\frac{1}{2}\int f^2(s)ds}, \qquad f \in S_d.$$

Heuristically the "paths" $X \in S'_d$ can be interpreted as the velocities of Brownian paths, as the d-dimensional Brownian motion is given by

$$B(t) = (\int_0^t X_1(s)ds, \ldots, \int_0^t X_d(s)ds),$$

X_i being the i_{th} component of the path X. One then considers the triple

$$[S_d] \subset L^2(S'_d) \subset [S^*_d]$$

and realizes the Feynman integrand as an element of $[S^*_d]$. More precisely the paths are modeled by

$$\gamma(s) = x - \sqrt{\hbar}\int_s^t X(\sigma)d\sigma := x - \sqrt{\hbar}(X, 1_{(s,t]}),$$

and the Feynman integrand for the free particle is realized as the distribution

$$I_0(x,t;x_0,t_0) = Ne^{\frac{i+1}{2}\int_0^t X(s)^2 ds}\delta(\gamma(0)-y),$$

(where N stands for normalization and $\delta(\gamma(0)-y)$ fixes the initial point of the path). The same technique allows one to handle more general potentials, such as those which are Laplace transform of bounded measures (see [210] for a detailed exposition).

Particularly interesting is the application of this formalism [39, 15, 215, 154–156, 158, 157, 159, 37, 35] to the study of the Chern-Simons topological field theory [46, 258, 259, 292] (for similar results by means of different methods see also [38, 61, 128, 151, 200, 197–199, 219]).

6.4 Poisson processes

An alternative mathematical definition of Feynman integrals has been proposed by V.P. Maslov and A.M. Chebotarev [222, 208] and further developed by Ph. Combe, R. Høegh-Krohn, R. Rodriguez, M. Sirugue, M. Sirugue-Collin, Ph. Blanchard [60, 86, 85, 84]. Some recent results and new applications have been proposed by V.N. Kolokoltsov [207, 206].

The main idea is the definition of the Feynman path integral for the solution of the Schrödinger equation in momentum representation:

$$\begin{cases} \frac{\partial}{\partial t}\tilde{\psi}(p) = -\frac{i}{2}p^2\tilde{\psi}(p) - iV(-i\nabla_p)\tilde{\psi}(p) \\ \tilde{\psi}(0,p) = \tilde{\phi}(p) \end{cases} \tag{6.18}$$

in terms of the expectation with respect to a Poisson process.

The potentials V which can be handled by this method are those belonging to the Fresnel class $\mathcal{F}(\mathbb{R}^d)$:

$$V(x) = \int_{\mathbb{R}^d} e^{ikx}d\mu_v(k), \qquad x \in \mathbb{R}^d.$$

In fact for any $\mu_v \in \mathcal{M}(\mathbb{R}^d)$, there exist a positive finite measure ν and a complex-valued measurable function f such that

$$\mu_v(dk) = f(k)\nu(dk).$$

One can assume that $\nu(\{0\}) = 0$ without loss of generality, because otherwise the condition can be fullfilled by a translation of the potential. In this case ν is a finite Lévy measure and one can consider the Poisson process having Lévy measure ν (see, e.g. [243], for these concepts). This process has almost surely piecewise constant paths. More precisely, a typical path P on the time interval $[0, t]$ is defined by a finite number of independent random jumps $\delta_1, ..., \delta_n$, distributed according to the probability measure ν/λ_ν, with $\lambda_\nu = \nu(\mathbb{R}^d)$, occurring at random times $\tau_1, ...\tau_n$, distributed according to a Poisson measure with intensity λ_ν.

Under the assumption that $\tilde{\phi}(p)$ is a bounded continuous function, it is possible to prove that the solution of the Cauchy problem (6.18) can be represented by the following path integral:

$$\tilde{\psi}(t,p) = e^{t\lambda_\nu}\mathbb{E}_p^{[0,t]}[e^{-\frac{i}{2}\sum_{j=0}^{n}(P_j,P_j)(\tau_{j+1}-\tau_j)}\prod_{j=1}^{n}(-if(\delta_j))\tilde{\phi}(P(t))],$$

where the expectation is taken with respect to the measure associated to the Poisson process and the sample path $P(\cdot)$ is given by

$$P(\tau) = \begin{cases} P_0 = p, & 0 \le \tau < \tau_1 \\ P_1 = p + \delta_1, & \tau_1 \le \tau < \tau_2 \\ \dots \\ P_n = p + \delta_1 + \delta_2 + \dots + \delta_n, & \tau_n \le \tau \le t. \end{cases} \tag{6.19}$$

The present approach has also been successfully applied to the study of relativistic quantum theory [1], to Klein-Gordon equation [86, 84], to Fermi systems [85] and to the solution of Dirac equation [208]. We refer to Kolokoltsov's book [207] for a detailed exposition of more general applications of this technique.

6.5 Further approaches and results

In recent years alternative mathematical definitions of Feynman path integrals have been proposed.

As an example we recall the work of Belokurov, Smolianov, Solov'ev and Shavgulidze [260, 266, 52–55], who introduce a regularized path integral which can be expressed in terms of a convergent power series in the coupling constant. This procedure allows a perturbative treatment of polynomial potentials.

In [18] an analytic operator-valued Feynman integral is defined for potentials given by a class of generalized signed measures described in terms of additive functions associated with Dirichlet forms.

Another interesting approach which has not been systematically developed yet makes use of nonstandard analysis [12]. The main idea is the extension of the expression

$$\left(2\pi i \frac{\hbar}{m}\frac{t}{n}\right)^{-\frac{dn}{2}} \int_{\mathbb{R}^{dn}} e^{\frac{i}{\hbar}S_t(x_0,\ldots,x_n)}\phi(x_0)dx_0\ldots dx_{n-1},$$

representing the finite n-dimensional approximation of the heuristic Feynman integral, from $n \in \mathbb{N}$ to a nonstandard hyperfinite infinite $n \in {}^*\mathbb{N}$. The result is just an *internal* quantity. For a suitable class of potentials, its standard part can be shown to exist and to solve the Schrödinger equation. Analogously the semiclassical approximation of the solution of the Schrödinger equation can be obtained by taking the parameter $\hbar$ as an infinitesimal quantity. Some interesting results can be found in [231, 232].

Besides Feynman integration theory, it is worthwhile to mention also Feynman operational calculus, that is an heuristic procedure which allows to define product and functions of noncommuting operators. It was developed by Feynman in 1951 [126] and applied to quantum electrodynamics. The study of this technique from a rigorous mathematical point of view is extensively described in Johnson and Lapidus' book [180] (see also [98, 99, 178, 179, 212, 213, 141–143]).

The number of applications of (heuristic) Feynman path integrals to quantum theory is really huge and it is impossible to give a complete description. Let us only mention some rigorous results concerning the solution of Dirac equation [20, 21, 168, 171, 232, 233, 295, 297, 296], the application to hyperbolic systems [273, 146], the mathematical approaches to supersymmetric Feynman path integrals [253].

Finally we should not forget the enormous predictive power of Feynman description of quantum dynamics when this is applied to quantum field theory [76, 174, 203, 238, 244, 288, 289, 298]. In this case however the gap between what can be heuristically calculated and what can be rigorously mathematically defined is particularly deep. Much work has still to be done...

Appendix A

Abstract Wiener Spaces

A.1 General theory

In the present section we give some elements of the theory of abstract Wiener spaces. For a more detailed treatment see for instance [149, 211].

In the following we shall denote by $(\mathcal{H}, \langle\ ,\ \rangle, \|\ \|)$ a real separable Hilbert space and by $(\mathcal{B}, |\ |)$ a real Banach space. Let us introduce some definitions.

Definition A.1. A Gaussian measure on a Banach space $(\mathcal{B}, |\ |)$ is a probability measure on the Borel σ-algebra on $\mathcal{B}$ such that for each $x \in \mathcal{B}^*$, the random variable $x : \mathcal{B} \to \mathbb{R}$ has a Gaussian distribution on $\mathbb{R}$.

Definition A.2. A cylinder set $Z \subset \mathcal{H}$ of a real separable Hilbert space $\mathcal{H}$ is a set of the form

$$Z = \{x \in \mathcal{H}, |Px \in F\}$$

with $P : \mathcal{H} \to \mathcal{H}$ is a projection operator on $\mathcal{H}$ with finite dimensional range, i.e. $P\mathcal{H} \equiv \mathbb{R}^n$ for some $n \in \mathbb{N}$, and $F \in \mathcal{B}(P\mathcal{H})$ is a Borel set in $P\mathcal{H}$.

In the following we shall denote by $\sigma(\mathcal{Z})$ the $\sigma-$algebra generated by all cylinder sets.

Definition A.3. A cylinder measure on $\mathcal{H}$ is a positive and finitely additive set function ν defined on the $\sigma-$algebra $\sigma(\mathcal{Z})$ of cylinder sets.

Let us consider the cylinder measure ν on $\mathcal{H}$, defined by its characteristic functional $\hat{\nu}(x) = e^{-\frac{1}{2}\|x\|^2}$. Equivalently, ν is given on the cylindrical sets of $\mathcal{H}$ by the following formula

$$\nu(\{x \in \mathcal{H}, |Px \in F\}) = (2\pi)^{-n/2} \int_F e^{-\frac{1}{2}\|Px\|^2} d(Px), \qquad F \in \mathcal{B}(P\mathcal{H}).$$

ν is called *standard Gaussian measure associated with $\mathcal{H}$.*

By Prokhorov's theorem [211] the following holds:

Theorem A.1. *Let $\mathcal{H}$ be infinite dimensional. Then the standard Gaussian measure associated with $\mathcal{H}$ is not $\sigma-$additive on $\sigma(\mathcal{Z})$.*

Definition A.4. A norm $|\ |$ (or a semi-norm) on $\mathcal{H}$ is called measurable if for every $\epsilon > 0$ there exist a finite-dimensional projection $P_\epsilon : \mathcal{H} \to \mathcal{H}$, such that for all $P \perp P_\epsilon$ one has:

$$\nu(\{x \in \mathcal{H}|\ |P(x)| > \epsilon\}) < \epsilon,$$

where P and P_ϵ are called orthogonal ($P \perp P_\epsilon$) if their ranges are orthogonal in $(\mathcal{H}, \langle\ ,\ \rangle)$.

One can easily verify that $|\ |$ is weaker than $\|\ \|$. Indeed there exists a $c \in \mathbb{R}^+$ such that

$$|x| \leq c\|x\|, \qquad \forall x \in \mathcal{H}. \tag{A.1}$$

In the following we shall denote with $\mathcal{B}$ the Banach space constructed as the completion of $\mathcal{H}$ in the $|\ |$-norm and by i the inclusion of $\mathcal{H}$ in $\mathcal{B}$, which is continuous by Eq. (A.1). Analogously, the dual map $i^* : \mathcal{B}^* \to \mathcal{H}^*$, which is given by restriction, i.e. $i^*(x) = x_{|\mathcal{H}}$, is continuous. Identifying $\mathcal{H} \equiv \mathcal{H}^*$ we have the following chain of densely embedded subspaces

$$\mathcal{B}^* \subset \mathcal{H} \subset \mathcal{B}.$$

The triple $(i, \mathcal{H}, \mathcal{B})$ is called an *abstract Wiener space.*

The following holds [149]:

Theorem A.2. *There exist a unique probability measure μ on the Borel $\sigma-$ algebra of $\mathcal{B}$, such that for any $y \in \mathcal{B}^*$:*

$$\int_{\mathcal{B}} e^{iy(x)} d\mu(x) = e^{-\frac{1}{2}\|y\|^2}. \tag{A.2}$$

Let us consider a particular kind of cylinder sets in $\mathcal{H}$. Given $y_1, \dots y_n \in \mathcal{B}^*$, and $F \in \mathcal{B}(\mathbb{R}^n)$, let $Z_F(y_1, \dots y_n)$ be the subset of $\mathcal{H}$

$$Z_F(y_1, \dots y_n) := \{x \in \mathcal{H} \,|\, (i^*y_1(x), \dots, i^*y_n(x)) \in F\}.$$

Analogously the subset of $\mathcal{B}$ defined as

$$\{x \in \mathcal{B} \,|\, (y_1(x), \dots, y_n(x)) \in F\}, \tag{A.3}$$

is called a cylinder set of $\mathcal{B}$. The following holds:

Theorem A.3. *The $\sigma-$algebra on $\mathcal{B}$ generated by the cylinder sets of the form (A.3) coincides with the Borel $\sigma-$algebra on $\mathcal{B}$. Moreover the Gaussian measure μ on $\mathcal{B}$ is an extension of the standard Gaussian measure ν on $\mathcal{H}$ in the sense that*

$$\mu(\{x \in \mathcal{B} | (y_1(x), \dots, y_n(x)) \in F\}) = \nu(\{x \in \mathcal{H} | (i^* y_1(x), \dots, i^* y_n(x)) \in F\}).$$

By theorem A.2, each element $y \in \mathcal{B}^*$ can be regarded as a random variable $n(y)$ on $(\mathcal{B}, \mu)$, which is normally distributed, with covariance $\|y\|^2$. More generally, given $y_1, y_2 \in \mathcal{B}^*$, one has

$$\int_{\mathcal{B}} n(y_1) n(y_2) d\mu = \langle y_1, y_2 \rangle. \tag{A.4}$$

Equation (A.4) and the density of $\mathcal{B}^*$ in $\mathcal{H}$ allow the extension of the map $n : \mathcal{B}^* \to L^2(\mathcal{B}, \mu)$ to a map $n : \mathcal{H} \to L^2(\mathcal{B}, \mu)$ (with an abuse of notation we denote the map n on $\mathcal{B}^*$ and its extension to $\mathcal{H}$ with the same symbol)[149]. Moreover, given a complete orthonormal system $\{e_i\}$ in $\mathcal{H}$, for any $h \in \mathcal{H}$ the sequence of random variables

$$\sum_{i=1}^n h_i n(e_i), \qquad h_i = \langle e_i, h \rangle,$$

converges in $L^2(\mathcal{B}, \mu)$, and by subsequences μ a.e., to the random variable $n(h)$.

Given an orthogonal projection P in $\mathcal{H}$, with

$$P(x) = \sum_{i=1}^n \langle e_i, x \rangle e_i$$

for some orthonormal $e_1, \dots, e_n \in \mathcal{H}$, let us define the stochastic extension $\tilde{P}$ of P on $\mathcal{B}$ as the random variable

$$\tilde{P}(\cdot) = \sum_{i=1}^n n(e_i)(\cdot) e_i.$$

More generally:

Definition A.5. Given a function $f : \mathcal{H} \to \mathcal{B}_1$, where $(\mathcal{B}_1, \| \, \|_{\mathcal{B}_1})$ is another real separable Banach space, we say that the stochastic extension $\tilde{f}$ of f to $\mathcal{B}$ exists if the functions $f \circ \tilde{P} : \mathcal{B} \to \mathcal{B}_1$ converge to $\tilde{f}$ in probability with respect to μ as P converges strongly to the identity in $\mathcal{H}$.

The following holds [149]:

Theorem A.4. *If $g : \mathcal{B} \to \mathcal{B}_1$ is continuous and $f := g|_{\mathcal{H}}$, then the stochastic extension of f is well defined and it is equal to g $\mu-$a.e.*

Analogously it is also possible to prove that given a self-adjoint trace class operator $B : \mathcal{H} \to \mathcal{H}$, the quadratic form on $\mathcal{H} \times \mathcal{H}$:

$$x \in \mathcal{H} \mapsto \langle x, Bx \rangle$$

can be extended to a random variable on $\mathcal{B}$, denoted again by $\langle \cdot , B \cdot \rangle$. Indeed for each increasing sequence of finite dimensional projectors P_n converging strongly to the identity, $P_n(x) = \sum_{i=1}^n e_i \langle e_i, x \rangle$ ($\{e_i\}$ being a complete orthonormal system in $\mathcal{H}$), the sequence of random variables

$$x \in \mathcal{B} \mapsto \sum_{i,j=1}^{n} \langle e_i, Be_j \rangle n(e_i)(x) n(e_j)(x)$$

is a Cauchy sequence in $L^1(\mathcal{B}, \mu)$. By passing if necessary to a subsequence, it converges to $\langle \cdot , B \cdot \rangle$ μ–a.e. Moreover the following holds:

Theorem A.5. *Let $B : \mathcal{H} \to \mathcal{H}$ be symmetric and trace class. Let us assume that the largest eigenvalue of B is strictly less than 1 (or, in other words, that $(I - B)$ is strictly positive). Let $y \in \mathcal{H}$.*
Then the random variable on $(\mathcal{B}, \mu)$

$$f(\,\cdot\,) = e^{n(y)(\,\cdot\,)} e^{\frac{1}{2}\langle \cdot , B \cdot \rangle}$$

is well defined, it is μ-summable and

$$\int_{\mathcal{B}} f d\mu = (\det(I - B))^{-1/2} e^{\frac{1}{2}\langle y,(I-B)^{-1} y \rangle},$$

where $\det(I - B)$ *denotes the Fredholm determinant of the operator* $I - B$.

Proof. By considering a complete orthonormal system $\{e_i\}$ made of eigenvectors of the operator B, b_i being the corresponding eigenvalues, the sequence of random variables

$$g_n : \mathcal{B} \to \mathbb{C}, \qquad x \mapsto g_n(x) = e^{\frac{1}{2}\sum_{i=1}^n b_i [n(e_i)(x)]^2},$$

converges to $g(x)$ μ-a.e..

On the other hand one has

$$\int_{\mathcal{B}} g_n(x) d\mu(x) = \prod_{i=1}^{n} \int \frac{e^{-\frac{1}{2}(1-b_i)x_i^2}}{\sqrt{2\pi}} dx_i = (\prod_{i=1}^{n}(1 - b_i))^{-1/2}$$

so that $\int_{\mathcal{B}} g_n d\mu$ converges, as $n \to \infty$, to $(\det(I - B))^{-1/2}$, where $\det(I - B)$ denotes the Fredholm determinant of $(I - B)$, which is well defined as B is trace class.

Moreover $0 \leq g_n \leq g_{n+1}$ for each n . It follows that, as $n \to \infty$,

$$\int_{\mathcal{B}} g_n d\mu \to \int_{\mathcal{B}} g d\mu = (\det(I - B))^{-1/2}.$$

By an analogous reasoning, for any $y \in \mathcal{H}$, the sequence of random variables f_n:

$$x \mapsto f_n(x) = e^{\sum_{i=1}^n y_i n(e_i)(x)} e^{\frac{1}{2}\sum_{i=1}^n b_i [n(e_i)(x)]^2}$$

where $y_i = \langle y, e_i \rangle$, converges μ–a.e. as n goes to ∞ to the random variable

$$f(\,\cdot\,) = e^{n(y)(\,\cdot\,)} e^{\frac{1}{2}\langle\,\cdot\,, B\,\cdot\,\rangle}$$

and

$$\int_{\mathcal{B}} f_n d\mu \to \int_{\mathcal{B}} f d\mu = (\det(I-B))^{-1/2} e^{\frac{1}{2}\langle y, (I-B)^{-1} y\rangle}. \tag{A.5}$$

(see [211, 191]). □

The last result of this section is a quasi-invariance property of the measure μ on $\mathcal{H}$.

Theorem A.6. [149, 211] *Let $(i, \mathcal{H}, \mathcal{B})$ be an abstract Wiener space. Let $y \in \mathcal{H}$. then for any $f \in L^1(\mathcal{B}, \mu)$, the following holds*

$$\int_{\mathcal{B}} f(x) d\mu(x) = \int_{\mathcal{B}} f(x+y)\rho(y,x) d\mu(x),$$

where ρ is given by

$$\rho(y,x) = e^{-\frac{1}{2}\|y\|^2 - n(y)(x)}.$$

A.2 The classical Wiener space

Let us consider the Sobolev space $\mathcal{H}^{1,2}([0,t], \mathbb{R}^d)$, i.e. the space of absolutely continuous functions $\gamma : [0,t] \to \mathbb{R}^d$ such that $\gamma(0) = 0$ and $\dot{\gamma} \in L^2([0,t])$, where $\dot{\gamma}$ denotes the distributional derivative of the function γ, endowed with the inner product

$$\langle \gamma_1, \gamma_2 \rangle = \int_0^t \dot{\gamma}_1(s)\dot{\gamma}_2(s) ds.$$

Let $(C_0([0,t], \mathbb{R}^d), |\cdot|)$ be the Banach space of continuous functions $\omega : [0,t] \to \mathbb{R}^d$ such that $\omega(0) = 0$, endowed with the supremum norm:

$$|\omega| := \sup_{s \in [0,t]} |\omega(s)|.$$

The following holds:

Proposition A.1. *$\mathcal{H}^{1,2}([0,t], \mathbb{R}^d)$ is dense in $C_0([0,t], \mathbb{R}^d)$ with respect to the supremum norm $|\cdot|$.*

The following result makes $(i, \mathcal{H}^{1,2}([0,t],\mathbb{R}^d), C_0([0,t],\mathbb{R}^d))$ an abstract Wiener space [211]

Theorem A.7. [148] *Let us consider the standard Gaussian measure* ν *associated with* $\mathcal{H}^{1,2}([0,t],\mathbb{R}^d)$. *Then the supremum norm* $|\cdot|$ *is a* ν-*measurable norm on* $\mathcal{H}^{1,2}([0,t],\mathbb{R}^d)$.

The space $(C_0([0,t],\mathbb{R}^d), \mathcal{B}(C_0([0,t],\mathbb{R}^d)), \mu)$ is called *classical Wiener space* and μ *Wiener measure*.

It is possible to prove (see for instance [211]) that the norm $\|i\|$ of the continuous embedding

$$i : \mathcal{H}^{1,2}([0,t],\mathbb{R}^d) \to C_0([0,t],\mathbb{R}^d)$$

is equal to $\|i\| = \sqrt{t}$.

In the theory of stochastic processes, Wiener measure is introduced by means of a different technique (see for instance [195]).

Let $(\Omega, \mathcal{F})$ be a measurable space, where Ω is a generic set and $\mathcal{F}$ a σ-algebra of subsets of Ω. A real valued *stochastic process* X is defined as a collection of random variables $X = \{X_t : 0 \leq t < \infty\}$ on $(\Omega, \mathcal{F})$ with values in $\mathbb{R}$.

The sample space $(\Omega, \mathcal{F})$ is equipped with a *filtration*, i.e. a non decreasing family $\{\mathcal{F}_t; t \geq 0\}$ of sub σ-algebras of $\mathcal{F}$ such that $\mathcal{F}_s \subseteq \mathcal{F}_t \subseteq \mathcal{F}$ for $0 \leq s < t < \infty$. The stochastic process X is called *adapted to the filtration* $\{\mathcal{F}_t\}$ if, for each $t \geq 0$, the random variable X_t is $\mathcal{F}_t$ measurable.

Definition A.6. A standard Brownian motion B is a continuous, adapted process $B = \{B_s, \mathcal{F}_s; 0 \leq s \leq t\}$, defined on a probability space $(\Omega, \mathcal{F}, \mathcal{P})$, with the properties that $B_0 = 0$ almost surely and for $0 \leq r < s$, the increment $B_s - B_r$ is independent of $\mathcal{F}_r$ and is normally distributed with variance $t - s$.

From the construction of the classical Wiener space

$$(C_0([0,t],\mathbb{R}^d), \mathcal{B}(C_0([0,t],\mathbb{R}^d)), \mu),$$

it is possible to recognize in it a concrete realization of the Wiener process, i.e. $\Omega = C_0([0,t],\mathbb{R}^d)$, $\mathcal{F} = \mathcal{B}(C_0([0,t],\mathbb{R}^d))$, $\mathcal{P} = \mu$ and $B_s(\omega) = \omega(s)$.

Bibliography

[1] Albeverio, S., Blanchard, P., Combe, P., Høegh-Krohn, R. and Sirugue, M. (1983). Local relativistic invariant flows for quantum fields, *Comm. Math. Phys.* **90**, 3, pp. 329–351.

[2] Albeverio, S., Blanchard, P. and Høegh-Krohn, R. (1980). Stationary phase for the Feynman integral and the trace formula, in *Functional integration* (Plenum, New York), pp. 341–361.

[3] Albeverio, S., Blanchard, P. and Høegh-Krohn, R. (1981). Feynman path integrals, the Poisson formula and the theta function for the Schrödinger operators, in *Trends in Applications of Pure Mathematics to Mechanics, Vol III* (Pitman, Boston), pp. 1–21.

[4] Albeverio, S., Blanchard, P. and Høegh-Krohn, R. (1982). Feynman path integrals and the trace formula for the Schrödinger operators, *Comm. Math. Phys.* **83**, 1, pp. 49–76.

[5] Albeverio, S., Boutet de Monvel-Berthier, A. M. and Brzeźniak, Z. (1995). Stationary phase method in infinite dimensions by finite dimensional approximations: applications to the Schrödinger equation, *Potential Analysis* **4**, pp. 469–502.

[6] Albeverio, S., Boutet de Monvel-Berthier, A. M. and Brzeźniak, Z. (1996a). The trace formula for Schrödinger operators from infinite dimensional oscillatory integrals, *Math. Nachr.* **182**, pp. 21–65.

[7] Albeverio, S. and Brzeźniak, Z. (1993). Finite-dimensional approximation approach to oscillatory integrals and stationary phase in infinite dimensions, *J. Funct. Anal.* **113**, 1, pp. 177–244.

[8] Albeverio, S. and Brzeźniak, Z. (1994). Feynman path integrals as infinite-dimensional oscillatory integrals: some new developments, *Acta Appl. Math.* **35**, pp. 5–27.

[9] Albeverio, S. and Brzeźniak, Z. (1995). Oscillatory integrals on Hilbert spaces and Schrödinger equation with magnetic fields, *J. Math. Phys.* **36**, 5, pp. 2135–2156.

[10] Albeverio, S., Brzeźniak, Z. and Haba, Z. (1998). On the Schrödinger equation with potentials which are Laplace transform of measures, *Potential Anal.* **9**, 1, pp. 65–82.

[11] Albeverio, S., Cattaneo, L., Di Persio, L. and Mazzucchi, S. (2007). A rigorous approach to the Feynman-Vernon influence functional and its applications. I, *J. Math. Phys.* **48**, 10, p. 102109.

[12] Albeverio, S., Fenstad, J., Høegh-Krohn, R. and Lindstrøm, T. (1986). *Non Standard Methods in Stochastic Analysis and Mathematical Physics* (Pure and Applied Mathematics 122, Academic Press, Orlando, FL).

[13] Albeverio, S., Guatteri, G. and Mazzucchi, S. (2003). Representation of the Belavkin equation via Feynman path integrals, *Probab. Theory Relat. Fields* **125**, pp. 365–380.

[14] Albeverio, S., Guatteri, G. and Mazzucchi, S. (2004a). Representation of the Belavkin equation via phase space Feynman path integrals, *Infin. Dimens. Anal. Quantum Probab. Relat. Top.* **7**, 4, pp. 507–526.

[15] Albeverio, S., Hahn, A. and Sengupta, A. (2004b). Rigorous Feynman path integrals, with applications to quantum theory, gauge fields, and topological invariants, in *Stochastic analysis and mathematical physics (SAMP/ANESTOC 2002),* (World Sci. Publishing, River Edge, NJ), pp. 1–60.

[16] Albeverio, S. and Høegh-Krohn, R. (1977). Oscillatory integrals and the method of stationary phase in infinitely many dimensions, with applications to the classical limit of quantum mechanics I, *Invent. Math.* **40**, 1, pp. 59–106.

[17] Albeverio, S., Høegh-Krohn, R. and Mazzucchi, S. (2008). *Mathematical theory of Feynman path integrals. An Introduction*, 2nd edn. (Springer, Berlin).

[18] Albeverio, S., Johnson, J. and Ma, Z. (1996b). The analytic operator-value Feynman integral via additive functionals of Brownian motion, *Acta Appl. Math.* **42**, 3, pp. 267–295.

[19] Albeverio, S., Khrennikov, A. and Smolyanov, O. (1999). The probabilistic Feynman-Kac formula for an infinite-dimensional Schrödinger equation with exponential and singular potentials, *Potential Anal.* **11**, 2, pp. 157–181.

[20] Albeverio, S., Khrennikov, A. and Smolyanov, O. (2001). Representation of solutions of Liouville equations for generalized Hamiltonian systems by functional integrals. (Russian), *Dokl. Akad. Nauk.* **381**, 2, pp. 155–159.

[21] Albeverio, S., Khrennikov, A. and Smolyanov, O. (2002). A local Liouville theorem for infinite-dimensional Hamilton-Dirac systems, *Russ. J. Math. Phys.* **9**, 2, pp. 123–139.

[22] Albeverio, S., Kolokoltsov, V. N. and Smolyanov, O. G. (1996c). Représentation des solutions de l'équation de Belavkin pour la mesure quantique par une version rigoureuse de la formule d'intégration fonctionnelle de Menski, *C. R. Acad. Sci. Paris Sér. I Math.* **323**, 6, pp. 661–664.

[23] Albeverio, S., Kolokoltsov, V. N. and Smolyanov, O. G. (1997). Continuous quantum measurement: local and global approaches, *Rev. Math. Phys.* **9**, 8, pp. 907–920.

[24] Albeverio, S. and Liang, S. (2005). Asymptotic expansions for the Laplace approximations of sums of Banach space-valued random variables, *Ann.*

Probab. **33**, 1, pp. 300–336.

[25] Albeverio, S. and Mazzucchi, S. (2004a). Generalized infinite-dimensional Fresnel integrals, *C. R. Acad. Sci. Paris* **338**, 3, pp. 255–259.

[26] Albeverio, S. and Mazzucchi, S. (2004b). Some New Developments in the Theory of Path Integrals, with Applications to Quantum Theory, *J. Stat. Phys.* **112**, 12, pp. 191–215.

[27] Albeverio, S. and Mazzucchi, S. (2005a). Feynman path integrals for polynomially growing potentials, *J. Funct. Anal.* **221**, 1, pp. 83–121.

[28] Albeverio, S. and Mazzucchi, S. (2005b). Feynman path integrals for the time dependent quartic oscillator, *C. R. Acad. Sci. Paris* **341**, 10, pp. 647–650.

[29] Albeverio, S. and Mazzucchi, S. (2005c). Feynman path integrals for time-dependent potentials, in G. D. Prato and L. Tubaro (eds.), *Stochastic Partial Differential Equations and Applications - VII* (Lecture Notes in Pure and Applied Mathematics, vol. 245, Taylor & Francis), pp. 7–20.

[30] Albeverio, S. and Mazzucchi, S. (2005d). Generalized Fresnel Integrals, *Bull. Sci. Math.* **129**, 1, pp. 1–23.

[31] Albeverio, S. and Mazzucchi, S. (2006). The time dependent quartic oscillator - a Feynman path integral approach, *J. Funct. Anal.* **238**, 2, pp. 471–488.

[32] Albeverio, S. and Mazzucchi, S. (2007a). Theory and applications of infinite dimensional oscillatory integrals, in *Stochastic Analysis and Applications. Proceedings of the Abel Symposium 2005 in honor of Prof. Kiyosi Ito* (Springer), pp. 73–91.

[33] Albeverio, S. and Mazzucchi, S. (2007b). The trace formula for the heat semigroup with polynomial potential, SFB-611-Preprint no. 332, Bonn.

[34] Albeverio, S. and Mazzucchi, S. (2008a). Infinite dimensional oscillatory integrals with polynomial phase function and the trace formula for the heat semigroup, Preprint of the University of Trento. UTM 721. To appear in *Astérisque.*

[35] Albeverio, S. and Mazzucchi, S. (2008b). A survey on mathematical Feynman path integrals: construction, asymptotics, applications, in B. Fauser, J. Tolksdorf and E. Zeidler (eds.), *Quantum Field Theory: Competitive Models* (Birkhauser, Boston), pp. 49–66.

[36] Albeverio, S. and Mazzucchi, S. (2009). An asymptotic functional integral solution for the Schrödinger equation with polynomial potential, Preprint of the University of Trento. To appear in *J. Funct. Anal.* http://dx.doi.org/10.1016/j.jfa.2009.02.005

[37] Albeverio, S. and Mitoma, I. (2009). Asymptotic Expansion of Perturbative Chern-Simons Theory via Wiener Space, *Bull. Sci. Math.*

[38] Albeverio, S. and Schäfer, J. (1995). Abelian Chern-Simons theory and linking numbers via oscillatory integrals, *J. Math. Phys.* **36**, pp. 2157–2169.

[39] Albeverio, S. and Sengupta, A. (1997). A mathematical construction of the non-Abelian Chern-Simons functional integral, *Commun. Math. Phys.* **186**, pp. 563–579.

[40] Albeverio, S. and Steblovskaya, V. (1999). Asymptotics of infinite-dimensional integrals with respect to smooth measures I, *Infin. Dimens. Anal. Quantum Probab. Relat. Top.* **2**, 4, pp. 529–556.

[41] Arnold, V. (1972a). Integrals of quickly oscillating functions and singularities of projections of Lagrange manifolds, *Funct. Analys. and its Appl.* **6**, pp. 222–224.

[42] Arnold, V. (1972b). Une classe caractéristique intervenant dans les conditions de quantification, in *Théorie des perturbations et méthodes asymptotiques* (Dunod. Paris), pp. 341–361.

[43] Arnold, V. (1981). Remarks on the stationary phase method and Coxeter numbers, in *Singularity Theory Selected Papers* (Cambridge Univerity Press. Cambridge), pp. 61–90.

[44] Arnold, V., Goryunov, V. V., Lyashko, O. V. and Vassiliev, V. A. (1993). *Singularity theory. I.*, Vol. 6 (Springer, Berlin).

[45] Arnold, V., Gusein-Zade, S. M. and Varchenko, A. (eds.) (1988). *Singularities of differentiable maps Vol II* (Birhäuser, Basel).

[46] Atiyah, M. (1990). *The Geometry and Physics of Knots* (Cambridge University Press, Cambridge).

[47] Azencott, R. (1982). *Formule de Taylor stochastique et développement asymptotique d'intégrales de Feynman. (French) [Stochastic Taylor formula and asymptotic expansion of Feynman integrals], Lecture Notes in Math.*, Vol. 921 (Springer, Berlin-New York), pp. 237–284.

[48] Azencott, R. and Doss, H. (1985). *L'équation de Schrödinger quand h tend vers zéro: une approche probabiliste. (French) The Schrödinger equation as h tends to zero: a probabilistic approach, Lecture Notes in Math.*, Vol. 1109 (Springer, Berlin), pp. 1–17.

[49] Balser, W. (1994). *From divergent power series to analytic functions* (LNM 1582, Springer Berlin-Heidelberg).

[50] Barchielli, A. (1993). On the quantum theory of measurements continuous in time, *Rep. Math. Phys.* **33**, pp. 21–34.

[51] Barchielli, A. and Belavkin, V. (1991). Measurements continuous in time and a posteriori states in quantum mechanics, *J. Phys. A.* **24**, 7, pp. 1495–1514.

[52] Belokurov, V. V., Solovev, Y. P. and Shavgulidze, E. T. (1995). New perturbation theory for quantum field theory: convergent series instead of asymptotic expansions, *Mod.Phys. Let. A* **19**, 39, pp. 3033–3041.

[53] Belokurov, V. V., Solovev, Y. P. and Shavgulidze, E. T. (1997a). A method for the approximate calculation of path integrals using perturbation theory with convergent series. I. *Theor. and Math. Phys.* **109**, 1, pp. 1287–1293.

[54] Belokurov, V. V., Solovev, Y. P. and Shavgulidze, E. T. (1997b). A method for the approximate calculation of path integrals using perturbation theory with convergent series. II. Euclidean quantum field theory, *Theor. and Math. Phys.* **109**, 1, pp. 1294–1301.

[55] Belokurov, V. V., Solovev, Y. P. and Shavgulidze, E. T. (1997c). Perturbation theory with convergent series for functional integrals with respect to the Feynman measure. (Russian), *Math. Surveys* **52**, 2, pp. 392–393.

[56] Ben Arous, G. (1988). Méthode de Laplace et de la phase stationnaire sur l'espace de Wiener. (French) [The Laplace and stationary phase methods on Wiener space], *Stochastics* **25**, 3, pp. 125–153.

[57] Ben Arous, G. and Castell, F. (1996). A probabilistic approach to semi-classical approximations, *J. Funct. Anal.* **137**, 1, pp. 243–280.

[58] Ben Arous, G. and Léandre, R. (1991). Décroissance exponentielle du noyau de la chaleur sur la diagonale. II. (French) [Exponential decay of the heat kernel on the diagonal. II], *Probab. Theory Related Fields* **90**, 3, pp. 377–402.

[59] Benatti, F., Ghirardi, G., Rimini, A. and Weber, T. (1987). Quantum mechanics with spontaneous localization and the quantum theory of measurement, *Nuovo Cimento B* **11**, 100, pp. 27–41.

[60] Blanchard, P., Combe, P., Sirugue, M. and Sirugue-Collin, M. (1987). Jump processes: an introduction and some applications in quantum theories, *Rend. Circ. Mat. Palermo.* **II**, 17, pp. 47–104.

[61] Broda, B. (1994). A path-integral approach to polynomial invariants of links. Topology and physics, *J. Math. Phys.* **35**, 10, pp. 5314–5320.

[62] Busch, P., Lahti, P. and Mittelstaedt, P. (1996). *The quantum theory of measurement*, 2nd edn. (Springer-Verlag, Berlin).

[63] Caldeira, A. and Leggett, A. (1983). Path integral approach to quantum Brownian motion, *Physica* **121 A**, pp. 587–616.

[64] Cameron, R. (1960). A family of integrals serving to connect the Wiener and Feynman integrals, *J. Math. and Phys.* **39**, pp. 126–141.

[65] Cameron, R. (1968). Approximation to certain Feynman integrals, *Journal d'Analyse Math.* **21**, pp. 337–371.

[66] Cameron, R. and Storvick, D. (1966). A translation theorem for analytic Feynman integrals, *Trans. Am. Math. Soc.* **125**, pp. 1–6.

[67] Cameron, R. and Storvick, D. (1968a). A Lindeløf theorem and analytic continuation for functions of several variables, with an application to the Feynman integral, in *Proc. Symposia Pure Mathem., XI* (Am. Math. Soc., Providence), pp. 149–156.

[68] Cameron, R. and Storvick, D. (1968b). An operator valued function space integral and a related integral equation, *J. Math. and Mech.* **18**, pp. 517–552.

[69] Cameron, R. and Storvick, D. (1970). An integral equation related to Schrödinger equation with an application to integration in function space, in C. Gunning (ed.), *Problems in Analysis, A Sympos. In Honor of S. Bochner* (Princeton Univ. Press, Princeton), pp. 175–193.

[70] Cameron, R. and Storvick, D. (1980). Some Banach algebras of analytic Feynman integrable functionals, in *Analytic functions, Kozubnik, 1979* (Lecture Notes in Mathematics, n 798, Springer-Verlag, Berlin), pp. 18–67.

[71] Cameron, R. and Storvick, D. (1983). A simple definition of the Feynman integral with applications, *Memoirs of the American Mathematical Society* **228**, pp. 1–46.

[72] Cartier, P. and DeWitt-Morette, C. (1995). A new perspective on functional integration, *J. Math. Phys.* **36**, 5, pp. 2237–2312.

[73] Cartier, P. and DeWitt-Morette, C. (1997). A rigorous mathematical foundation of functional integration, in *Functional integration (Cargése, 1996). NATO Adv. Sci. Inst. Ser. B Phys., 361* (Plenum, New York), pp. 1–50.

[74] Cartier, P. and DeWitt-Morette, C. (2000). Functional integration, *J. Math. Phys.* **41**, 6, pp. 4154–4187.

[75] Cartier, P. and DeWitt-Morette, C. (2006). *Functional integration: action and symmetries* (Cambridge University Press, Cambridge).

[76] Chang, S. J. (1990). *Introduction to quantum field theory* (World Scientific, Singapore).

[77] Chazarain, J. (1974). Formula de Poisson pour les Variétés Riemmaniennes, *Invent. Math.* **24**, pp. 65–82.

[78] Chernoff, P. (1968). Note on product formulas for operator semigroups, *J. Functional Analysis* **2**, pp. 238–242.

[79] Chernoff, P. (1974). Product formulas, nonlinear semigroups and addition of unbounded operators, *Mem. Am. Math. Soc.* **140**.

[80] Chiu, C. B., Misra, B. and Sudarshan, E. C. G. (1977). Time evolution of unstable quantum states and a resolution of Zeno's paradox, *Phys. Rev. D* **16**, 2, pp. 520–529.

[81] Chung, D. (1991). Conditional analytic Feynman integrals on Wiener spaces, *Proc. AMS* **122**, pp. 479–488.

[82] Chung, L. and Zambrini, J. (2001). *Introduction to random time and quantum randomness* (Monographs of the Portuguese Mathematical Society. McGraw-Hill, Lisbon).

[83] Cohn, D. (ed.) (1980). *Measure Theory* (Birkhäuser, Boston).

[84] Combe, P., Høegh-Krohn, R., Rodriguez, R., Sirugue, M. and Sirugue-Collin, M. (1978). Feynman path integrals and Poisson processes with piecewise classical paths, *J. Math. Phys.* **23**, 2, pp. 405–411.

[85] Combe, P., Høegh-Krohn, R., Rodriguez, R., Sirugue, M. and Sirugue-Collin, M. (1980). Poisson Processes on Groups and Feynman Path Integrals, *Commun. Math. Phys.* **77**, pp. 269–288.

[86] Combe, P., Høegh-Krohn, R., Rodriguez, R., Sirugue, M. and Sirugue-Collin, M. (1981). Generalized Poisson processes in quantum mechanics and field theory, *Phys. Rep.* **77**, pp. 221–233.

[87] Cruzeiro, A. and Zambrini, J. (1991). Feynman's functional calculus and stochastic calculus of variations, in *Stochastic analysis and applications (Lisbon, 1989)* (Progr. Probab., 26, Birkhuser Boston, Boston, MA), pp. 82–95.

[88] Daubechies, I. and Klauder, J. R. (1982). Constructing measures for path integrals, *J . Math. Phys.* **23**, 10, pp. 1806–1822.

[89] Daubechies, I. and Klauder, J. R. (1983). Measures for more quadratic path integrals. *Lett. Math. Phys.* **7**, 3, pp. 229–234.

[90] Daubechies, I. and Klauder, J. R. (1984). Quantum-mechanical path integrals with Wiener measure for all polynomial Hamiltonians. *Phys. Rev. Lett.* **52**, 14, pp. 1161–1164.

[91] Daubechies, I. and Klauder, J. R. (1985). Quantum-mechanical path integrals with Wiener measure for all polynomial Hamiltonians II. *J. Math. Phys.* **26**, pp. 2239–2256.

[92] Davies, E. (1976). *Quantum Theory of Open Systems* (Academic Press. London).

[93] Davies, I. M. (1998). Laplace asymptotic expansions for Gaussian functional integrals, *Electron. J. Probab.* **3**, 13, pp. 1–10.

[94] Davies, I. M. and Truman, A. (1982a). Laplace asymptotic expansions of conditional Wiener integrals and generalized Mehler kernel formulas, *J. Math. Phys.* **23**, 11, pp. 2059–2070.

[95] Davies, I. M. and Truman, A. (1982b). Laplace expansions of conditional Wiener integrals and applications to quantum physics, in *Stochastic processes in quantum theory and statistical physics (Marseille, 1981)* (Lecture Notes in Phys., 173, Springer, Berlin), pp. 40–55.

[96] Davies, I. M. and Truman, A. (1983). On the Laplace asymptotic expansion of conditional Wiener integrals and the Bender-Wu formula for x^{2N}-anharmonic oscillators, *J. Math. Phys.* **24**, 2, pp. 255–266.

[97] de Verdière, Y. C. (2003). Singular Lagrangian manifolds and semiclassical analysis, *Duke Math. J.* **116**, 2, pp. 263–298.

[98] DeFacio, B., Johnson, J. and Lapidus, M. (1992). Feynman's operational calculus as a generalized path integral, in C. S. et al. (ed.), *Stochastic Processes. A Festrischritf in Honour of Gopinath Kallianpur* (Springer-Verlag. New York), pp. 51–60.

[99] DeFacio, B., Johnson, J. and Lapidus, M. (1997). Feynman's operational calculus and evolution equations, *Acta Applicandae Mathematicae* **47**, pp. 155–211.

[100] DeWitt-Morette, C. (1972). Feynman's path integral. Definition without limiting procedure, *Commun. Math. Phys.* **28**, pp. 47–67.

[101] DeWitt-Morette, C. (1974). Feynman path integrals, I. Linear and affine techniques, II. The Feynman – Green function, *Commun. Math. Phys.* **37**, pp. 63–81.

[102] DeWitt-Morette, C. (1984a). Feynman path integrals, *Acta Phys. Austr. Suppl.* **XXVI**, pp. 101–170.

[103] DeWitt-Morette, C. (1984b). Feynman path integrals: from the prodistribution definition to the calculation of glory scattering, in *Stochastic methods and computer techniques in quantum dynamics (Schladming, 1984)*, Vol. XXVI (Springer, Vienna), pp. 101–170.

[104] DeWitt-Morette, C., Maheshwari, A. and Nelson., B. (1977). Path integration in phase space, *General Relativity and Gravitation* **8**, 8, pp. 581–593.

[105] Diosi, L. (1988). Continuous quantum measurements and Ito formalism, *Phys. Lett. A* **129**, 8, pp. 419–423.

[106] Dirac, P. (1933). The Lagrangian in quantum mechanics, *Phys. Zeitschr. d. Sowjetunion* **3**, 1, pp. 64–72.

[107] Dirac, P. (1958). *The Principles of Quantum Mechanics* (Clarendon Press, Oxford).

[108] Doss, H. (1980). Sur une Résolution Stochastique de l'Equation de Schrödinger à Coefficients Analytiques, *Commun. Math. Phys.* **73**, pp. 247–264.

[109] Dreyfus, T. and Dym, H. (1978). Product formulas for the Eigenvalues of a Class of Boundary Value Problems, *Duke Math. J.* **45**, 1, pp. 15–37.

[110] Duistermaat, J. (1974). Oscillatory integrals, Lagrange immersions and unfoldings of singularities, *Comm. Pure Appl. Math.* **27**, pp. 207–281.

[111] Ekeland, I. (1990). *Convexity Methods in Hamiltonian Mechanics* (Springer Verlag, Berlin).

[112] Ellis, A. and Rosen, J. (1981). Asymptotic analysis of Gaussian integrals II. Manifold of mimimum points, *Commun. Math. Phys.* **82**, pp. 153–181.

[113] Ellis, A. and Rosen, J. (1982). Asymptotic analysis of Gaussian integrals I: Isolated minimum points, *Trans. Amer. Math. Soc.* **273**, pp. 447–481.

[114] Elworthy, D. and Truman, A. (1984). Feynman maps, Cameron-Martin formulae and anharmonic oscillators, *Ann. Inst. H. Poincaré Phys. Théor.* **41**, 2, pp. 115–142.

[115] Erdélyi, A. (1956). *Asymptotic expansions* (Dover Publications, Inc., New York).

[116] Exner, P. (1985). *Open quantum systems and Feynman integrals* (Reidel, Dortrecht).

[117] Faria, M. D., Oliveira, M. J. and Streit, L. (2005). Feynman integrals for nonsmooth and rapidly growing potentials, *J. Math. Phys.* **46**, 6, pp. 063505, 14 pp.

[118] Faria, M. D., Potthoff, J. and Streit, L. (1991). The Feynman integrand as a Hida distribution, *J. Math. Phys.* **32**, pp. 2123–2127.

[119] Fedoryuk, M. (1993). *Asymptotic analysis* (Springer-Verlag, Berlin).

[120] Feynman, R. and Hibbs, A. (1965). *Quantum mechanics and path integrals* (McGraw-Hill, New York).

[121] Feynman, R. and Vernon, F. (1963). The theory of a general quantum system interacting with linear dissipative system, *Annals of phyics* **24**, pp. 118–173.

[122] Feynman, R. P. (1948). Space-time approach to non-relativistic quantum mechanics, *Rev. Mod. Phys.* **20**, pp. 367–387.

[123] Feynman, R. P. (1949a). Space-time approach to quantum electrodynamics, *Phys. Rev.* **76**, pp. 769–789.

[124] Feynman, R. P. (1949b). The theory of positrons, *Phys. Rev.* **76**, pp. 749–759.

[125] Feynman, R. P. (1950). Mathematical formulation of the quantum theory of electromagnetic interaction, *Phys. Rev.* **80**, pp. 440–457.

[126] Feynman, R. P. (1951). An operator calculus having applications in quantum electrodynamics, *Phys. Rev.* **84**, pp. 108–128.

[127] Friedman, C. (1972). Semigroup product formulas, compression and continuous observation in quantum mechanics, *Indiana Univ. Math. J.* **21**.

[128] Fröhlich, J. and King, C. (1989). The Chern-Simons Theory and Knot Polynomials, *Comm. Math. Phys.* **126**, pp. 167–199.

[129] Fujiwara, D. (1980). Remarks on convergence of Feynman path integrals, *Duke Math. J.* **47**, pp. 559–600.

[130] Fujiwara, D. (1990). The Feynman path integral as an improper integral over the Sobolev space, in *Journées Équations aux Dérivées Partielles (Saint Jean de Monts, 1990), Exp. No. XIV* (École Polytech. Palaiseau).

[131] Fujiwara, D. (1991). The stationary phase method with an estimate of the remainder term on a space of large dimension, *Nagoya Math. J.* **124**, pp. 51–67.

[132] Fujiwara, D. (1993). Some Feynman path integrals as oscillatory integrals over a Sobolev manifold, in *Functional analysis and related topics,1991 (Kyoto)* (Lecture Notes in Math., 1540, Springer, Berlin), pp. 39–53.

[133] Fujiwara, D. (1999). *Mathematical Methods for Feynman Path Integrals. (Japanese)* (Springer, Tokyo).

[134] Fujiwara, D. and Kumano-go, N. (2005). Smooth functional derivatives in Feynman path integrals by time slicing approximation, *Bull. Sci. math* **129**, pp. 57–79.

[135] Fujiwara, D. and Kumano-go, N. (2006). The second term of the semi-classical asymptotic expansion for Feynman path integrals with integrand of polynomial growth, *J. Math. Soc. Japan* **58**, 3, pp. 837–867.

[136] Fujiwara, D. and Kumano-go, N. (2008a). Feynman Path Integrals and Semiclassical Approximation, *RIMS KoKyuroku Bessatsu* **B5**, pp. 241–263.

[137] Fujiwara, D. and Kumano-go, N. (2008b). Phase space Feynman path integrals via piecewise bicharacteristic paths and their semiclassical approximations, *Bull. Sci. math.* **132**, pp. 313–357.

[138] Gatarek, D. and Gisin, N. (1991). Continuous quantum jumps and infinite-dimensional stochastic equation, *J. Math. Phys* **32**, 8, pp. 2152–2157.

[139] Gaveau, B. and Schulman, L. S. (1986). Explicit time-dependent Schrödinger propagators, *J. Phys. A* **19**, 10, pp. 1833–1846.

[140] Gelfand, I. and Vilenkin, N. (1964). *Generalized Functions. IV Applications of Harmonic Analysis* (Academic Press, New York).

[141] Gill, T. (1981). Time-ordered operators, I. Foundations for an alternate view of reality, *Trans. Amer. Math. Soc.* **266**, pp. 161–181.

[142] Gill, T. (1983). Time-ordered operators, II. *Trans. Amer. Math. Soc.* **279**, pp. 617–634.

[143] Gill, T. and Zachary, W. (1987). Time-ordered operators and Feynman-Dyson algebras, *J. Math. Phys.* **28**, pp. 1459–1470.

[144] Golubitsky, M. (1978). An Introduction to Catastrophe Theory and Applications, *SIAM Mathematical Review* **20**, pp. 352–387.

[145] Grabert, H., Inomata, A., Schulman, L. S. and Weiss, U. (eds.) (1993). *Path integrals from meV to MeV: Tutzing 1992* (World Scientific Publishing Co., Inc., River Edge, NJ).

[146] Grosche, C. (1996). *Path integrals, hyperbolic spaces, and Selberg trace formulae* (World Scientific Publishing Co., Inc, River Edge, NJ).

[147] Grosche, C. and Steiner, F. (1998). *Handbook of Feynman path integrals* (Springer Tracts in Modern Physics, 145. Springer-Verlag, Berlin).

[148] Gross, L. (1962). Measurable functions on Hilbert spaces, *Trans. Amer. Mat. Soc.* **105**, pp. 375–390.

[149] Gross, L. (1965). Abstract Wiener Spaces, in 5th *Berkeley Symp. Math. Stat. Prob.*, pp. 31–42.

[150] Grothaus, M., Khandekar, D., da Silva, J. and Streit, L. (1997). The Feynman integral for time-dependent anharmonic oscillators, *J. Math. Phys.* **38**, 6, pp. 3278–3299.

[151] Guadagnini, E. (1993). *The link invariants of the Chern-Simons field theory, de Gruyter Expositions in Mathematics*, Vol. 10 (Walter de Gruyter & Co., Berlin).

[152] Gutzwiller, M. C. (1990). *Chaos in classical and quantum mechanics* (Springer-Verlag, New York).

[153] Gutzwiller, M. C., Inomata, A., Klauder, J. R. and Streit, L. (eds.) (1986). *Path integrals from meV to MeV* (Bielefeld Encounters in Physics and Mathematics, VII. World Scientific Publishing Co., Singapore).

[154] Hahn, A. (2001). *Chern-Simons theory on* $\mathbb{R}^3$ *in axial gauge*, Ph.D. thesis, Bonn University.

[155] Hahn, A. (2004a). Chern-Simons theory on $\mathbb{R}^3$ in axial gauge: a rigorous approach, *J. Funct. Anal.* **211**, 2, pp. 483–507.

[156] Hahn, A. (2004b). The Wilson loop observables of Chern-Simons theory on $\mathbb{R}^3$ in axial gauge, *Commun. Math. Phys.* **248**, 3, pp. 467–499.

[157] Hahn, A. (2005a). An analytic approach to Turaevs shadow invariants, *SFB 611 preprint 233, Bonn* .

[158] Hahn, A. (2005b). Chern-Simons models on $S^2 \times S^1$, torus gauge fixing, and link invariants I, *J. Geom. Phys.* **53**, 3, pp. 275–314.

[159] Hahn, A. (2005c). White noise analysis, quantum field theory, and topology, in *Stochastic analysis: classical and quantum* (World Sci. Publ. Hackensack, NJ), pp. 13–30.

[160] Hardy, G. H. (1963). *Divergent series* (Oxford University Press, London).

[161] Helffer, B. (1988). *Semi-classical analysis for the Schrödinger operator and applications* (Lecture Notes in Mathematics, 1336. Springer-Verlag, Berlin).

[162] Hida, T. (2001). White noise approach to Feynman integrals, *J. Korean Math. Soc.* **38**, 2, pp. 275–281.

[163] Hida, T., Kuo, H., Potthoff, J. and Streit, L. (1993). *White Noise. An infinite-dimensional calculus* (Mathematics and its Applications, 253. Kluwer Academic Publishers Group, Dordrecht).

[164] Hörmander, L. (1971). Fourier integral operators I, *Acta Math.* **127**, 1, pp. 79–183.

[165] Hörmander, L. (1983). *The Analysis of Linear Partial Differential Operators. I. Distribution Theory and Fourier Analysis* (Springer-Verlag, Berlin/Heidelberg/New York/Tokyo).

[166] Hudson, R. L. and Parthasarathy, K. R. (1984). Quantum Ito's Formula and Stochastic Evolutions, *Commun. Math. Phys.* **93**, pp. 301–323.

[167] Hurd, N. (1983). *Geometric quantization in action* (D. Reidel, Dortrecht).

[168] Ichinose, T. (1982). Path integral for the Dirac equation in two space-time dimensions, *Proc. Japan Acad. Ser. A Math. Sci.* **58**, 7, pp. 290–293.

[169] Ichinose, T. (1997). On the formulation of the Feynman path integral through broken line paths, *Comm. Math. Phys.* **189**, 1, pp. 17–33.
[170] Ichinose, T. (2003). Convergence of the Feynman path integral in the weighted Sobolev spaces and the representation of correlation functions, *J. Math. Soc. Japan* **55**, 4, pp. 957–983.
[171] Ichinose, T. and Tamura, H. (1984). Propagation of a Dirac particle. A path integral approach, *J. Math. Phys.* **25**, 6, pp. 1810–1819.
[172] Ito, K. (1961). Wiener integral and Feynman integral, in *Proc. Fourth Berkeley Symp. Math. Stat. and Prob.*, Vol. II (Univ. California Press, Berkeley), pp. 227–238.
[173] Ito, K. (1967). Generalized uniform complex measures in the Hilbertian metric space with their application to the Feynman path integral, in *Proc. Fifth Berkeley Symposium on Mathematical Statistics and Probability*, Vol. II.1 (Univ. California Press, Berkeley), pp. 145–161.
[174] Itzykson, C. and Zuber, J. (1980). *Quantum field theory* (McGraw-Hill International Book Co., New York).
[175] Johnson, G. (1982). The equivalence of two approaches to Feynman integral, *J. Math. Phys.* **23**, pp. 2090–2096.
[176] Johnson, G. and Kallianpur, G. (1993). Homogeneous chaos, *p*-forms, scaling and the Feynman integral, *Trans. Amer. Math. Soc.* **340**, 2, pp. 503–548.
[177] Johnson, G. and Kim, J. G. (2000). A dominated-type convergence theorem for the Feynman integral, *J. Math. Phys.* **41**, 5, pp. 3104–3112.
[178] Johnson, G. and Lapidus, M. (1986). Generalized Dyson series, generalized Feynman diagrams, the Feynman integral and Feynman's operational calculus, *Mem. Amer. Math. Soc.* **62**, 351, pp. 1–78.
[179] Johnson, G. and Lapidus, M. (1988). Noncommutative operations on Wiener functionals and Feynman's operational calculus, *J. Funct. Anal.* **81**, 1, pp. 74–99.
[180] Johnson, G. and Lapidus, M. (2000). *The Feynman integral and Feynman's operational calculus* (Oxford University Press, New York).
[181] Johnson, G. and Skoug, D. (1970a). Operator-valued Feynman integrals of certain finite-dimensional functionals, *Proc. Am. Math. Soc.* **24**, pp. 774–780.
[182] Johnson, G. and Skoug, D. (1970b). Operator-valued Feynman integrals of finite-dimensional functionals, *Pac. Journ. of Math.* **34**, pp. 415–425.
[183] Johnson, G. and Skoug, D. (1971a). Feynman integrals of non-factorable finite-dimensional functionals, *Pac. Journ. of Math.* **37**, pp. 303–317.
[184] Johnson, G. and Skoug, D. (1971b). An operator valued function space integral: A sequel to Cameron and Storvick's paper, *Proc. Am. Math. Soc.* **27**, pp. 514–518.
[185] Johnson, G. and Skoug, D. (1973). A Banach algebra of Feynman integrable functionals with application to an integral equation formally equivalent to Schrödinger's equation, *J. Funct. Analysis* **12**, pp. 129–152.
[186] Johnson, G. and Skoug, D. (1974). A function space integral for a Banach

space of functionals on Wiener space, *Proc. Am. Math. Soc.* **43**, pp. 141–148.

[187] Kac, M. (1949). On distributions of certain Wiener functionals, *Trans. Amer. Math. Soc.* **65**, pp. 1–13.

[188] Kac, M. (1951). On some connections between probability theory and differential and integral equations, in *Proc. Second Berkeley Symp.* (Univ. California Press, Berkeley), pp. 189–215.

[189] Kac, M. (1980). *Integration in function spaces and some of its applications* (Accademia Nazionale dei Lincei, Pisa).

[190] Kallianpur, G. and Bromley, C. (1984). Generalized Feynman integrals using analytic continuation in several complex variables, in *Stochastic analysis and applications* (Adv. Probab. Related Topics, 7, Dekker, New York), pp. 217–267.

[191] Kallianpur, G., Kannan, D. and Karandikar, R. (1985). Analytic and sequential Feynman integrals on abstract Wiener and Hilbert spaces, and a Cameron Martin formula, *Ann. Inst. H. Poincaré, Prob. Th.* **21**, pp. 323–361.

[192] Kallianpur, G. and Oodaira, H. (1978). Freĭdlin-Wentzell type estimates for abstract Wiener spaces, *Sankhyā Ser. A* **40**, 2, pp. 116–137.

[193] Kallianpur, G. and Üstünel, A. S. (1992). Distributions, Feynman integrals and measures on abstract Wiener spaces, in *Stochastic analysis and related topics (Silivri, 1990)* (Birkhuser. Boston (MA)), pp. 237–284.

[194] Kanasugi, H. and Okada, H. (1975). Systematic Treatment of General Time-Dependent Harmonic Oscillator in Classical and Quantum Mechanics, *Progr. Theoret. Phys.* **16**, 2, pp. 384–388.

[195] Karatzas, I. and Shreve, S. (1988). *Brownian motion and stochastic calculus* (Springer-Verlag).

[196] Kato, T. (1966). *Perturbation Theory for Linear Operators* (Springer).

[197] Kauffman, L. (1995). Functional integration and the theory of knots, *J. Math. Phys.* **36**, 5, pp. 2402–2429.

[198] Kauffman, L. (1997). An introduction to knot theory and functional integrals, in *Functional integration (Cargèse, 1996)* (NATO Adv. Sci. Inst. Ser. B Phys., 361, Plenum, New York), pp. 247–308.

[199] Kauffman, L. (1998). Witten's integral and the Kontsevich integral, in *Particles, fields, and gravitation (Łódź, 1998)* (AIP Conf. Proc., 453, Amer. Inst. Phys., Woodbury, NY), pp. 368–381.

[200] Kauffman, L. (2004). Vassiliev invariants and functional integration without integration, in *Stochastic analysis and mathematical physics (SAMP/ANESTOC 2002)* (World Sci. Publishing, River Edge, NJ), pp. 91–114.

[201] Kelvin, L. (1887). *Phil. Mag.* **23**, pp. 252–255.

[202] Khandekar, D. C. and Lawande, S. V. (1975). Exact propagator for a time-dependent harmonic oscillator with and without a singular perturbation, *J. Math. Phys.* **16**, 2, pp. 384–388.

[203] Kinoshita, T. (1990). *Quantum Electrodynamics* (World Scientific, Singapore).

[204] Klauder, J. (1974). *Functional techniques and their application in quantum field theory* (Colorado Assoc. Univ. Press).

[205] Kleinert, H. (2004). *Path integrals in quantum mechanics, statistics, polymer physics, and financial markets*, 3rd edn. (World Scientific Publishing Co., Inc., River Edge, NJ).

[206] Kolokoltsov, V. N. (1999). Complex measures on path space: an introduction to the Feynman integral applied to the Schrödinger equation, *Methodol. Comput. Appl. Probab.* **1**, 3, pp. 349–365.

[207] Kolokoltsov, V. N. (2000). *Semiclassical analysis for diffusions and stochastic processes* (Lecture Notes in Mathematics, 1724. Springer-Verlag, Berlin).

[208] Kostantinov, A., Maslov, V. and Chebotarev, A. (1990). Probability representations of solutions of the Cauchy problem for quantum mechanical equations, *Russian Math. Surveys* **45**, 6, pp. 1–26.

[209] Kumano-go, N. (2004). Feynman path integrals as analysis on path space by time slicing approximation, *Bull. Sci. Math.* **128**, 3, pp. 197–251.

[210] Kuna, T., Streit, L. and Westerkamp, W. (1998). Feynman integrals for a class of exponentially growing potentials, *J. Math. Phys.* **39**, 9, pp. 4476–4491.

[211] Kuo, H. (1975). *Gaussian Measures in Banach Spaces* (Lecture Notes in Math., Springer-Verlag Berlin-Heidelberg-New York).

[212] Lapidus, M. (1987). The Feynman-Kac formula with a Lebesgue-Stieltjes measure and Feynman's operational calculus, *Stud. Appl. Math.* **76**, 2, pp. 93–132.

[213] Lapidus, M. (1996). The Feynman integral and Feynman's operational calculus: a heuristic and mathematical introduction, *Ann. Math. Blaise Pascal* **3**, 1, pp. 89–102.

[214] Lascheck, A., Leukert, P., Streit, L. and Westerkamp, W. (1993). Quantum Mechanical Propagators in terms of Hida Distribution, *Rep. Math. Phys.* **33**, pp. 221–232.

[215] Leukert, S. and Schäfer, J. (1996). A Rigorous Construction of Abelian Chern-Simons Path Integral Using White Noise Analysis, *Reviews in Math. Phys.* **8**, pp. 445–456.

[216] Malgrange, B. (1995). Sommations des séries divergentes, *Expo. Math.* **13**, pp. 163–222.

[217] Malgrange, B. and Ramis, J. P. (1992). Fonctions Multisommables, *Ann. Inst. Fourier* **42**, 1-2, pp. 353–368.

[218] Mandrekar, V. (1983). Some remarks on various definitions of Feynman integrals, in K. J. Beck (ed.), *Lectures Notes Math.* (Springer. Berlin), pp. 170–177.

[219] Manoliu, M. (1998). Abelian Chern-Simons theory. II. A functional integral approach, *J. Math. Phys.* **39**, 1, pp. 207–217.

[220] Maslov, V. P. (1971). Characteristics of pseudo-differential operators, in *Proc. Internat. Congress Math. (Nice 1970) vol 2* (Gautier-Villars, Paris), pp. 290–294.

[221] Maslov, V. P. (1972). *Théorie des perturbations et méthodes asymptotiques* (Dunod, Paris).

[222] Maslov, V. P. and Chebotarev, A. M. (1979). Processus à sauts et leur application dans la mécanique quantique, in S. A. et al. (ed.), *Feynman path integrals* (Springer Lecture Notes in Physics 106), pp. 58–72.

[223] Maslov, V. P. and Fedoryuk, M. V. (1981). *Semiclassical approximation in quantum mechanics* (Reidel Publishing Co., Dordrecht-Boston, Mass.).

[224] Mazzucchi, S. (2008). Feynman path integrals for the inverse quartic oscillator, *J. Math. Phys.* **49**, 9, p. 093502 (15 pages).

[225] Mensky, M. (1993). *Continuous Quantum Measurements and Path Integrals* (Taylor & Francis, Bristol and Philadelphia).

[226] Mensky, M. (2000). *Quantum measurements and decoherence. Models and phenomenology* (Fundamental Theories of Physics, 110. Kluwer Academic Publishers, Dordrecht).

[227] Milnor, J. (1963). *Morse theory* (Princeton University Press).

[228] Milnor, J. (1965). *Topology from the differentiable viewpoint* (The University Press of Virginia, Charlottesville).

[229] Misra, B. and Sudarshan, E. C. G. (1975). The Zeno's paradox in quantum theory, *J. Math. Phys.* **10**, 4, pp. 756–763.

[230] Muldowney, P. (1987). *A general theory of integration in function spaces, including Wiener and Feynman integration* (Pitman Research Notes in Mathematics Series, 153. Longman Scientific & Technical, Harlow; John Wiley & Sons, Inc., New York).

[231] Nakamura, T. (1991). A nonstandard representation of Feynman's path integrals, *J. Math. Phys.* **32**, 2, pp. 457–463.

[232] Nakamura, T. (1997). Path space measures for Dirac and Schrödinger equations: nonstandard analytical approach, *J. Math. Phys.* **38**, 8, pp. 4052–4072.

[233] Nakamura, T. (2000). Path space measure for the $3+1$-dimensional Dirac equation in momentum space, *J. Math. Phys.* **41**, 8, pp. 5209–5222.

[234] Nelson, B. and Sheeks, B. (1982). Path integration for velocity-dependent potentials, *Comm. Math. Phys.* **84**, 4, pp. 515–530.

[235] Nelson, E. (1964). Feynman integrals and the Schrödinger equation, *J. Math. Phys.* **5**, pp. 332–343.

[236] Nevanlinna, F. (1919). Zur Theorie der asymptotischen Potenzreihen, *Ann. Acad. Sci. Fenn. (A)* **12**, 3, pp. 1–81.

[237] Olver, F. W. J. (1974). *Asymptotics and special functions.* (Computer Science and Applied Mathematics. Academic Press, New York-London).

[238] Parisi, G. (1988). *Statistical Field Theory* (Addison-Wesley, Redwood City, Cal.).

[239] Pechukas, P. and Light, J. C. (1966). On the Exponential Form of Time-Displacement Operators in Quantum Mechanics, *J. Chem. Phys.* **44**, 10, pp. 3897–3912.

[240] Peres, A. (1980). Zeno paradox in quantum theory, *Am. J. Phys.* **48**, 11, pp. 931–932.

[241] Piterbarg, V. I. (1996). *Asymptotic methods in the theory of Gaussian pro-*

cesses and fields (Translations of Mathematical Monographs, 148. American Mathematical Society, Providence, RI).

[242] Piterbarg, V. I. and Fatalov, V. R. (1995). The Laplace method for probability measures in Banach spaces, *Russian Math. Surveys* **50**, 6, pp. 1151–1239.

[243] Protter, P. (1990). *Stochastic Integration and Differential Equations* (Applications of Mathematics 21, Springer-Verlag).

[244] Ramis, J. P. (1993). Divergent series and asymptotic theories, *Bull. Soc. Math. France, Panoramas et Syntheses* .

[245] Reed, M. and Simon, B. (1973). *Methods of Modern Mathematical Physics I. Functional Analysis* (Academic Press, New York).

[246] Reed, M. and Simon, B. (1975). *Methods of Modern Mathematical Physics II. Fourier Analysis, Self-Adjointness* (Academic Press, New York).

[247] Rezende, J. (1984). Quantum systems with time dependent harmonic part and the Morse index, *J. Math. Phys.* **25**, 11, pp. 3264–3269.

[248] Rezende, J. (1985). The method of stationary phase for oscillatory integrals on Hilbert spaces, *Comm. Math. Phys.* **101**, pp. 187–206.

[249] Rezende, J. (1993). A note on Borel summability, *J. Math. Phys.* **34**, 9, pp. 4330–4339.

[250] Rezende, J. (1994). Feynman integrals and Fredholm determinants, *J. Math. Phys.* **35**, 8, pp. 4357–4371.

[251] Rezende, J. (1996). Time-Dependent Linear Hamiltonian Systems and Quantum Mechanics, *Lett. Math. Phys.* **38**, pp. 117–127.

[252] Robert, D. (1987). *Autour de l'approximation semi-classique. (French) [On semiclassical approximation]* (Progress in Mathematics, 68. Birkhauser Boston, Inc., Boston, MA).

[253] Rogers, A. (1987). Supersymmetric path integration, *Phys. Lett. B* **193**, 1, pp. 48–54.

[254] Rudin, W. (1974). *Real and Complex Analysis* (McGraw-Hill, New York).

[255] Sa-yakanit, V., Sritrakool, W. and Berananda, J.-O. (eds.) (1989). *Path integrals from meV to MeV* (World Scientific Publishing Co., Inc., Teaneck, NJ).

[256] Schilder, M. (1966). Some asymptotic formulas for Wiener integrals, *Trans. Amer. Math. Soc.* **125**, pp. 63–85.

[257] Schulman, L. S. (1981). *Techniques and applications of path integration* (John Wiley & Sons, Inc., New York).

[258] Schwartz, A. S. (1978). The partition function of a degenerate quadratic functional and Ray-Singer invariants, *Lett. Math. Phys* **2**, pp. 247–252.

[259] Schwartz, A. S. (1979). The partition function of a degenerate functional, *Comm. Math. Phys.* **67**, 1, pp. 1–16.

[260] Shavgulidze, E. T. (1996). Solutions of Schrödinger equations with polynomial potential by path integrals, in *Path integrals: Dubna '96* (Joint Inst. Nuclear Res.), pp. 269–272.

[261] Silva, J. L. and Streit, L. (2004). Feynman integrals and white noise analysis, in *Stochastic analysis and mathematical physics (SAMP/ANESTOC 2002)* (World Sci. Publ., River Edge, NJ), pp. 285–303.

[262] Simon, B. (1979). *Trace ideals and their applications* (London Mathematical Society Lecture Note Series 35. Cambridge U.P., Cambridge).

[263] Simon, B. (2005). *Functional integration and quantum physics*, 2nd edn. (AMS Chelsea Publishing, Providence, RI).

[264] Skoug, D. (1974). Partial differential Systems of generalized Wiener and Feynman integrals, *Portug. Math.* **33**, pp. 27–33.

[265] Smolyanov, O. and Shavgulidze, E. (1990). *Path integrals. (Russian)* (Moskov. Gos. Univ., Moscow).

[266] Smolyanov, O. and Shavgulidze, E. (2003). Feynman formulas for solutions of infinite-dimensional Schrödinger equations with polynomial potentials. (Russian), *Dokl. Akad. Nauk.* **390**, 3, pp. 321–324.

[267] Sokal, A. D. (1980). An improvement of Watson's theorem on Borel summability, *J. Math. Phys.* **21**, 2, pp. 261–263.

[268] Sommerfeld, A. (1954). *Optics. Lectures on theoretical physics, Vol. IV* (Academic Press Inc., New York).

[269] Stokes, G. G. (1857). *Camb. Phil. Trans.* **10**, pp. 106–128.

[270] Streit, L. (2003). *Feynman paths, sticky walls, white noise* (World Sci. Publ. River Edge (NJ)), pp. 105–113.

[271] Sugita, H. and Taniguchi, S. (1998). Oscillatory integrals with quadratic phase function on a real abstract Wiener space, *J. Funct. Anal.* **155**, 1, pp. 229–262.

[272] Sugita, H. and Taniguchi, S. (1999). A remark on stochastic oscillatory integrals with respect to a pinned Wiener measure, *Kyushu J. Math.* **53**, 1, pp. 151–162.

[273] Takeo, F. (1989). Generalized vector measures and path integrals for hyperbolic systems, *Hokkaido Math. J.* **18**, 3, pp. 497–511.

[274] Taniguchi, S. (1998a). On the exponential decay of oscillatory integrals on an abstract Wiener space, *J. Funct. Anal.* **154**, 2, pp. 424–443.

[275] Taniguchi, S. (1998b). Oscillatory integrals with quadratic phase function on a real abstract Wiener space, *J. Funct. Anal.* **155**, 1, pp. 229–262.

[276] Taniguchi, S. (1998c). A remark on stochastic oscillatory integrals with respect to a pinned Wiener measure, *Kyushu J. Math.* **53**, 1, pp. 151–162.

[277] Thaler, H. (2003). Solution of Schrödinger equations on compact Lie groups via probabilistic methods, *Potential Anal.* **18**, 2, pp. 119–140.

[278] Thaler, H. (2005). The Doss Trick on Symmetric Spaces, *Letters in Mathematical Physics* **72**, pp. 115–127.

[279] Trotter, H. (1959). On the product of semigroups of operators, *Proc. Amer. Math. Soc.* **10**, pp. 545–551.

[280] Truman, A. (1976). Feynman path integrals and quantum mechanics as $\hbar \to 0$, *J. Math. Phys.* **17**, pp. 1852–1862.

[281] Truman, A. (1977). Classical mechanics, the diffusion (heat) equation, and the Schrödinger equation, *J. Math. Phys.* **18**, pp. 2308–2315.

[282] Truman, A. (1978). The Feynman maps and the Wiener integral, *J. Math. Phys.* **19**, pp. 1742–1750.

[283] Truman, A. (1979). The polygonal path formulation of the Feynman path integral, in S. Albeverio (ed.), *Feynman path integrals*, *Springer Lecture*

Notes in Physics, Vol. 106 (Springer).

[284] Truman, A. and Zastawniak, T. (1999). Stochastic PDE's of Schrödinger type and stochastic Mehler kernels—a path integral approach, in *Seminar on Stochastic Analysis, Random Fields and Applications (Ascona, 1996)* (Birkhäuser, Basel), pp. 275–282.

[285] Truman, A. and Zastawniak, T. (2001). Stochastic Mehler kernels via oscillatory path integrals, *J. Korean Math. Soc.* **38**, 2, pp. 469–483.

[286] Watling, K. D. (1992). Formulae for solutions to (possibly degenerate) diffusion equations exhibiting semi-classical asymptotics, in *Stochastics and quantum mechanics (Swansea, 1990)* (World Sci. Publ., River Edge, NJ), pp. 248–271.

[287] Watson, G. N. (1911). A Theory of Asymptotic Series, *Philosophical Transaction of the Royal Society of London, ser A* **CCXI**, pp. 279–31.

[288] Weinberg, S. (2005a). *The quantum theory of fields. I. Foundations* (Cambridge University Press, Cambridge).

[289] Weinberg, S. (2005b). *The quantum theory of fields. II. Modern applications* (Cambridge University Press, Cambridge).

[290] Wiegel, F. W. (1986). *Introduction to path-integral methods in physics and polymer science* (World Scientific Publishing Co., Singapore).

[291] Wiener, N. (1976). *Norbert Wiener. Collected Works.* (MIT Press, Cambridge).

[292] Witten, E. (1989). Quantum field theory and the Jones polynomial, *Commun. Math. Phys.* **121**, pp. 353–389.

[293] Yajima, K. (1996). Smoothness and Non-Smoothness of the Fundamental Solution of Time Dependent Schrödiger Equations, *Commun. Math. Phys.* **181**, pp. 605–629.

[294] Zambrini, J. C. (2001). Feynman integrals, diffusion processes and quantum symplectic two-forms, *J. Korean Math. Soc.* **38**, 2, pp. 385–408.

[295] Zastawniak, T. J. (1988). Path integrals for the telegrapher's and Dirac equations; the analytic family of measures and the underlying Poisson process, *Bull. Pol. Acad. Sci. Math.* **36**, 5-6, pp. 341–356.

[296] Zastawniak, T. J. (1989a). The nonexistence of the path-space measure for the Dirac equation in four space-time dimensions, *J. Math. Phys.* **30**, 6, pp. 1354–1358.

[297] Zastawniak, T. J. (1989b). Path integrals for the Dirac equation—some recent developments in mathematical theory, in *Stochastic analysis, path integration and dynamics (Warwick, 1987)* (Pitman Res. Notes Math. Ser., 200, Longman Sci. Tech., Harlow), pp. 243–263.

[298] Zinn-Justin, J. (1993). *Quantum field theory and critical phenomena*, 2nd edn. (Oxford University Press, New York).

Index